Tibetan mastiffs were first imported into England over 150 years ago, and only three decades ago to North America. During the Chinese Cultural Revolution these remarkable high altitude livestock guardian dogs were nearly annihilated, but they have recovered and are now in high demand, some for *very* high prices among the *nouveau riche* of modern China. Today, thousands of these dogs are found around the world, promoted by many breeders, raised as pets, guardians and faithful companions. Some have scored high marks at international dog shows. Interest in Tibetan mastiffs and related dogs—their history, breeding, temperament, function and future—has never been as high.

This is a book of discovery of the exotic and relatively rare breeds of big dogs from Tibet and the Himalayas: the Tibetan mastiff, the rare KyiApso (the 'bearded' or 'shaggy' Tibetan mastiff), the Himalayan mountain dog, and the least known Sha-kyi (Tibetan hunting dog).

Research on Tibetan dogs is contentious. This book challenges some of the conventional wisdom about the big dogs with evidence showing how some big dog fanciers have gotten it wrong. It questions the notion that there were *gigantic* dogs in history, an idea that has inspired some modern breeders to create enormous critters, mistakenly evoking a mythical past—and much more.

Simply a 'must read' for all big dog owners and admirers.

*I've just completed a wonderful morning reading through your canine charivari... it's a cracking book. It is patently both a labour of love and a work of exhaustive scholarship. You've done a magnificent job.* **Charles Allen**, author of popular historical books on the Himalayas.

*You do know that the s**t is going to hit the fan when you publish, right? You're going to tick off a <u>lot</u> of people!! Personally, I think that could be a good thing for the breed if it gets some folks thinking...* **A Tibetan dog breeder** (anonymous).

## About the Author

Don Messerschmidt, award-winning author and anthropologist, is a Himalayan specialist and an authority on Tibetan mastiffs (TMs) from the 'Roof of the World'. He has spent over four decades studying them in their natural setting and their places of origin (and has scars to show for it). Dr Messerschmidt has raised, trained, shown, bred and photographed Tibetan mastiffs. His long-time companion TM, the International Champion Saipal Baron of Emodus (aka 'Kalu'), is well known in the bloodlines of many fine TMs in North America and Europe.

Don's writings about Tibetan dogs have appeared in *Dog World*, *Rangelands* and *ECS Nepal* magazines, and in *The Himalayan Times* and various dog club newsletters.

In this book, the author describes his lifelong enchantment with the big dogs and his quest for and discovery of them in old writings, as well as in travels on the Tibetan plateau and in the high pastures of the Himalayan borderlands.

Other books by the author:

*Against the Current: The Life of Lain Singh Bangdel*; Bangkok 2004.

*Anthropologists at Home in North America: Methods and Issues in the Study of One's Own Society*; Cambridge, 1981, 2010.

*Development Studies: Essays*; Bibliotheca Himalayica, Series IV, Vol, I; Kathmandu; 1995.

*Fr Moran of Kathmandu*; Bangkok 1997; new edition forthcoming.

*Gurungs of Nepal: Conflict and Change in a Village Society*; Warminster 1976.

*Muktinath: Himalayan Pilgrimage, Cultural and Historical Guide*; Kathmandu 1992; new edition forthcoming.

# BIG DOGS OF TIBET AND THE HIMALAYAS

## *A Personal Journey*

Don Messerschmidt

Orchid Press

Don Messerschmidt
BIG DOGS OF TIBET AND THE HIMALAYAS
A Personal Journey


ORCHID PRESS
PO Box 1046,
Silom Post Office,
Bangkok 10504, Thailand

*www.orchidbooks.com*


Copyright ©, Orchid Press 2010
Protected by copyright under the terms of the International Copyright Union: all rights reserved. Except for fair use in book reviews, no part of this publication may be reproduced in any form or by any means, electronic or mechanical, including photocopying, recording, or by any information storage or retrieval system without prior permission in writing from the copyright holder.


Front cover graphic design by Hans D. Messerschmidt, <www.themodcorp.com>

Author's contact Email: <dmesserschmidt@gmail.com>

ISBN: 978-974-524-130-5

**Dedication**

*For Jay N. Singh of Kathmandu,
Nepal's pre-eminent Tibetan mastiff expert.*

*And, remembering Ann Rohrer,
America's Tibetan mastiff pioneer.*

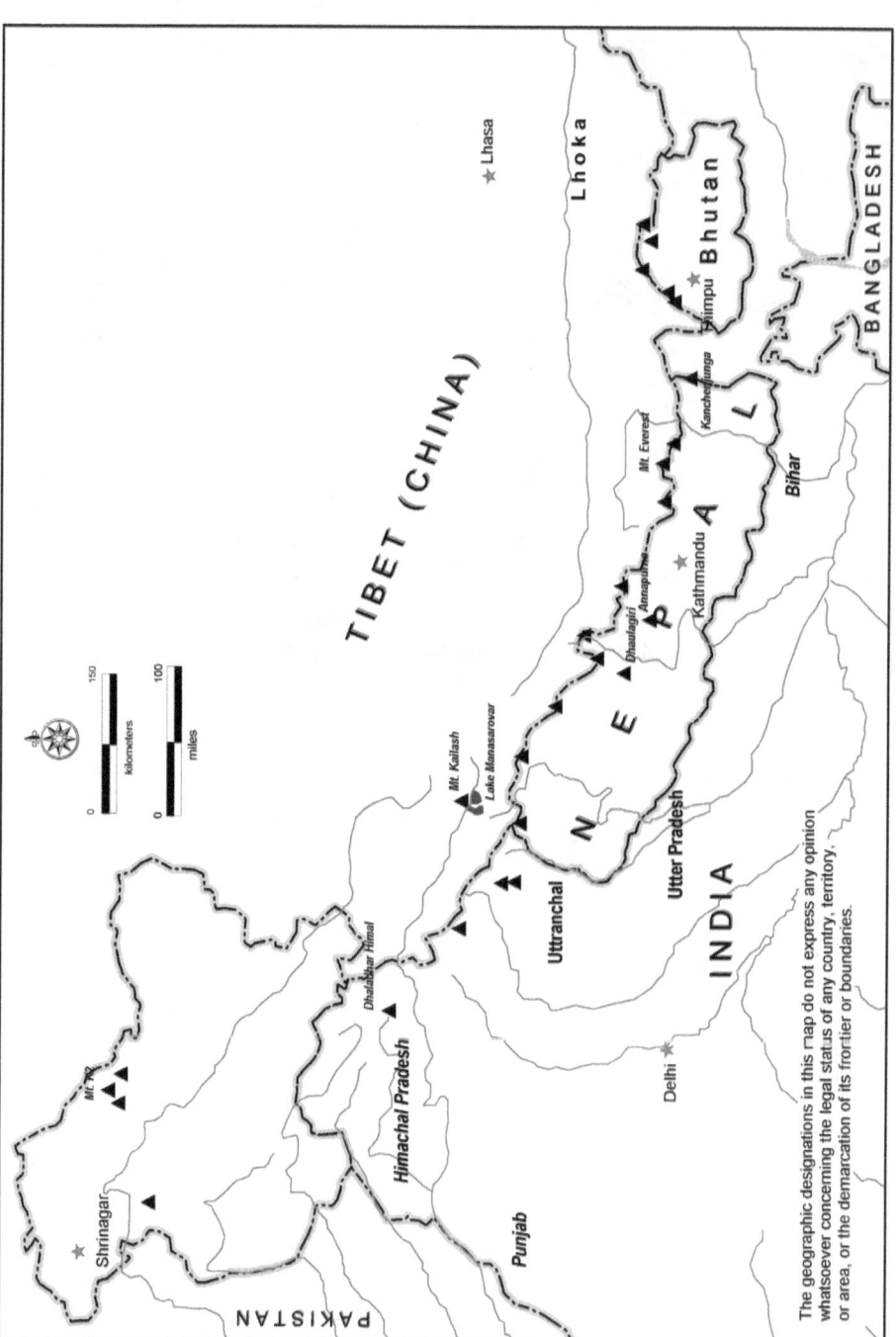

Map 1. Nepal and surrounding regions.

# Contents

| | |
|---|---|
| List of Illustrations | viii |
| Sources of Plates | x |
| Preface and Acknowledgments | xi |
| Chapter 1: Forget Marco Polo! | 1 |
| Chapter 2: The Noblest of the Party | 19 |
| Chapter 3: First Encounters | 52 |
| Chapter 4: Saipal's Kalu | 82 |
| Chapter 5: The Shaggy KyiApso | 106 |
| Chapter 6: Across Tibet | 133 |
| Chapter 7: Summing Up | 161 |
| Endnotes | 178 |
| Appendices | |
| 1. 'Dogs from the Roof of the World' (1937), by Hon. Mrs. Eric Bailey | 203 |
| 2. 'Role of the dog in Tibetan nomadic society' (1963), by Robert Ekvall | 208 |
| 3. Adaptive Traits of Tibetan Mastiffs: Natural vs. Cultural, Landrace vs. Purebred, by Don Messerschmidt | 218 |
| 4. 'Coat Color Genetics in Tibetan Mastiffs' (2004, rev. 2008), by Charles W. Radcliffe and Matthew J. Taylor | 221 |
| 5. The Tibetan Mastiff in Popular Fiction | 227 |
| Bibliography | 228 |
| Index | 247 |

# ILLUSTRATIONS

## Plates

1.1. Tomb guard dog (Tibetan mastiff?), clay figurine from the Han Dynasty. ... 5
1.2. Messers Marco Polo, Nicolo Polo and Maffeo Polo, on return to Venice after 26 years' absence. ... 11
2.1. An old illustration of a Tibetan mastiff. ... 25
2.2. Charm against dog-bite. ... 26
2.3. A village woman of gTsang (Tsang) tying up the watch-dog in front of her house. ... 27
2.4. Takkar had to be held by two men. ... 32
2.5. Takkar anchored. ... 34
2.6. The famous Tibetan mastiff named 'Bhotean', imported into England from Tibet in early 1900s. ... 37
2.7. Tendrup' sitting in front of Hugh Ruttledge and other climbers on the British Everest Expedition of 1936. ... 41
3.1. A typical do-kyi (tied watchdog) of northern Tibet and Nepal. ... 56
3.2. Cream colored (white) Himalayan mountain dog in a shepherd camp of Nepal's Lamjung Himal. ... 59
3.3. The author with Amjo, a Himalayan mountain dog on trek in Nepal. ... 63
3.4. Lata, the Tibetan mastiff of the Khampas, in Kesang Camp on the Nepal-Tibet border. ... 72
4.1. Jay N. Singh with the bitch, Bounty, in Kathmandu, 1983. ... 84
4.2. Tü-bo in Germany, the half brother of Saipal Baron (Kalu). ... 85
4.3. Kalu (Saipal Baron) with the author's son, Hans, 1983. ... 87
4.4. Kalu (Saipal Baron) with the author's daughter, Liesl, 1983. ... 87
4.5. Kalu's pups in 1983. ... 88
4.6. Bhalu, son of Saipal Baron (Kalu) with Elliott Johnson in Canada, 1986. ... 102
4.7. The strange halo of Bhalu, son of Saipal Baron (Kalu) in Canada, 1986. ... 104
5.1. The Dalai Lama's Do-Kyi-Apso (KyiApso); photo by Mrs Irma Bailey. ... 107
5.2. Captain Graham's 'Great Dog of Tibet', from a newspaper clipping. ... 109
5.3. Kinnauri kutta (beardedTibetan mastiff) of the north Indian Himalaya, with a man from Kinnaur. ... 112
5.4. High pasture 'dog show' with shepherds and dogs in the rain, in the Dhauladhar Himalayas of north India. ... 113
5.5. Bearded Tibetan mastiff (KyiApso) from Humla, far western Nepal Himalayas. ... 118

5.6 and 5.7. Thumdru, KyiApso import from Tibet. 122
5.8. Apo, a KyiApso. 123
6.1. Example of gigantism in a Tibetan mastiff. 143
6.2. KyiApso in the Lhasa dog market. 149
6.3. Tibetan mastiff with wool ruff, in the Lhasa dog market. 150
6.4. Tibetan mastiff (golden/cream color) from the Mt Kailash area of Western Tibet, at Gangaling Kennel, Lhasa. 151
6.5. Tibetan mastiff at Gangaling Kennel, Lhasa. 152
6.6. Tibetan hunting dog, early 20th century. 156
6.7. Tibetan nomads with hunting dogs. 157
6.8. Tibetan hunting dog. 158
6.9. 'Shikari' Himalayan mountain dog. 159
6.10. Tibetans with yaks and Tibetan mastiff guardian dog. 160
7.1. Tibetan mastiff by the English painter Edwin Landseer (1802–1873). 170
7.2. HAGAR by Chris Browne. 177

## Tables

Table 1. Passages most often borrowed from Polo about Tibetan dogs. 9
Table 2. Kirkpatrick and plagiarized passages. 17
Table 3. The major distinctions described for each type of Tibetan mastiff. 142

## Maps

Map 1. Nepal and surrounding regions (map by Ramesh Shrestha). vi
Map 2. Nepal (map by Ramesh Shrestha). 46
Map 3. Nepal Himalaya (map by Alton Byers). 75
Map 4. Eastern Dolpa and Mustang Districts, West Nepal (map by Alton Byers). 76
Map 5. Northern Nepal and Southern Tibet (map by Ramesh Shrestha). 134

# Sources of Plates

1.1. After *The Animal in Chinese Art*, The Arts Council of Great Britain and the Oriental Ceramic Society, London (1968, fig. 303).

1.2. Frontispiece in Yule, Henry, *The Book of Marco Polo the Venetian* (translated and edited with notes by Henry Yule, 3rd ed., rev.), London: Henri Cordier (1903). Reprinted in 2 vols. by J. Murray, London (1975).

2.1. Unknown.

2.2. L.A. Waddell, *The Buddhism of Tibet or Lamaism*, 2nd ed., 1934, p.405.

2.3. David Snellgrove and Hugh Richardson, *A Cultural History of Tibet*, 1968, following p.160).

2.4.; 2.5. Sven Hedin, *My Life as an Explorer*, Garden City and New York: Garden City Publ. Co. (1925, pp.496-7.).

2.6. Unknown.

2.7. Ruttledge, Hugh, *Everest: The Unfinished Adventure*, London: Hodder and Stoughton (1937), with permission.

3.1.-3.4., 4.1-4.5, 4.7., 5.4., 5.5., 6.2.-6.5, 6.8., 6.10. The author

4.6. Bruce Morrison.

5.1., 6.6. Irma Bailey, 'Dogs from the roof of the world,' *American Kennel Gazette* (1937, v.54, n.3).

5.2. Old newspaper clipping. Accessed at www.tmcamerica.org (2001).

5.3. Christian Ehrich.

5.6, 5.7. Dan'l Taylor.

5.8. Judy Steffel.

6.1. Li Qian, *China's Tibetan Mastiff*, Beijing: Foreign Language Press (2006).

6.7., 6.9. M.C. Goldstein and C.M. Beall, 1990, *Nomads of Western Tibet*, Berkeley and Los Angeles: University of California Press (1990).

7.1. Unknown.

7.2. Hagar © King Features Syndicate, with permission.

# PREFACE AND ACKNOWLEDGMENTS

*There are no books in Tibet about dogs and even if one does make an attempt, it would be considered silly.*
Jigme Wangchuk Taring, 'Tibetan dogs' (1981)[1]

## A Book of Discovery

I didn't set out to write a book, but the more I learned about big Tibetan dogs and as my admiration and respect for them increased, the more I realized that, silly or serious, everyone loves a good dog story. This is a dog story, a book of discovery, travel and research and of a long personal journey, a life lived in association with some of the biggest dogs in the highest place on earth, the Tibetan and Himalayan 'Roof of the World.'

The written record of large Asian dogs in the West began with Marco Polo in the late thirteenth century CE. As a schoolboy, I was inspired by Marco Polo's *Travels*, but I never saw any of the dogs he mentioned until 1964 when, as a Peace Corps Volunteer in Nepal, I went trekking and mountaineering in the north of the country along the Tibet border. There I met them, face-to-face. I was attacked, but unhurt (*that* time). I was impressed. I was enamored. I was hooked. And, while studying for advanced degrees in Anthropology, I returned to Asia looking for them. For the next four decades, after completing my doctorate and while pursuing other research, teaching, consulting and various writing assignments, I intermittently documented the big dogs in their natural environments and cultural settings; that is, in their places of origin. This book tells the story.

Much that is written about the big dogs of Tibet is based on faulty assumption and myth. This book presents the facts about these dogs, but it also addresses the myths that have grown up about them—especially myths about their origins and their size. In her *Short History of Myth*, Karen Armstrong defines myth as a form of "make-believe". "It is a game", she says, "that transfigures our fragmented, tragic world, and helps us to glimpse new possibilities by asking 'what if?'" More to the point, the anthropologist E.B. Tylor wrote well over a century ago that "Myths of observation are inferences from observed facts, which take the form of positive assertions, and they differ principally from the inductions of modern science in being much more generally crude and erroneous." A myth may go so far, Tylor says, as to become historical tradition and thereby "be all the more puzzling" for the hint of scientific truth that it may contain. This is what puzzles observers—the grain of truth that sometimes overwhelms incontrovertible evidence to the contrary. There are grains of truth in every myth, but the time has come to sort them out about the big Tibetan dogs.

## Dog Lore

There are few more fascinating subjects than Tibetan-Himalayan history and culture, including canine lore. When I began looking for the big dogs in the mountains, I did so as ethnographers have always done when conducting studies in foreign places—on foot, with rucksack, pen, notebook and camera. I soon found the dogs both "good to think or to imagine", as the anthropologist Eugenia Shanklin has put it; and, good to *be* with.[2] In my studies I examined and documented their roles in the cultures, economies, rituals and beliefs of the mountain people. During the 1960s when I began to hear reports that the Chinese were systematically annihilating dogs in Tibet, I was disturbed. I set a goal to find and describe them in their natural setting along the Himalayan-Tibetan borderlands and on the Tibetan plateau. I traveled long and high in my quest and, over time, I met, petted and hugged many big dogs, and was attacked by a few. My goal has always been to understand the big dogs and preserve their legacy.

I have encountered virtually every known breed and type of high Asian dog, and have owned several. One of the first was an intelligent and agile Himalayan mountain dog from Helambu, north of Kathmandu. I named him 'Amjo Gipu' ('Long Ears' in the Sherpa language; though, in retrospect, his ears were not exceptionally long). Some years later, I acquired a black-and-tan Tibetan mastiff named 'Kalu' (née Saipal Baron) from Nepal's preeminent dog authority, Jay N. Singh of Kathmandu. Kalu's heritage was traceable directly back to Tibet through the Khumbu region over the Nangpa La (pass) on his dam's side and to Kyirong, the 'Valley of Happiness' in southwestern Tibet on his sire's side.

My focus is on big dogs, but there are, of course, some very fine and popular small dog breeds of Tibetan origin. Nonetheless, as the contemporary adventurer and writer Michel Peissel reminds us, "The immense popularity in the West of Tibetan lapdogs... makes one forget that Tibet is primarily the home of large guard dogs."[3] This book reflects my long-standing affection for the largest of them, as working dogs, guardians, and companions.

With Kalu I became closely involved in Tibetan mastiff breeding, training and showing, and not least, in writing about them over almost half a century. For awhile a rambunctious, roly-poly Tibetan terrier named Momo (a Tibetan 'dumpling') lived at my house. He too, like Kalu before him, often sat at my feet while I wrote parts of this book. I have also raised some fine companion dogs of mongrel heritage, some with obvious big Tibetan dog genes. Along the way I became a life member of the Nepal Kennel Club, a co-founder and first vice president of the American Tibetan Mastiff Association (ATMA), an honorary member of the American Tibetan KyiApso Club, and an acknowledged authority on the dogs.[4]

## Acknowledgments

Many people have participated in the preparation of this book, with information, hospitality, and other help. They include Tibetans, Nepalese, Bhutanese, Chinese, Indians, Europeans, Australians and Americans. Among them are dog breeders, kennel club members and officers, dog show judges, show dog owners and pet fanciers, historians, cynologists, anthropologists and geographers, livestock and pasture experts, geneticists and physiologists, linguists, Tibetologists and other Himalayan/Tibetan area specialists, livestock herders (nomads) and villagers, Buddhist and Bönpo nuns and monks (including reincarnate lamas) and eminent religious scholars, development workers, literary scholars, librarians, writers, Tibetan partisans (the famous Khampa guerillas), and Tibetan expatriates living far from their homeland.

For *all* of you who have helped in one way or another with your intimate or passing knowledge of the dogs, your supportive or controversial opinions, your impartialities or biases, your pros and cons..., you have *my sincere thanks and appreciation!*

Frankly, I do not recall everyone who helped, but here are those whom I remember (in no particular order): A.J.S. Grewal, Alton Byers, Andrew Manzardo, Ann Rohrer, Bruce Morrison, Camille Richard, Charles Allen, Charlie Radcliffe, Chökyi Nyima (Rinpoche), Chris Cluett, Christian Ehrich, Dan James, Daniel Taylor, Diana Quinn, Dorjé and Lodenla, Ellen Coon, Eric Holliday, Eva Garcia, Gabrielle Gruenwald, Hans Messerschmidt, Hedy Nouc, Himmat Singh Sekhon, Jay N. Singh, Jennifer Ide, Judith Pettigrew, Judy Steffel, Juliette Cunliffe, Kareen Messerschmidt, Karma Wangyal, Kristina Sherling, Lance Hart, Liesl Messerschmidt, Linda Bennett, Linda Smith, Liisa Sarakontu, Marcia Calkowski, Mary Fischer, Melvyn C. Goldstein, Naresh Gurung, Paul Smethurst, Paula Allen, Peter Moran, Phillip Sponenberg, Philippe Touret, Ramesh Shrestha, Richard Eichhorn, Rinchen Tsognyi, Robert B. Ekvall, Robyn Allwright, Ron Bombliss, Saralouise Anderson, Sharon Hepburn, Shyam Tulachan, Susan Elworthy, Ted Riccardi, Thinley Chogyal, Victor Mair, and Wangchuk Dorjé. If you helped out, and are not on the list, please accept my apologies for my bad memory.

I have accessed many secondary sources including a wealth of books old and new, in private collections and in public and international libraries. Some essential sources have been difficult to obtain, and many early travelers' books are not readily available in hard copy outside such austere repositories as the Library of Congress and the British Library. Various book archives on the Internet have been invaluable to me (e.g., books.google.com, amazon.com, gutenberg.com, and others), along with university and community libraries and inter-library loan services. Some writings about the dogs are in languages that I neither read nor speak, and a number of interviews have challenged my comprehension of the Asian languages that I do speak. For some foreign materials I have obtained translations from friends and colleagues. I thank you all.

Note that I have found the World Wide Web (Internet sites and postings) to be a valuable resource *except* when it comes to what appears on *some* Tibetan dog breeders' websites. I have a healthy skepticism for much that professional breeders write about their dogs, primarily because there is so much fiction and fancy (mythology) being perpetrated online in the guise of historical 'fact' and scientific 'truth.' In my opinion, the web is overloaded with romanticized beliefs and myths about Tibetan dogs. I detect a disturbing trend towards blatant exaggeration by some breeders whose inspiration and aspirations appear to me, frankly, to be based on economic self-interest (to sell pups) and for the purpose of bolstering their egos. There are, of course, notable exceptions to my generalizations, and I recognize them. And, ultimately, when the negatives are weighed against all the honest folks out there who own, raise, train, breed, study, write about, love and brag about, are seen walking their fine dogs in the neighborhood park, write about and discuss them on the Internet, and who sometimes sell big Tibetan dogs, then the Tibetan dog fancy seems to right itself. Those who seek what is best for these dogs are to be complimented.

A number of the essays, stories and anecdotes in the book were requested or volunteered by interested colleagues, or reprinted whole or in part from other sources. Soliciting their contributions, however, was not always smooth. When one early contributor resented my suggestions for following the style and format that I set for this book (my editorial prerogative) it was withdrawn in a huff with the demand that I neither quote nor cite anything that individual has written. Wow. Chalk it up to experience. Dog doo happens. That disturbed soul shall remain unnamed, unquoted and unreferenced in these pages. *Woof!*

Despite a few minor setbacks and some long delays, once I had the topics for the book mapped out, I shared drafts with other writers and dog experts. They thoughtfully helped me by pointing out a few errors and questions of interpretation, and suggested some points of clarification. Their inputs have been very helpful, though we have not always agreed on some points.

Much of this book was written in Kathmandu where I have lived for many years. It has taken me almost a quarter century to complete it. I began the research in the early 1980s, although some of my observations on the big dogs of high Asia go back to the early 1960s when I first lived and traveled the Himalayas as an American Peace Corps Volunteer. The writing was delayed off and on over the decades while I raised a dog-loving family and worked as an international development consultant, and as a writer and editor, bringing bread and rice to the table and fresh bones and rawhide chews to the dog at my feet. Along the way, I wrote articles and essays on Tibetan mastiffs for such publications as *Dog World* (USA), *Rangelands* magazine, *The Himalayan Times* (Kathmandu), Kathmandu's *ECS Nepal* magazine (for which I served several years as editor), the *ATMA Gazette*, the *TMCA Sentinel* and the *TKC Newsletter* (of the American Tibetan Mastiff Association, the Tibetan Mastiff Club of America, and the Tibetan KyiApso Club, respectively). All my publications about Tibetan dogs are listed in the Bibliography.

Lastly, the reader must be aware that all that I have written in this book and anything I have left out or ignored is my responsibility alone—true or false, 'spot on' or in error, fact or fancy.

## A Note on Sources and Terminology

I have made every effort to credit the sources of all quotations and illustrations in the book, including those in the Annexes. In some instances, however, my inquiries seeking permission to use them were met with total silence. Some sources simply eluded me despite many attempts at making contact. For the most part, however, I have received the proper permission to use specific materials. If, in the end, however, a photographer, author, or copyright holder remains unacknowledged, I apologize for the oversight.

Note that the page numbers given in some references and citations may vary by the edition, version or translation of the work cited.

Regarding the use of foreign terms (in Tibetan, Nepali and other languages), for the most part I have presented them as they are pronounced by English speakers, not necessarily as scholars might prefer to spell them (with a few exceptions in direct quotations).

# 1

# Forget Marco Polo!*

Every school child knows the story. It comes down to us across seven centuries that a young Venetian lad named Marco Polo traveled to China with his father and an uncle, and spent over two decades as an honored guest and sometimes personal emissary of the Great Khan, Kublai, the grandson of Genghis Khan. Later, when Polo published the account of his travels, he described dogs in Tibet as large as asses (or donkeys, depending on the translation).

When I was boy, I remember reading Polo's *Travels* with awe and wonder. I don't remember which of the many translations it was, but I know that it inspired me. It still does. It caught my fancy then, and again years later when I met big dogs in high Himalayan pastures for the first time. Those dogs and Polo's story ultimately encouraged me to study the place of dogs in the cultures of Tibet and the Himalayas, and this book is the result of that quest. My own tale of discovery begins with Marco Polo.

## Introduction

Marco Polo was born in 1254 in Korčula, a town in Venetian Dalmatia (now Croatia) on the Mediterranean coast. In those days, Venice was "a kind of gateway to the world, you could say," as Elizabeth Kostova has put it.[1] It was a gateway that Polo threw wide open for his generation, with repercussions lasting down to the present time. Marco, his father Nicolo and his uncle Maffeo Polo were fortunate in their timing, for trade between the Mediterranean and East Asia flourished during their century as never before. The century beginning with 1245 CE was a time that historians call the "Mongol Peace," a period of openness for overland travel and commerce by Europeans.

Polo lived a long and venturesome life (seven decades) before he died in Venice in 1324. The stories of his journeys along the Silk Road through Persia to Cathay, through some or all of the countries now known as Iran, Iraq, Afghanistan, Pakistan, India, China and Mongolia, are available to us in many versions and translations. The first edition of his original writings,

---

*A shorter, earlier version of this chapter was published as 'Big Dogs of Tibet and «The Marko Polo Effect»'in *PADS Newsletter* (2009, No.17), from the *Proceedings of the First International Conference on 'Aboriginal Dog Breeds as Part of Biodiversity and of the Cultural Heritage of Humankind'* held in Almaty, Kazakhstan, September 10-15, 2007, sponsored by the Russian Branch of the Primitive and Aboriginal Dog Society (PADS).

often referred to simply as Polo's *Travels*, was published in 1298 as *Le divisament dou monde* (*The Description of the World*). It was written in Old French, the language of commerce in his day. Between 1310 and 1320, Polo reworked the book and reissued it in Italian.

The text of Polo's *Travels* was first dictated to Rustichello da Pisa, who heard the fascinating story of Polo's travels to Cathay while the two were incarcerated together in a prison in Genoa. Rustichello was a thirteenth-century romance writer; thus, the book has been described as "a product of an observant merchant and a professional romancer."[2] The Italian name of the book, *Il Milione* ('*The Million*'), reflects the Polo family nickname, '*Emilione,*' and Marco's personal nickname: '*Messer Marco Milione.*' Because Marco and his father and uncle had sewn gemstones up in their clothing for safekeeping on the long journey home from the East, they were considered to be "millionaires."[3] Other editions of the book (in translation) have followed over the centuries. The earliest that we have access to today is the so-called *Z Codex*, dated 1400 CE.[4]

In its day, Polo's *Travels* was a popular story. It was written for a European audience that knew little or nothing of the Orient. It was read by the educated classes, translated into many languages, printed and distributed widely, an immense achievement given that the printing press was not invented until 1440, over a century after Polo's death. The significance of Polo's book down through the centuries inspired one translator and editor, Henry Yule, to call it one of the greatest books of all time.

## Dogs "the size of asses" and "mastiffs as big as donkeys"

In a short chapter 'Of the Province of Thebeth' (Tibet), Polo briefly mentions big dogs that were used to hunt wild yak and musk deer. He describes them (in some translations) as "dogs the size of asses" and (in others) as "mastiffs as big as donkeys." We do not know for sure, however, if they or *any* of the dogs he mentions in the book were what we call today "Tibetan mastiffs." Nonetheless, many mastiff and molossus dog historians, breeders and fanciers of today refer uncritically to Polo's account as authoritative.

In a different chapter, Polo describes (in translation) the "mastiff dogs" used in large hunting parties of 10,000 men and 5000 dogs, in pursuit of various game animals. "It is indeed beautiful," he says, "to see the speed of these dogs and the hunters, for when the prince goes out with his barons, boars and other animals are running on every side, and the dogs pursuing."[5] There is no indication that these particular dogs were of *Tibetan* origin or anything like what are called Tibetan mastiffs.

And in yet another part of the book, he mentions a large breed of dog encountered in "Cuiju Province"—dogs "so fierce and bold that two of them together will attack a lion." Tibetan mastiff fanciers sometimes cite this passage together with the comparison of dogs to asses or donkeys to justify a modern quest to produce a breed of giant *super*-mastiffs for show and for sale. Once again, it is impossible from Polo's description to tell the breed of dog he is referring to. It may well *not* have been a Tibetan mastiff, if for no other reason than where he was talking about. The western province that Polo calls Cuiju is, in fact, Guizhou (sometimes spelled Kwei-chau), situated east of Yunnan, bordering Hunan. That's a long ways from both physical and ethnographic Tibet, and so low in elevation that highland dogs would have had difficulty thriving or surviving.[6]

Problems of interpretation arise when one assumes something other than the facts. In their otherwise excellent book, *The Tibetan Mastiff* (1989), for example, Ann Rohrer and Cathy Flamholtz mistakenly identify Kwei-chau (Cuiju/Guizhou) as Szechuan, a mountainous province bordering southeastern Tibet. Then they rather wishfully assume that this was where "Polo first encountered Tibetan traders and their dogs,"[7] further promoting the belief in Polo as some sort of early expert on big Tibetan dogs.

This is not the only time that Polo is misrepresented in writings about Tibetan dogs. Such errors occur and are repeated frequently in books and articles, and on the Internet. I call this phenomenon '*The Marco Polo Effect*'—the repetitive misinterpretation of what Polo reputedly saw or did, where he went (or not) and what he said or meant. The Marco Polo Effect holds true *any* time someone quotes, misquotes or repeats any source from the past, or a translation of a source, or another translation of that, with or without attribution, in order to establish the putative history of something (such as a dog breed). It is part of the process of myth-making—Karen Armstrong's make-believe, E.B. Tylor's "crude and erroneous" inferences and puzzles—based on assumptions 'said to be true,' then using them to support or promote questionable propositions.[8]

## Forget Marco Polo!

If you have read anything about Tibetan mastiffs you have probably encountered some version of Polo's famous 'first' description of them as big as donkeys. With that as a starter in my quest to document the dogs in their native habitat, I have tried to ascertain how much Marco Polo *really knew* about the dogs of Tibet, if he *really saw* the same dogs that we know today as Tibetan mastiffs, and *how accurate* modern accounts are when referring to them based on his travelogue. It turns out that a good deal of what Polo wrote about big dogs (and other subjects) is puzzling, confused and shrouded in mystery. In the end, it is my opinion that much of what is attributed today to dear Marco by big dog aficionados suffers from the Marco Polo Effect—as uncritically exaggerated and grossly misrepresented.

If that's the case, should we forget him? Not easily. To abandon the first European to write about the big dogs of Tibet is both radical and controversial. Tibetan mastiff and other large breed fanciers have long looked to Marco Polo for inspiration about their dogs. It makes a great story. But we should be very careful, for much of what Polo described is probably not about Tibetan mastiffs *per se*. Consider this:

How much did Marco Polo write about the big dogs of Tibet?
— *Very little; one sentence.*

Did he actually visit the high plateau of Tibet proper?
— *All indications are that he did not.*

Did he observe Tibetan mastiffs serving their masters as livestock guardians or as monastery compound guards?
— *Probably not; there's no evidence.*

Were they the size of donkeys?
— *No, for reasons explained below.*

Did he really mean that the thousands of dogs he describes traveling with Kublai Khan's troops and hunting parties were 'Tibetan mastiffs' as we know them?
— *Unlikely.*

Did he call *any* dogs "mastiffs"
— *No.*

Does he deserve credit for being the first European to introduce us to the big dogs of Tibet?
— *Yes. Undoubtedly.*

For centuries, the legendary big dogs of Tibet and the Himalayas have been described with awe and apprehension by adventurers and explorers, mountaineers and missionaries, anthropologists and tourists, diplomats and spies, in journals and travelogues, encyclopedias, historical tomes, dog books and magazine articles. The dogs that have come to be known as Tibetan mastiffs were known for their size and sound, protective instincts and guardianship, faithfulness and ferocity. We also know that there is more than one sort of big dog on the so-called 'Roof of the World,' including livestock guardians and hunting dogs. We know that some of them, especially the big guardian dogs, invariably leave a strong impression (and sometimes scars) on those who encounter them.

While Marco was the first to write about them, the Polos were not the first to hear about or to have 'discovered' the big dogs. They are preceded in the literature by occasional and obscure references to several earlier encounters. One speculative suggestion dates to the Han Dynasty (206 BCE to 221 CE) in the form of a glazed clay image of a dog, usually interpreted as a tomb guard, which some think "looks like" a Tibetan mastiff.[9] A more well-known reference is to an often cited Chinese chronicler's account from 1121 BCE that tells of a large Tibetan dog "called the 'Ngao'...," given as a gift

# 1. Forget Marco Polo!

from a western province to the Emperor Wuwang (or Wou-Wang).[10]

Nor were the Polos the first adventurous travelers from the West to travel the Silk Road and other routes to the East. Arab traders had gone to the East by sea for at least a thousand years, and Mongols, Turks and others, including European merchants, monks and friars had also traveled the overland route to and from China well before the Polos. Between the seventh and eleventh centuries CE, Jewish merchants called Radhanites traded in Asia. In the twelfth century Rabbi Benjamin ben Jonah of Tudela went east from Navarre as far as China and later described his travels in his *Book of Travels*. And, in the mid-thirteenth century, a few years before Polo, Pope Innocent IV sent a Franciscan monk, Plano Carpini, as a peace envoy to the Mongols. He was followed a few years later by Friar William of Rubruck. There were others, but we know little about them or about the dogs that they undoubtedly encountered.[11]

Less than a quarter century after Polo one Odoric of Pordenone (Italy) (1286?–1331) went east by sea, visiting what is now India and Indonesia before landing in China in 1324. Odoric stayed in China for three years where he met various European missionaries of the Franciscan order. In 1327 he set out to return overland by way of Tibet, and may have been the first European to visit Lhasa (though this is disputed by some scholars). Odoric left no formal written record, but the substance of his travels and observations in East Asia were incorporated into the works of other writers of his time. Of dogs, however, he wrote nothing of significance.

A few years later, Ibn Battuta, a Moroccan Muslim whose full name was Muhammad ibn 'Abd Allah Ibn Battuta (or Batuta, Battutah) also went east as far as China. His travels lasted thirty years and covered 75,000 miles on three continents, for which he has been described as "one of the greatest travelers of the Middle Ages, perhaps of all ages." One would expect to find reference to the dogs he encountered; but, alas, as a Muslim, he was averse to dogs and mentions them only rarely in his book, such as when he notes they were eaten by the Chinese.[12]

Some of Polo's contemporaries, predecessors and followers had vivid imaginations

Plate 1.1. Tomb guard dog (Tibetan mastiff?), clay figurine from the Han Dynasty.

and wrote accounts that were both fanciful and vague; but big dogs tend not to be mentioned. Ancient geographers were more interested in exotic places and peoples. Traders focused their attentions on economic goods and markets. Missionaries wrote about pagans and spreading Christianity. Dogs were not typically in their sights nor on their itineraries, though some were certainly at their heels at times.[13]

It was left to Marco Polo, almost by default, to be 'the first' to tell the outside world about the big dogs of the East.

## What Are We to Believe?[14]

Throughout the literature on Tibetan mastiffs, it is commonplace to refer to something ostensibly said by Marco Polo without paying much attention to its veracity. Is that good enough? If so, read no more; skip ahead to the next chapter. If not, then consider this—Polo's writings have been used and abused for a very long time to shore up questionable claims about the origin, history, functions, size, sound and temperament of Tibetan mastiffs. The practice has even jumped breeds, and occurs in the literature on other so-called 'mastiff-type' dogs, including Wolfhounds and Saint Bernards.

I am not saying that some of the quotations from Polo that we often see are not found in one or another translation of his book, but rather that they are largely misinterpreted, often contradictory and frequently repeated imperfectly, unquestioningly and *ad nauseam* by contemporary armchair writers without checking, without proof. At this point I am reminded of what that great sleuth, Agatha Christie's Miss Marple, once said about truth seeking: "The truth is, you see, that most people... are far too trusting... They believe what is told them. I never do. I'm afraid I always like to prove a thing for myself."[15]

The 'truth' of the matter, the proof, is undoubtedly buried in two of the earliest Polo manuscripts, written in Old French and Italian. The problem is that we do not have the originals to fall back on, and subsequent versions, translations and translations of translations, in English (and other European languages), tend to suffer from being "refracted through the prism of translation," to borrow the writer Doug Brown's phrase.[16]

The following examples are my take on how Polo has been misappropriated. Some are minor; others change the whole meaning of what he had to say. Each describes an attribute of the big dogs of Tibet—but *which* big dogs?

### On their size

When reading the many references to Marco Polo in print and on the Internet, he is typically attributed with describing dogs *the size of...*, *as tall as...*, *as large as...*, *as big as...*—*a donkey* or *donkeys*, or *an ass* or *asses...* (depending on the translation). Furthermore, he is also either quoted or paraphrased as having *encountered..., recorded..., described..., claimed to have seen..., happened upon..., told of seeing..., wrote of seeing..., came across..., met with..., was introduced to...,* or *wrote about...*—Tibetan mastiffs.[17]

*Which is it?* The term 'Tibetan mastiff' (or an equivalent term in Italian or Old French) was unknown to Polo; to him the canines of which he wrote were simply large dogs (but with no sure dimensions). The term 'Tibetan mastiff' was not popularized until five centuries later, in the 1800s, when this inappropriate moniker was rather blithely inserted into the English language discourse to describe big Tibetan dogs.[18]

It is fine to use descriptive terms, but some of the words chosen today to describe what Marco Polo wrote about long ago are very different from the meaning that translators have gotten out of the early versions of his famous travel book. Nowhere does he actually say he *saw* the dogs *in Tibet* (nor Tibetan wild asses or *kiang*, nor tame donkeys or asses in Tibet with which to compare them). As early as 1880 Thomas Wright accused Polo of hyperbole over the donkey/ass size comparison, noting that while "a few other (more recent) travelers" mention large dogs, their accounts "do not convey an idea of the same magnitude." Even the great nineteenth century mastiff expert, M.B. Wynn, questioned Polo's reference to asses, which he took to mean "the smaller breeds of the domestic ass." Summing up, a recent but anonymous observer on the Internet has recently said "Surely, the breed was never quite *that* large."[19]

*How* large? Compared to *what?* Some breeders and Tibetan mastiff fanciers interpret Polo to be comparing the dogs with the small domestic donkey, the *Equus asinus* of Tibet. Others think he meant the large Tibetan wild ass, *Equus hemionus kiang.*

*Which is it?* Probably *neither,* since there is no conclusive evidence that he ever saw Tibetan donkeys or wild asses with which to make such a comparison. He was very familiar, however, with the relatively small domesticated burro (also called 'ass,' *Equus asinus*) common in Mediterranean countries. They were more likely his reference point. *Why?* Because he was, after all, writing for European readers who knew nothing about Tibetan donkeys or asses, domestic or wild, but a lot about those little donkey/ass critters close to home.[20]

### On their loud voice

Polo's statement about the dog of Tibet "as tall as a donkey" is often appended "with a voice like a lion" or "with a voice as powerful as a lion."

*Where did that lion voice come from?* It's a puzzle, twice over. For one, although the Tibetan mastiff does have a uniquely loud voice or bark that some may think *sounds* lion-like, the comparison to a "voice like a lion" does not appear in Polo's chapter 'On the Province of Thebeth.' Furthermore, there are no lions (*leone* to Polo, in Italian) in Tibet or China (other than, perhaps, gifts from abroad to some royal menagerie). This fact makes the lion-voice comparison spurious. There is a statement elsewhere in Polo's *Travels*, however, that some big dogs are strong enough to hunt lions. Some current accounts rather loosely splice the two comments together to *sound* like Tibetan mastiffs.

Thomas Wright correctly notes that there are "imaginary and grotesque" representations of lion statuary in China, undoubtedly borrowed from the *Singha*, or 'lion,' of India's Hindu mythology. Wright also got it right when he wrote: "when our author [Polo] speaks of lions in China, as living animals, he undoubtedly means tigers." Even this is puzzling, as tigers do not roar like lions, nor are they exceptionally loud in any other way. The most parsimonious explanation is that Polo (or, rather, his translators) used the term "lion" simply because it was *more familiar to his European audience* than the Asian tiger.[21]

Consider now the passages most often borrowed (and distorted) from Polo about Tibetan dogs. Compare the following excerpts (Table 1, opposite) from three of the most respected and authentic translations (and note how little Polo actually says about dogs):[22]

### *On their primary function(s)*

The various terms for the canines mentioned here and there in Polo's *Travels* are translated as "messenger dogs," "sentry dogs" and "dogs of war" (in Kublai Khan's army). It is remarkable that *nowhere* does Polo describe true *guardian dogs*, 'tied dogs' (*do-kyi*), 'door dogs' (*go-kyi*) or 'loose dogs' (*yun-kyi*) used by Tibetan nomads to protect livestock or to guard monastery compounds. References to their specific functions as *messenger, sentry* or *war dogs* are derived from his sketchy discussion of the Khan's army found elsewhere in Polo's travelogue, quite apart from his description of Tibet. He undoubtedly exaggerated their roles with the army for effect, since much writing about armies then and now is intended more to impress than to inform. (Did Kublai Khan's army really include a "contingent of 30,000 Tibetan Mastiffs," as one recent writer claims?)[23] It is highly unlikely that dogs with the bulk of Tibetan mastiffs, as we know them, could have functioned well as "dogs of war," as they would have had great difficulty keeping up with the Khan's mounted soldiers over long distances. The most reasonable Tibetan mastiff function on the list is that of sentry.

In retrospect, Polo may have meant some other large, more lithe, faster, short-haired dog, of which there are several types in the territories in and around the high regions of central and south central Asia and the Himalayas. Had Polo, in fact, *met or observed* Tibetan nomads on the high plateau, first hand—if he had actually *seen* the big dogs in action—he would certainly have described them more correctly as rather fierce and territorial livestock guardians, for which landrace Tibetan mastiffs are primarily bred and kept.

**From Thomas Wright (1880):**

Here are found the animals that produce the musk, and such is the quantity, that the scent of it is diffused over the whole country... They are called *gudderi*\* in the language of the natives, and are taken with dogs...

**They have dogs of the size of asses, strong enough to hunt all sorts of wild beasts, particularly the wild oxen**, which are called *beyamini*,\* and are extremely large and fierce...

**From Henry Yule (1903):**

...in this country there are many of the animals that produce musk, which are called in the Tartar language Gudderi. Those rascals have great numbers of large and fine dogs, which are of great service in catching the musk-beasts, and so they procure great abundance of musk...

**They have mastiff dogs as big as donkeys, which are capital at seizing wild beasts...** They have also sundry other kinds of sporting dogs...

**From Ronald Latham (1958):**

...The country abounds with animals that produce musk, which in their language are called *gudderi*. They are so plentiful that you can smell musk everywhere... The rascally natives have many excellent dogs, who catch great number of these animals; so they have no lack of musk...

**They have mastiffs as big as donkeys, very good at pulling down game, including wild cattle, which are plentiful here and of great size and ferocity.** They also have a great variety of other hunting dogs, besides excellent lanner and saker falcons, good fliers and apt for hawking...

---

\* *gudderi* refers to the musk deer, and *beyamini* refers to the yak

Table 1. *Passages most often borrowed from Polo about Tibetan dogs.*

### On their fearlessness

Tibetan mastiffs are often described as hunters of all sorts of "wild beasts" including "oxen" (*beyamini,* yak), "animals that produce musk" (*gudderi,* musk deer), and "lions" (*leone,* lion or tiger). There are those mythical lions again!

There is no doubt that the big dogs Polo heard about were both fearsome and fearless. While there are big dogs in Tibet and elsewhere in China and Mongolia that *are* used for hunting, Tibetan mastiffs only rarely (I won't say never) serve this purpose, for they are neither light enough nor fast enough to chase down big game on the open plateau as a normal routine. The usual hunting and sporting dog types of Tibet are quite different (see Plates 6.6-6.8, pp.156-8). If he was talking about the latter, which he probably was, then this statement does not belong in a description of Tibetan mastiffs. Landrace Tibetan mastiffs are only occasionally used for hunting, according to scattered accounts. In those cases they most often assist nomads tangentially during a hunt. Robert Ekvall describes them as "accomplices in the subsidiary subsistence techniques of hunting." Being an accomplice in a subsidiary activity implies a partner, helper or associate.[24]

In Henry Yule's translation of the section in Polo's *Travels* about the various wonders of Cuiju Province (Guizhou or Kwei-chau) "a large breed of dogs" is described "so fierce and bold that two of them together will attack a lion."[25] As already noted these, too, are not likely to have been Tibetan mastiffs, for Guizhou is a long ways from the high plateau, far to the southeast of Tibet between Yunnan and Hunan. Furthermore, there were no lions in Tibet or China; the nearest Asiatic lions (*Panthera leo persica*) were far away in southwest Asia, and once ranged from India to Macedonia. (In Asia today, lions are limited to a few hundred animals in Gir Forest National Park in Gujarat, in western India.) As noted earlier, I suspect Marco Polo used the image of the lion because it was an animal familiar to his European readers.

### On their companionship

The companionship of Tibetan mastiffs is a common theme, and more than one breeder and aficionado believes that Polo was so taken with the breed that he had his own Tibetan mastiff to accompany him on his travels. This spurious assumption is based on an illustration printed in Henry Yule's 1903 translation of Polo's *Travels*. It shows the three Polos approaching the gates of Venice after twenty-six years absence. (See Plate 1.2, p.11) On arrival, they were initially refused admittance to the family mansion because they were unrecognizable and had been presumed dead. The Polos in the illustration are shown dressed in Tartar clothing. In the foreground is a large black dog on a leash that looks remarkably like a Tibetan mastiff. Many people apparently assume that this *proves* that Marco Polo brought one back to Venice with him.

*Plate 1.2. Messers Marco Polo, Nicolo Polo and Maffeo Polo, on return to Venice after 26 years' absence. This illustration was used as the Frontispiece to Yule's 1903 translation of Polo's* Travels, *as instructed by the book's editor, Cordier, to the illustrator, Signor Quinto Cenni, of Milan, Italy. The image showing a big dog like a Tibetan mastiff was created for the 1903 book, and is purely the result of editorial and illustrator license.*

But, *it doesn't!* Check the facts.

First, the full story of the Polos' return to Venice was not recorded until the sixteenth century, by Giovanni Battista Ramusio. (It is not in Polo's *Travels*, except as inserted in translators' notes long after the fact.) Ramusio based his account on a story handed down by family members. There is *no dog* in what Ramusio wrote about their arrival.

Second, on further researching the subject it is apparent that the sketch is comparatively recent. It does not date to Marco Polo's time. It was created on commission by an Italian artist, Signor Quinto Cenni of Milan, for the frontispiece to the 1903 Yule-Cordier edition of *The Book of Marco Polo the Venetian*, from a design drawn by the editor (Henri Cordier). The commission is described by Miss Amy Frances Yule, daughter of Henry Yule, in her Preface to the book. By the late 1800s, when Yule was working on his translation, Tibetan mastiffs were well known in Europe. Adding one to this renowned edition of the book was a nice touch.

Even the Tibetologist Robert Ekvall, in his otherwise authoritative 1963 article on the 'Role of the dog in Tibetan nomadic society' (reprinted in Appendix 2, pp.208-17), got it wrong when he states in a note that the illustration showing "the Tibetan mastiff which Marco Polo brought to Italy with him" was taken from the *original* Italian edition. Unlike Ekvall, and various adventurers in Tibet such as Sven Hedin, Andrew Wilson, Himalayan mountaineers, and other nineteenth and twentieth century travelers who adopted local dogs and wrote fondly of their companionship, Marco Polo *never mentions one as* his traveling companion.[26]

### And other stupendous features

One exaggerated account of the Tibetan mastiff ostensibly found "in the travel notes of Marco Polo" states that this big dog was "as sturdy as black bear, as agile as leopard and as clever as huntsman"...! *Where* did Marco Polo say this?

In another account, which speaks of "people [who] cannot tell true stories from myths," it is written that "it is not a make-up [to say] that a Tibetan mastiff can defeat three wolves." And more: that "People once even saw a two-meter long tiger-like Tibetan mastiff in a Tibetan pasturing area."

All of this is blatant hyperbole, and none of it is from Marco Polo.[27]

## More Questions to Marco Polo

If, by now, you haven't begun to doubt even a little of Marco Polo's very brief and minor role in defining the Tibetan mastiff all those centuries ago, here are some more puzzles to ponder.

### Did Marco Polo go to Tibet?

I have already indicated that he probably did not. Scholars generally agree that Polo may have gotten *close* to the Tibetan plateau during his travels in western China, but that he did not visit Tibet proper. Polo's description of Tibet, its culture and its big dogs is exceedingly thin and undoubtedly based on hearsay, probably from traders, pilgrims or other travelers he met elsewhere in the Khan's China.

Some scholars even question if Polo even reached China! It has been pointed out, for example, that much of Polo's vocabulary is Persian, not Chinese. This has been taken to imply that he may not have traveled farther east than Persia. Others have noted that Polo omits describing certain Chinese cultural items and practices that surely would have intrigued him and would prove he had actually been there—e.g., tea, chopsticks, the custom of foot-binding and the Great Wall. None of the claims *against* his traveling as far as China is conclusive, however, and for every argument against it there are counter arguments *for* it. Polo's staunch defenders point out that there are good reasons why he deigned not to describe some things. They also point out that he mentions many other thoroughly Chinese inventions and socio-cultural peculiarities, conclusively demonstrating an intimate knowledge of the place and the culture.

Besides Persian, other languages also come into Polo's account. Why, then, if he traveled through western China to Tibet, doesn't he use Tibetan names for the animals he mentions in association with the big dogs there? For example, Polo refers to yak as *beyamini,* an obvious corruption of the term *brahmini* or *brahminy* ('bull') from Sanskrit and Hindi. The Tibetan term is *yak*. For musk deer he uses the term *gudderi,* which Henry Yule interprets to be derived from the Mongol term *kuderi,* for musk deer. The Tibetan term is *lau* or *lawa*.[28]

Part of the answer to this puzzle is addressed by Stephen Haw in his definitive narrative in support of Marco Polo having gone to China. Haw points out that the Tibetan country that Polo claims to have penetrated to a distance of five days' walk was not Tibet at all, but parts of western China far to the east of the high plateau, at a considerably lower elevation.[29]

Nor is Polo mentioned in any official Chinese accounts from Kublai Khan's court. The Chinese were typically quite thorough about keeping records of people with the reputed fame and stature of Marco Polo. Recent scholarship suggests, however, that during the fifteenth century official Chinese records of knowledge about or relations with outsiders were systematically destroyed by the xenophobic leaders of the time.[30]

### *Did Marco Polo exaggerate?*
Most likely, yes. His descriptions of some places, cultural practices and strange things seen or heard about are often distorted by superlatives (even his own nickname: 'Messer Marco Milione,' the Venetian 'millionaire'). Paul Smethurst, a Polo expert, reports that the family mansion in Venice was known as the Corte del Milione, which suggests, he says, that Polo was prone to exaggeration, to "talking big."[31]

*If so, why?* Scholars point out that books of his day were often "misleading and sentimental." Popular travel writing pandered to the expectations of audiences who delighted in reading about faraway places, strange and marvelous creatures and customs and other wonders to behold. Marco Polo followed the literary conventions of his time, though to a lesser degree than some of his contemporaries. He set out to both inform and entertain, so much so that, as one Polo expert has said, his "delight in the diversity of nature and social customs...is palpable... It is peppered with wonders and marvels, with fabulous cities, Pygmies, exotic plants and birds, ornate palaces, wild beasts, vast deserts, rivers of gems, beautiful women, fine silks, spices, miracles, legends, cannibalism, matriarchal societies, and bizarre marriage customs and funeral rites."[32]

The exuberance, exaggeration and hyperbole of Polo's *Travels* moved Henry Yule to call it "a great book of puzzles."[33]

*Can we trust the translations of Polo?*
Not comfortably. Contemporary references to Marco Polo are largely based on copies and translations, and translations of translations. These include each translator's own interpretations on various subjects, and the introduction into the translations of terms familiar to their readers, but which Polo never used (such as "Tibetan mastiff" and "donkey"). As Paul Smethurst has noted, "There are no 'original' copies as such—all existing manuscripts in various libraries and museums across the world are of copies and translations."[34] Because these are all that the many translators have had to rely on, we must wonder about their veracity.

*Do we sometimes seek what we want to hear?*
Undoubtedly. Yes. It is human nature to find what we *want* to see and hear in various accounts, then pass it along to impress others, pursuing myth and fancy and ignoring facts to the contrary. Inflated egos and economics often seem to get in the way of caution. Italo Calvino says it well in his postmodernist novelistic account of Marco Polo's interviews with Kublai Khan, where he writes the following truism as part of an imaginary conversation: "I speak and speak," Marco says, "but the listener retains only the words he is expecting... It is not the voice that commands the story: It is the ear."[35]

Samuel Johnson, that great man of eighteenth-century English literature, attributed this trait in writers to negligence rather than deliberate fabrication, according to James Boswell, his equally famous biographer.[36]

In short, we often read and write and hear what we are looking for, despite or in ignorance of the truth.

*Misusing Polo to prove a point*
There is one more example of the Marco Polo Effect where Polo's observations have been used to 'prove' that Tibetan mastiffs are true molosser canines. There are many Tibetan dog aficionados (on the Internet and elsewhere) who claim that the Tibetan mastiff is a molosser, a breed of stalwart hunting dog associated with ancient Greece and Assyria and vicinity. Molosser advocates, themselves, have difficulty defining what a molosser is, however,[37] and the notion that the Tibetan mastiff is a molosser has recently been put down as spurious by the Norwegian breeder and dog origin researcher, Kåre Konradsen (and others). The idea of a molosser connection comes from that fact that some writers compare Polo's descriptions of the big Tibetan breed with the famous Assyrian bas reliefs of molossus-type dogs. The prominent nineteenth-century expert on the history of the European mastiff, M.B. Wynn, for example, rather too casually connects what he calls the "Asiatic mastiff" of "Thibet" to those Assyrian images.[38] Konradsen calls this "The Tibetan Theory" then rather irreverently and in the true fashion of the dogged cynic he writes that the "Tibetan theory advocates

claim that the dog on the plate from Niniveh, dated about 580 BCE, is a Tibetan Mastiff brought down from the mountains, just because it has a bushy tail carried like the modern Tibetan Mastiff. Yeah, and the moon is a cheese, because it look like a cheese. What about the other bas reliefs? Perfectly Mastiff typed dogs without a bushy tail carried over the back? Tibetan Mastiff mutants?"[39]

Konradsen's discussion also scotches the notion that Tibetan mastiffs accompanied Alexander the Great's army across Persia to the Indus River (though molossus-type dogs from Alexander's homeland probably did). Simple logic should have put this notion to rest long ago, if for no other reason than the region from Persia to India (crossing present day Iran, Afghanistan and Pakistan) is far too hot for Tibetan mastiffs to survive.

## The 'Nepaul Dog' and Yet Another Historic 'Effect'

Marco Polo is not the only author to have been so carelessly rubbished in print. It happens often, especially since the rise of the Internet and its uncritical and encapsulating effect on modern communications (sometimes called the 'Wiki Effect,' after Wikipedia, the name of the Internet's free content collaborative public encyclopedia, at www.wikipedia.org).

Polo-like 'effects' are also found with the translations and replications of other old writings. Even the Bible has not been spared.[40] The misappropriation of an author's intent may occur any time quotations are extracted and used in other writings without considering how true they are to the original. Translators and others who borrow important passages from previous writings are often guided by more recent or modern agendas and expectations. (It has even happened to me.)[41] Hence, it seems a bit dodgy to consider much of what we read in translation of older works, particularly those on important historical, social, philosophical or theological themes, to be entirely faithful to the author's intent.[42]

On the theme of big Himalayan/Tibetan dogs, another intriguing case comes out of writings that date back over 150 years ago. The misuse here comes from two mid-nineteenth century books that plagiarize and distort the earlier work by Colonel William Kirkpatrick, specifically his description of the so-called 'Nepaul dog.' (I am tempted to call what happened next: 'The Kirkpatrick Effect.')

Col Kirkpatrick was a British officer with the East India Company. In 1793 he trekked overland into Nepal on a brief visit, followed eighteen years later by publication of his *An Account of the Kingdom of Nepaul: Being the Substance of Observations Made During a Mission to that Country in the Year 1793* (1811). In it he describes many facets of Nepalese life, including descriptions of the forests of the Terai lowlands, Himalayan cattle, and the 'Nepaul dog.' Within a few decades Kirkpatrick's book was out of print and scarce, a fact that may have encouraged subsequent travelers to Nepal to publish their own accounts of the kingdom while leaning heavily on Kirkpatrick's original observations. In July 1852, for example, an articled entitled 'Nepaul' appeared in *Blackwood's Edinburgh Magazine* (of Scotland). There, the magazine's editors begin by expressing delight in describing Kirkpatrick's writing as "antiquated and cumbrous in form." But, then, they register their disgust at the blatant plagiarism of it by two British colonial officers: Orfeur Cavenagh in *Rough Notes of the State of Nepal* (1851) and Thomas Smith in *Narrative of a Five Years' Residence at Nepaul* (1852).[43]

According to the Blackwoods, both Cavenagh and Smith borrowed heavily from Kirkpatrick, even rewriting passages from Kirkpatrick's account as if their own—acts of misappropriation that the Blackwoods describe as "astounding examples of impudence in print," "clumsy" appropriations, and "ridiculous blunders." Thomas Smith, for example, wrote about "enormous fruit-trees which are to be found in the Terai" although his source (Kirkpatrick) only mentioned the lowland "forests" (no fruit-trees). Both Cavenagh and Smith also bungled Kirkpatrick's distinction between Himalayan and English cattle. But it was their rewriting of Kirkpatrick's reference to the "Nepaul dog" as "a native of upper and lower Tibets" (*sic*), which he described as "about the size of an English bull-dog," that caused the Blackwoods to express utter exasperation (complete with the parenthetical remark), in this statement: "Nepaul cows discussed and dismissed, we pass on to dogs, and find (this is really too bad) the rival captains, Cavenagh and Smith, both helping themselves at the expense of defunct Kirkpatrick."

Here (p.17) is the original passage from Kirkpatrick (on the left), alongside the two plagiarized versions (each with original spellings, which sometimes differ: e.g., Nepaul *vs*. Nepal and Thibet *vs*. Tibet). Neither Cavenagh nor Smith bother to cite Kirkpatrick as their source:

The only kind words the Blackwoods have about these two brazen word thieves is this comment about Smith's comparison of the Nepaul dog to the English Newfoundland. "Here we find Smith coming out victoriously with an original idea," the Blackwoods say. "Having, we may suppose, during his five years' residence in Nepaul, had frequent opportunities of contemplating the canine species in all the various phases of their interesting existence, he ventures authoritatively to correct the portrait sketched by Kirkpatrick, and copied by Cavenagh. The Nepaul dog does bear greater resemblance to a badly bred Newfoundland, both in appearance and size, than to a bull dog. So, for once, the (otherwise) plagiarist is not only original, but accurate."

The story of Kirkpatrick and of those who leaned so heavily on him (without attribution) is not as heavy as what has been done with Marco Polo's writings. Nonetheless, it is another remarkable example of misappropriating someone else's writings.

| Kirkpatrick's original (1811): | Cavenagh's version (1851): | Smith's version (1852): |
|---|---|---|
| The animal known in Bengal by the name of the Nepaul dog, is, properly speaking, a native of the upper and lower Tibets, from whence they are brought to Nepaul. It is a fierce, surly creature, about the size of an English bull-dog, and covered with thick long hair. | The dog, generally known as the Nepal dog, is also, properly speaking, a native of Thibet. It is a fierce, surly creature, about the size of an English bull-dog, and covered with thick long hair. | This dog, which is known in Bengal by the name of the Nepaul dog, is, properly speaking, a native of the Upper and Lower Tibets, whence it is usually brought to Nepaul. It is a fierce and surly creature, about the size of an English Newfoundland, and covered with thick long hair. |

Table 2. *Kirkpatrick and plagiarized passages.*

## A Reprieve

With all this copy-cat bashing, of people who misuse Marco Polo and others to stake out a position or make a point or claim to understand the history of something (trees, cows, dogs, donkeys, or whatever) or a place (Nepal, Tibet) about which they otherwise know very little, we are reminded that even when we know better, we do not always act better.

*What is it about the subject of dogs that causes some scientists,* including respected historians, anthropologists, geneticists, animal behaviorists, well-established dog breeders, and others, *to fall into the trap of self-deception?*

Dr Gustavo Aguirre, a veterinary scientist at Cornell University, has penned a reasonable answer to the question. Although he is speaking from the perspective of dog genome studies, what he says has a larger meaning to which we can all relate: that even the most scientific among us mess things up at times when we neglect to pause and think carefully and critically about what we know or assume, including what we say or write that is largely speculative. "Most scientists who talk about dogs," Aguirre says, "have their scientist hat and their dumb hat. And whenever they start talking about dogs, they put on their dumb hat. They say things that as scientists they have to know can't possibly be right."[44]

It is to Samuel Clemens (Mark Twain), however, that I give the last word—"*Get your facts first, and then you can distort them as much as you please.*"[45]

That said, we can now proceed to see what other travelers across Tibet and through the Himalayas have written more recently and far more authoritatively than Marco Polo, or Kirkpatrick, Cavenagh and Smith.

# 2

# The Noblest of the Party

Not until five centuries after Polo did descriptions of the big dogs begin to show up again in the writings European travelers in the Himalayas and Tibet. As I set out on my own quest to document the characteristics of these dogs and their roles in the mountain cultures, I scoured the travel literature for descriptions of the dogs. Some of that literature is exceedingly obscure and rare, dating as far back as the early eighteenth century. Many early adventurers described brief but fearsome encounters. Others went at great lengths to extol the dogs' virtues and, in some instances, their companionship. Each account reveals a little more about the temperament and importance of dogs in local life, especially among nomads. In many of the early writings, Marco Polo's brief remark is remembered. For example, the British explorer Laurence Austine Waddell wrote in *Lhasa and Its Mysteries* (1906): "The watch-dogs chained up at the doors of the houses gave us a fierce reception. They are huge Tibetan mastiffs—'the mastiff dogs' of which Marco Polo writes, 'as big as donkeys…'."[1]

## Introduction

Perhaps the earliest mention of big Tibetan dogs in the travel literature since Polo is by the Italian Jesuit Fr Ippolito Desideri who encountered them on his travels in Tibet ("Thibet") between 1712 and 1727. "Many of the Thibettan dogs are uncommon and extraordinary," he begins. "They are black with rather long glossy hair, very big and sturdily built, and their bark is most alarming. One or two are always chained at the entrance of every house, and a stranger would run great risks if no servant came to his help. Merchants who are traveling with many laden animals find two of these dogs sufficient to guard their whole party."[2]

Desideri goes on to say that "These beasts are always well fed, especially with meat to make them strong, and much milk to increase the hoarseness of their bark. They wear large collars ornamented with stiff red fur, so they seem to have a flame round their necks, which added to their natural ferocity increases the fear they inspire."

Half a century later George Bogle saw big dogs while on a trade mission to Tibet in 1774. He describes them as being "of the shepherd breed, the same kind with those called Nepal dogs." And a few years after that, Samuel Turner, a British colonial officer and diplomat encountered huge herds of yaks in the mountains with very large dogs guarding them at night from wolves. And in "Bhootan" he also saw big dogs of a ferocious disposition confined in cages in the courtyard of a Bhutanese royal: "The mansion stood upon the right; on the left was a row of wooden cages, containing a number of huge dogs, tremendously fierce, strong, and noisy. They were natives of Tibet; and whether savage by nature, or soured by confinement, they were so impetuously furious, that it was unsafe, unless the keepers were near, even to approach their dens."[3]

Near the end of the nineteenth century, Captain William John Gill described his visit to a community where there were some impressive animals. "The chief had a huge dog kept in a cage on the top of the wall at the entrance," he wrote. "It was a very heavily built black-and-tan, the tan of a very good colour; his coat was rather long, but smooth; he had a bushy tail, smooth tan legs, and an enormous head that seemed out of proportion to the body, very much like that of a bloodhound in shape with overhanging lips. His bloodshot eyes were very deep-set, and his ears were flat and drooping. He had tan spots over the eyes, and a tan spot on the breast. He measured four feet from the point of his nose to the root of the tail, and two feet ten inches in height at the shoulder. He was three years old, and was of the true Tibetan breed."[4]

Count Bela Széchenyí was a well-traveled Hungarian aristocrat with a special attraction to big Tibetan dogs. In his remarkable narrative of travels published in 1882, Széchenyí describes two types. One was a black mastiff with brown markings over the eyes and on the legs (a black-and-tan). He thought it looked like a Newfoundland, "only broader and stronger, with a fiery eye, shortish head with broad forehead and very strong, large teeth." He described them as "able to chew the strongest bones to fragments. The tail is curled and muzzle has large dewlaps, the colour is black but shaded in various places, including under the tail, to chocolate or brown, the eyebrows also have brown markings. The body hair is so dense that only the lower parts of the legs are visible. The coat has a bear-like appearance."

The other type was pure black, with long shaggy hair around its head. This suggests that his may be the first written description of what we call, today, the Tibetan KyiApso—the shaggy or bearded Tibetan mastiff. The evidence, however, is inconclusive; he may have been describing the unkempt mane on a shorter-coated dog.

Later, in the vicinity of Lake Kokonor in Qinghai Province, the Count purchased three Tibetan dogs that he eventually took back to Europe. The one he named 'Dianga' was quite wild. "It took two weeks before he allowed me to touch him in spite of feeding him myself by hand. At one such attempt he bit right through my hand but with luck, only through the fleshy part. I also had to pay a great deal of money for constant damage he did to other people's livestock. At his first sight of a buffalo [sic] herd he leapt onto the back of one of the beasts and scattered the rest in all directions."[5]

## 2. The Noblest of the Party

Count Széchenyí ends his story on a sad note. He had to shoot Dianga after it was severely injured in a fight with his other dogs. And, "Though I was able to treat the gaping wound on his thigh, it later on got infected and I had to put him out of his misery." He then donated the skull to the National Museum of Budapest.

Széchenyí's traveling companion was Gustav Kreitner, an Austrian military officer and historiographer. Kreitner was impressed by what he had heard about Tibetan dogs. He anticipated seeing them from the moment he and the Count arrived in China. And, "True enough," he says, "they deserved all the praise, because they looked very much like the most beautiful Newfoundland. Their heads were larger and more imposing due to the manelike ruff around their necks. It gives these dogs a sort of wild, impressive, look. To make them look even more magnificent, the owners often adorned them with collars made from reddish yak hair, woven like a wreath. The colour of these dogs varies from black to light brown but black is the dominating colour. Also they are mostly very dangerous beasts and are kept on chains around the house. Their deep bark makes everything shake."

Kreitner then makes an interesting observation: "Strangely," he notes, "whilst attacking their prey, they constantly wag their tails." Then as a sort of postscript he adds: "They are used to guard sheep, goat and yak herds against intruders and also to keep order within the herd."[6]

By the mid-nineteenth century the exploration of the Himalayas and Tibet was in full swing, and more explorers, researchers and diplomats published their journals and travelogues about the wonders, including big canines, that they saw. Some even adopted dogs for their companionship and protection. While most writers describe fearful encounters, a few of descriptions of the nomads' dogs are complementary and show genuine affection.

### The Rigors of Entering and Crossing Tibet

What was it like trekking over the mountains and across the vast Tibetan highlands? Before such modern conveniences as cars, trains and airplanes, travel to and through Tibet was a demanding and laborious experience and, at times, quite challenging and dangerous. It was a 'trek' in the truest sense of the word, arduous cross country travel. While traversing the high Tibetan plateau, pilgrims, missionaries, foreign adventurers and explorers were typically accompanied by porters, cooks, local guides and armed guards, for often enough, besides the typical peasants and nomads, traders and beggars, monks and nuns, the travelers encountered bands of armed robbers. Wise travelers sought safety in numbers and for both security and companionship they joined convoys of slow, lumbering yaks loaded with trade goods. Travel across the vast expanse of Tibet could take many months, since it was sometimes necessary to take circuitous routes to find pasture for the livestock and to avoid areas controlled by brigands.

In his classic book *Seven Years in Tibet* (1952), Heinrich Harrer tells of a harrowing experience that he and his companion Peter Aufschnaiter had when they met up with armed robbers while crossing Tibet. Harrer expressed great fear of the Khampas. In the 1960s, after the Chinese took control of Tibet and the Dalai Lama had fled to India, some of those same bandit-types from Kham became guerrilla fighters against the Chinese occupation of their homeland. They were often accompanied by large Tibetan mastiffs.[7]

Trekking across Tibet was all the more difficult due to the extreme altitudes, often well above 15,000 ft (*c.* 4500 m). Winter months were most difficult, with a dry, biting wind, along with ice, snow and extreme cold. Foreign travelers to Tibet prior to modern times trekked overland from the west or north, or from the south across the Himalayas, on foot, traversing some of the world's highest mountain passes, snowbound most of the year, open only briefly in summer. For many foreign travelers Lhasa was the goal, a medieval outpost at 12,000 ft (*c.* 3650 m) that for a long time was considered the world's highest city.[8]

One of the first Europeans to cross the Himalayas into Tibet was António Andrade, a Portuguese Jesuit priest, in 1624. He trekked from north India into western Tibet and, although dogs didn't interest him, his travelogue reveals for us some of the tremendous difficulties encountered by outsiders. Disguised as Hindu pilgrims, Andrade and his companion Manuel Marques ascended northward along a dangerous path beside India's roaring Alaknanda River to the sacred source of the Ganges at Badrinath, then over the 18,390 ft (5605 m) Mana Pass into Tibet. Andrade and Marques suffered many hardships: raging rivers, deep snow, biting cold and sheer cliffs where they risked being "dashed to pieces" with a single misstep. "But, we had constantly before our eyes the example of [Hindu pilgrims] who were braving all these difficulties to honour their gods...," writes Andrade. He found Tibet to be austere and barren, "intersected by a vast range of mountains... where there are neither human habitations, nor trees, nor grass; nothing, in fact, but rocks covered with snow..." The two Jesuits survived on roasted corn and were convinced that "only people in robust health can possibly endure such a journey, even under its most favourable circumstances."[9]

When travelers such as these finally arrived at Tibetan towns, fortresses, monasteries or nomad camps, they carried *tsampa* (roasted barley flour, the staple food), and occasionally purchased or bartered for yak, sheep and goat meat to eat. Tibet teemed with exotic wildlife, including the wild ass (*kiang*), blue sheep (*na*), antelope (*chiru*) and many smaller animals. Hunting is discouraged by the Buddhist precept against killing, but the nomads have always hunted to sustain themselves, their families and their dogs. Vegetables and fruits were virtually unknown until recently and then, as now, rice was imported from Nepal, India and lowland China.

After Father Andrade's visit, the Belgiun Jesuit Albert D'Orville and his Austrian companion Fr John Grueber arrived. It was 1661, and they may have been the first Europeans to enter Lhasa. Forty years later, in 1707, two Capuchin Fathers, Francois de Tours and Giuseppe d'Ascoli, arrived to establish a mission in Lhasa. They were followed by Domenico di Fano. In 1711 the Capuchin mission was temporarily closed for lack of funds. When it was briefly re-established a few years later it was attended by Fr Cassiano Beligatti and Fr Prefect Della Penna. These priests' interests were focused on spreading the Gospel, so we find nothing about dogs in their periodic reports back to Rome.[10]

In the following century, a long list of (mostly European) travelers entered Tibet. Among them are a few who ultimately became well known for their scientific or diplomatic endeavors. Besides Desideri and Bogle, already mentioned, the eccentric intellectual Thomas Manning is considered to have been the first Englishman to enter Lhasa, in 1811. Thirty-three years later, in 1844, L'Abbe Evariste Huc and Joseph Gabet, Lazarist priests, arrived. Their *Travels in Tartary, Thibet, and China* is one of the more interesting reads.

By the mid-nineteenth century, the prospect of scientific research in a virtually unknown land was attracting more adventurers to Tibet. In the 1840s, the botanist Joseph Dalton Hooker entered Tibet after crossing a high pass from eastern Nepal near Mount Kanchenjunga, which at 28,169 ft (8586 m) is the world's third highest peak. Hooker was one of the first great plant hunters and a friend of Charles Darwin. His description of big Tibetan dogs in *Himalayan Journals* (1854) is one of the most eloquent in the literature. Hooker had an eye for detail and was adept at recording what he saw, both natural and cultural. For awhile a large dog accompanied him in his travels, but it was tragically swept away in a raging river after falling from a bridge. "For many days I missed him by my side on the mountain, and by my feet in camp," he writes. "He had become a very handsome dog, with glossy black hair, pendent triangular ears, short muzzle, high forehead, jet-black eyes, straight limbs, arched neck, and a most glorious tail curling over his back."

Hooker's writings provide us with a rich word picture of Tibetan traders, their livestock and their dogs. Amidst the flurry of sheep, goats, yaks and colorfully dressed men, women and children, the mind's eye can clearly see the majestic mastiff he describes, etched against the skyline on the trail above him. Hooker records that "On the 18th November, I left Mywa Guola, and continued up the river to the village of Wallanchoon or Walloong, which was reached in six marches." There, he goes on, "A change in the population accompanies that in the natural features of the country, Tibetans replacing the Nepalese who inhabit the lower region.

I daily passed parties of ten or a dozen Tibetans, on their way to Mywa Guola, laden with salt; several families of these wild, black, and uncouth-looking people generally traveling together..."[11] Then, as now, salt from the shallow lake shores of western Tibet was an important item of trade in the lower Himalayas.

To Hooker, the Tibetans he met were "singularly picturesque" with their colorful dress and "odd appearance", accompanying their livestock along the mountain trails. What follows is an image that I carry with me on every Himalayan trek, hoping to see it for myself one fine day. It is, perhaps, Hooker's most oft-quoted passage.

"First comes a middle-aged man or woman," he begins, "driving a little silky black yak, grunting under his load of 260 lbs of salt, besides pots, pans, and kettles, stools, churn, and bamboo vessels, keeping up a constant rattle; and perhaps, buried amongst all, a rosy-cheeked and lipped baby, sucking a lump of cheese-curd. The main body follow in due order, and you are soon entangled amidst sheep and goats, each with its two little bags of salt; beside these stalks the huge, grave, bull-headed mastiff, loaded like the rest, his glorious bushy tail thrown over his back in a majestic sweep, and a thick collar of scarlet wool round his neck and shoulders, setting off his long silky coat to the best advantage; he is decidedly the noblest looking of the party, especially if a fine and pure black one, for they are often very ragged, dun-coloured, sorry beasts."

To Hooker, the big dog "seems rather out of place, neither guarding nor keeping the party together, but he knows that neither yaks, sheep, nor goats, require his attention; all are perfectly tame, so he takes his share of work as salt-carrier by day, and watches by night as well. The children bring up the rear, laughing and chatting together; they, too, have their loads, even to the youngest that can walk alone."

In 1895, A.E. Brehm summed the essence of such scenes, describing the Tibetan mastiff as "a magnificent, beautiful, large animal of really awe-inspiring appearance."[12]

Over the next half century, Tibet was visited more frequently by scientists and scholars, mostly for exploration, and some for espionage. The orientalist and diplomat William Woodville Rockhill's *The Land of the Lamas* (1891), *Diary of a Journey through Mongolia and Thibet...* (1894) and *The Life of the Buddha* (1894) are good reads. The Bengali explorer, 'Pundit' Sarat Chandra Das, wrote *A Journey to Lhasa and Central Tibet* (1902), which includes several references to dogs. In one place, Das describes some dogs he met on "a narrow pathway [that] led up to the gateway, near which were chained two fierce watchdogs (*do khyi*), who barked furiously and strained at their chains as we passed. The Yamdo [Amdo] dogs, I had heard, were famous throughout Tibet for their size and fierceness, and these certainly justified the reputation given them."[13]

Others who wrote of their travels in Tibet include Captain Hamilton

*Plate 2.1. An old illustration of a Tibetan mastiff.*

Bower (a spy), A. Henry Savage Landor (an explorer), and two Frenchmen, Gabriel Bonvalot and Prince Henri d'Orleans (dedicated adventurers).[14] One of the most noteworthy visitors was Sir Charles Bell, a British diplomat whose scholarly *The People of Tibet* (1928) and *The Religion of Tibet* (1931) provide us with considerable insight into Tibet of a century ago.

The first anthropologist to visit Lhasa was William Montgomery McGovern, who wrote *To Lhasa in Disguise* (1924). The anthropologist who has had the most to say about the big dogs was Robert Ekvall. His 1963 study of the 'Role of the dog in Tibetan nomadic society' stands out as the best of the otherwise rare scholarship on Tibetan dogs (reprinted here in Appendix 2, pp.208-17).[15]

The Italian archaeologist and Tibetologist Giuseppe Tucci also explored much of Tibet and the Himalayas, as did the British geologists Sidney G. Burrard and Horace H. Hayden. There was also L.A. Waddell, a British explorer and, after him, the naturalist Francis (Frank) Kingdon-Ward. Each wrote extensively about their research, the land, the people and, occasionally, of dogs.[16] Kingdon-Ward's writings, in particular, are among the most eloquent and fascinating of all.

In Waddell's *The Buddhism of Tibet or Lamaism, with its Mystic Cults, Symbolism and Mythology, and in its Relation to Indian Buddhism* we find a number of protective charms against various afflictions and misfortunes

*Plate 2.2. Charm against dog-bite.*

described; e.g., the plague, demons, birds of prey, dog bite, and even one for killing one's enemy. About the Tibetan charm against dog bite Waddell provides this detailed description: "The huge Tibetan mastiffs are let loose at night as watch-dogs, and roaming about in a ferocious state are a constant source of alarm to travellers, most of whom therefore carry the following charm against dog-bite. It consists of a picture of a dog fettered and muzzled by a chain, terminated by the mystic and all-powerful thunderbolt-sceptre; and it contains the following inscribed Sanskrit *mantras* and statements: 'The mouth of the blue dog is bound beforehand! *Omriti-sri-ti swāhā! Omriti-sri-ti swāhā!*' And this is repeated along the body of the dog, followed by:—*Om Vajra ghana kara kukuratsa sal sal nan marya smugs smugs kukuratsa khathamtsa le tsa le mun mun sar sar rgyug kha tha ma chhu chhinghchhang maraya rakkhya rakkhya!* (It is) fixed! fixed! *Om smege smege bhum bhummu!*"[17]

## Early Women Travelers and Big Dogs in Tibet

Not all of the early European travelers through the Himalayas and Tibet were men. Several stalwart women also made the trip, some of whom also wrote about the dogs they met. The first European missionary woman to visit Tibet was Annie R. Taylor, in 1892.

Three decades later, in 1924, the renowned French-Belgian adventuress, anarchist and spiritualist, Alexandra David-Neel, became the first European woman to enter Lhasa. She wrote *Tibetan Journey* (1926) and *My Journey to Lhasa* (1927), in which she describes several interesting encounters with dogs. In the latter there's a scene where she and Yongden, her traveling companion, were prevented from approaching a house: "At the *chukpo's* (richman) house five big dogs were allowed to attack us freely. While I worked hard with my iron-spiked staff to keep the beasts away," she writes, "Yongden endeavoured to express his request loudly enough to be heard above the furious barking of animals. Nobody answered at first. Then a young woman appeared on the flat roof and asked us many questions, without ordering the dogs to leave us. Yongden answered with angelic patience, while I continued the fight around him... We were not allowed to enter the house!"[18]

Other Western women explorers of note were Isabella Bird (Bishop), Nina Mazuchelli, Luree Miller, Susie Carson Rijnhart, and Fanny Bullock Workman. Isabella Bird was once saved from a dog attack by the fearless women of the tents. In her 1894 book *Among the Tibetans*, she writes: "Each of the fifty tents was visited: at every one a huge, savage Tibetan mastiff made an attempt to fly at me, and was pounced upon and held down by a woman little bigger than himself."[19]

By far the most dog-oriented of the women travelers was Irma Bailey (sometimes identified as Honourable Mrs Eric Bailey), the wife of the diplomat (and former spy against the Russians in Central Asia), Colonel Frederick ('Eric') Bailey. She accompanied him to Lhasa and lived there several years during the first decades of the last century. Mrs. Bailey brought several Tibetan dogs back to the UK, where she was instrumental in establishing the Tibetan Breeds Association. In 1937, in a famous and oft-cited article 'Dogs from the roof of the world' in *The American Kennel Gazette*, she writes: "In 1928, we imported five of these mastiffs. The best one was undoubtedly Tomtru (meaning 'young bear'), a village dog. Then there was Rakpa... a fine red dog which was given first prize for Foreign Dogs at the Kensington Show some years ago, and was also a winner at Crufts and the Kennel Club shows..."[20] (Irma Bailey's article is reprinted here as Appendix 1, (pp.203-7)

Plate 2.3. *A village woman of gTsang (Tsang) tying up the watch-dog in front of her house.*

"Speaking of names," she notes, "mastiffs in Tibetan are called Do-Kyi, which means 'a dog you can tie up.' Do-Kyi are kept by every nomad and sheep- or yak-herd to guard the tents..." When approaching a nomad encampment, she goes on, "... the first sign of life is usually the barking of dogs. On this, the owners come out their black yak-hair tents and inspect the cause of the alarm. They, or more usually their children, then see that all dogs fastenings are secure, and often hold the dogs down while it strains to reach the stranger. Although fierce, as the result of being tied up from puppyhood, these mastiffs are very affectionate and good tempered with the people they know, and one often sees the smallest children handling and calling them off from their attempts to attack the intruder with perfect ease and safety. The mastiffs, and sometimes the hunting dogs also, wear a large fluffy collar of wool dyed bright red. This red collar can usually be seen in pictures of scenery by Tibetan artists."

"Tibetan mastiffs," she goes on, "are usually black in color with tan points. One of the high officials of His Holiness the Tashi Lama once had some entirely black ones, of which he was very proud, but tan markings are more usual. Not frequently red dogs are found in a litter. The dog is very heavily built with a thick coat. The head is particularly heavy and the flews so pendant that the red of the eye is conspicuous. This, in a country like Tibet, subject to dust, wind and glare, often leads to diseased eyes owing to dirt and lack of care and cleanliness. The whole head is large and heavy, but the heads of females are notably smaller and lighter than those of males. The Tibetans especially admire a deep-voiced bark."

## Big Dog Attributes

The size, ferocity and bark of the big dogs are the three attributes attracting the most attention in much of the written record. We know how their size has been exaggerated since first described by Marco Polo and how, based on that, some modern breeders want to turn them into gigantic critters. According to contemporary description out of China, a few designer Tibetan mastiffs have been documented as over 35 inches (90 cm) in height and one has been described as being over six and a half feet (2 m) in length.[21]

Their aggressiveness has been highlighted for centuries by writers like Thomas Hungerford Holdich who describes dogs "whose ferocity even to this day is a by word amongst travelers in Tibet." While visiting southeastern Tibet in the early 1900s, the naturalist Frank Kingdon-Ward witnessed "The menace of dogs, savages, desperate from being kept tied up, and hungry..." Sometimes he was hesitant to go into villages where he could hear their agitated barking, though he was pleased to note that "the people themselves welcomed us whenever we cared to go."[22]

The Italian archaeologist Giuseppe Tucci described the role of big watchdogs: "The locked house is guarded by mastiffs, great tawny or black sheepdogs, something between a wolf and a bear, which leap towards an unknown caller with such rage and savage growls that they seem about to break the heavy chains which always secure them. To make them more fearsome and ferocious looking, their owners put great collars on them, made with the long hair from a yak, dyed red."[23]

The British adventurer John Baptist Lucius Noel, who sneaked illegally into Tibet in 1913, wrote the following about his encounter with big dogs after being repulsed by Tibetan authorities in his attempt to get close to Mount Everest. "... We were obliged to turn back and take a lower level route by the plain, hiding ourselves and passing villages at night, making detours to avoid the dogs that, hearing every movement of men and animals within the distance of a mile, barked and howled. These Tibetan mastiffs, found in all the Dok-pa [nomad] shepherds' encampments, are magnificent yet savage animals. Attacking, they show fangs like wolves, and their ferocious appearance is heightened by the immense scarlet ruffs of yak hair which the Tibetans place round their necks like Elizabethan collars. These dogs are prized for their ferocity, and the mark of their breeding is the depth of their bark. The Tibetans say their bark must ring like a well-made gong. Once I had to shoot one of these mastiffs that rushed out and bit one of my men badly."[24]

That dark bark has inspired some worthy prose. Reginald Fleming Johnston, for example, characterizes the bark of a Tibetan mastiff as "most peculiar: not sharp and crisp like that of most European dogs, but with a sepulchral and 'far away' sound as if each dog kept his own ghost in his stomach and it was only that ghost that barked." Another oft-quoted description compares the dog's voice to "the sound of a well-made copper gong."[25]

The sound of their booming bark travels far and wide, instills fear, encourages caution, and cannot be mistaken. The naturalist Lawrence Swan has observed that from a helium or hydrogen balloon sailing over the high Himalayas in absolute silence (a popular sport during the 1980s) "the pilot can hear dogs barking and people talking." He goes on to explain "that sound from the surface moves upward aided by an echo from the earth and travels better vertically than it does horizontally."[26]

In some instances, the bark takes on almost supernatural qualities. J.B.L. Noel writes of phantom dogs barking on Mount Everest during the ill-fated British expedition of 1924. One of the climbers described the condition in which he found the Sherpas after a serious accident on the mountain. "They were crying like little children," he begins. "They had completely lost their nerve. Alone for four days, they had exhausted their food, and for the last twenty-four hours had had nothing to eat. They were obsessed with fears of the demons of the mountain. They told us they had heard the barking of the phantom guard dogs of Chomolungma at night, and that they were terrified. They never expected to be rescued... Their sahibs had to come to save them, otherwise surely they would have perished."[27]

Years earlier, William Moorcroft and George Trebeck, in their 1841 *Travels in the Himalayan Provinces*, told of Hindu pilgrims visiting the sacred cave of Amarnath near the Tibet border in northwest India who "assert that they can hear the barking dogs in Tibet."[28] Their account, along with Swan's explanation and the trepidation of the Sherpas, reflect both the resonant and the mythical qualities sometimes attributed to the Tibetan mastiff's famous *Woof!* Is their sound exaggerated by the altitude? By the mind? Or by the mountain deities?

## "A Source of Constant Amusement"

Two intrepid gentlemen-explorers who wrote long and inspiring accounts about dogs were the English traveler Andrew Wilson and the Swedish explorer Sven Anders Hedin in the mid- and late-nineteenth century, respectively. Both men adopted dogs to accompany them, and their lives were richly rewarded and buoyed up by canine companionship. They gave their bushy-tailed charges intriguing names and wrote about them with genuine feelings of affection.

After trekking around western Tibet in the mid-1800s, Andrew Wilson wrote, "On the roof of every house there was a ferocious Tibetan mastiff, roused to the highest pitch of excitement by our arrival, and desiring nothing better than that some stranger should intrude upon his domain." The ones he saw were both fine and noble, mean and fearful at the same time. Early in one day's journal, "I met the finest Tibetan mastiff which I saw in all the Himáliya," he says. "It was a sheep-dog, of a dark colour, and much longer and larger than any of the ferocious guardians of Shipki. While we were talking to the shepherd who owned it, this magnificent creature sat watching us, growling and showing his teeth, evidently ready to fly at our throats at a moment's notice; but whenever I spoke of purchase, it at once put a mile of hills between us, and no calls of its master would induce it to come back. It seemed at once to understand that it was being bargained for, and so took steps to preserve its own liberty; but it need not have been so alarmed, for the shepherd refused to part with it on any terms."[29]

At one point in the mountains Wilson found his path "enlivened by flocks of sheep, some laden with salt, and by very civil shepherds from Kúlú and Bussahir. The usual camping-ground was occupied by large flocks." It was there that Wilson acquired 'Djeóla.' Here's the story.

"I purchased from the Kúlú shepherds," he begins, "a wonderful young dog called Djeóla, a name which, with my Indian servants and the public in general, very soon got corrupted into Julia. This animal did not promise at first to be any acquisition. Though only five or six months old, it became perfectly furious on being handed over to me and tied up. I fastened it to my tent-pole, the consequences of which was that it tore the drill, nearly pulled the tent down, hanged itself until it was insensible, and I only got sleep after somehow it managed to escape. I recovered it, however, next morning; and after a few days it became quite accustomed to me and affectionate. Djeóla was a source of constant amusement. I never knew a dog in which there was so fresh a spring of strong simple life. But the curious thing is, that it had all the appearance of a Scotch collie, though considerably larger than any of these animals. Take a black-and-tan collie, double its size, and you have very much what 'Julia' became after he had been a few months in my possession... The only differences were that the tail was thicker and more bushy, the jaw more powerful, and he had large dew claws upon his hind feet. Black dogs of this kind are called *sussa* by the Tibetans, and the red species, of which I had a friend at Pú, are *mustang*..."[30]

He goes on to say that although Djeóla "was most savage on being tied up and transferred to a new owner, there was nothing essentially savage, rude, brutish or currish in its nature... It not only became reconciled to me, but watched over me with an almost ludicrous fidelity, and never got entirely reconciled even to my servants. The striking of my tent in the morning was an interference with its private property to which it strongly objected, and if not kept away at that time, it would attack the *bigarries* engaged. I also found, on getting to Kashmir, that it regarded all sahibs as suspicious characters to be laid hold of at once; but, fortunately, it had a way of seizing them without doing much damage, as it would hold a sheep, and the men it did seize were good-natured sportsmen. It delighted in finding any boy among our *bigarries* that it could tyrannise over, but never really hurt him. It was very fond of biting the heels of yaks and horses, and then thinking itself ill-treated when they kicked... "

Wilson intended to take Djeóla back to England, but when it could not be conveniently arranged he left the dog with a friend in India.

A bit further in his narrative, he comes up with an effective way to ward off the advances of aggressive mountain dogs. "Not far from the top of the pass," he writes, "...the power of the sunbeams in the rarefied atmosphere, and of their reflection from the vast sheets of pure white snow, was something tremendous. I had on blue goggles to protect my eyes..." With the goggles on, he discovered that the big dogs were afraid of them: "The fiercest dog in the Himáliya will skulk away terrified if you walk up to it quietly in perfect silence with a pair of dark-coloured goggles on, and as if you meditated some villany; but to utter a word goes far to break the spell."

## "His Joy was Boundless... His Attacks were Dart-Like"

The Swedish explorer, Sven Hedin, also expressed great admiration for some of the dogs he met in his travels, especially those that attached themselves to his entourage from time to time during his many years of exploration in Central Asia, China, Mongolia and Tibet. He wrote about them in every book he published. He was especially observant when it came to the nomads' dogs, and was enamored with their skill as livestock guardians. The following account demonstrates his affection for the big dogs that joined his caravan while traveling near Mt. Mustagh-Ata (north of K2) in the Pamir mountains. One called 'Yollchi,' which Hedin defines to mean "Him who was picked up on the road," was a special friend. From his sparse description, Yollchi sounds more like a shaggy KyiApso than the traditional Tibetan mastiff. But, no matter, it's the story that counts.[31]

"In the hurry of my departure...," Hedin writes, "I had forgotten one thing—namely, a watch-dog, to lie outside the yurt at night. The oversight was made good in a curious way. On February 25th..., a big Kirghiz dog, yellow and long-haired, came and joined himself of his own accord to our troop. He followed us faithfully throughout the whole of the journey to Kashgar, and kept grim *karaol* (watch) outside the tent every night."

Later in the travelogue, while traveling in snow, Hedin shares Yollchi's ecstacy and acrobatic antics in the snowdrifts. "He rolled over and over in the snow, thoroughly cooling his thick, shaggy hide. One moment he would playfully catch up a mouthful of snow, the next he would race off swift as an arrow ahead of the caravan. The creature was half wild when he joined us; and I never succeeded in making him properly tame."

Sven Hedin then describes how the dog's wildness and previous relationships with humans affected his behavior: "Having been reared among the Kirghiz, he could never by any bribe be induced to come inside my tent. For the Kirghiz are Mohammedans, and look upon the dog as an unclean animal. The very dust off his feet would pollute the inside of a tent. I tried my best to wean Yollchi from such superstitious notions. But, do what I would, I could not get him past the tent door—neither by fair means nor foul. He had never once in his life set foot inside a tent, and

Plate 2.4. Takkar had to be held by two men.

## 2. The Noblest of the Party

obviously had made up his mind that he had no manner of business in such a place."

Hedin's initial encounters were often fraught with danger, but it seems that final departures from his canine companions were personally far more difficult for him. The following story is from the final pages of Hedin's autobiography, *My Life as an Explorer* (1925).

"We met a Tibetan, on a white horse," the tale begins, "followed by a big, ragged watchdog, black, with two white spots. [We] bought the horse for eighty-six rupees, and the dog for two. The dog was of the 'Takkar' race; so we called him Takkar. He was absolutely savage, and as ferocious as a wolf. The Tibetans helped us to tie a rope around his neck, leaving two long ends, by which Kunchuk and Sadik led him between them, to prevent him from biting."

After crossing a high pass, Hedin's party entered a meadow and made camp among other tents, horsemen and herds. There "the sight of [a] white horse seemed to make him happy," he writes. They needed a watchdog, but were worried that Takkar might run off if unleashed at night. Therefore, "To prevent Takkar from giving us the slip, we thought of tying a pole to his neck, in such a way that he could not gnaw it to pieces. But as soon as one of the men approached him, he rushed up, with bared fangs and bloodshot eyes, determined to get at the throat of his tormentor. The men therefore threw a thick *vojlok* blanket over the dog, and four men sat on him, while the others fastened the pole to his neck with a thick rope. Then they anchored the pole in the ground, and Takkar was moored. The operation over, he tried to hurl himself at the men, who made off in all directions. 'He will be nice to have around the house,' I thought."

Takkar was irreconcilable: "No one could go near him. If one of us even came out of a tent, he would bark himself hoarse. But he was most furious at Kunchuk, who had bought him…"

It is obvious that Hedin liked dogs and cherished their company on his otherwise lonely treks across the vast expanses of high Asia. A bit further on in his biography, he writes of a fascinating incident that occurred shortly after making camp in a quiet spot. It begins with "Takkar… , as usual, tied before my tent. I seated myself, entered the day's events in my diary, and sketched the panorama. It was a clear evening. Spring-like winds were blowing across the plain… All of a sudden, the big dog came up to me, looked at me steadily. 'Well, what do you want?' I asked. He leaned his head to one side and started to scratch my arm with his forepaws. I took his ragged head in my hands and patted him. We understood each other. He began to howl and whine with delight, leaped at me, and seemingly implied: 'Ah, come and play with me, instead of sitting there alone and sulky.' I unfastened the ropes and knots around his neck, and freed him from the nasty pole that had weighed him down since the

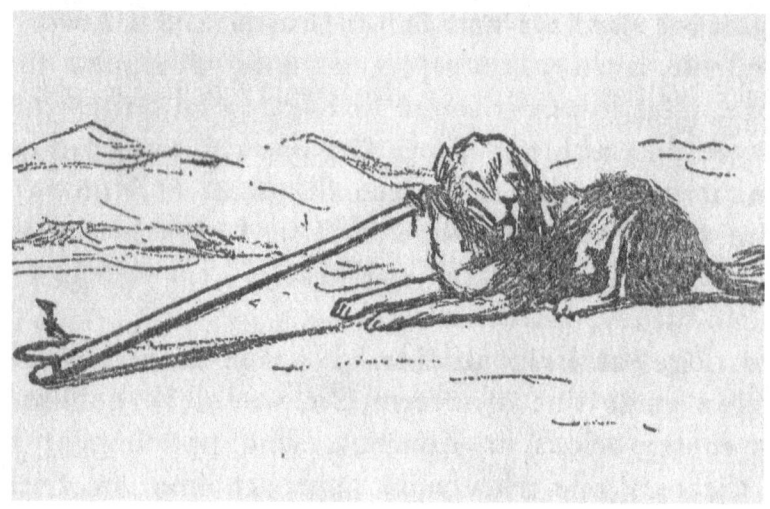
*Plate 2.5. Takkar anchored.*

day he became our captive. He stood motionless. Finally, I wiped away the dust-clots in the corners of his eyes. Now his joy was boundless. He shook himself so that the dust flew, and nearly upset me with his playful leaps. He cavorted and danced, howled and barked, and seemed both proud and happy at the confidence I had shown in restoring his freedom. Then he darted off over the plain like an arrow. 'Now he will run back to his former master,' thought I. But no; he came back at top speed, within a minute, and gave Baby Puppy [another of Hedin's dogs] a push, the impact of which caused the little dog to make several revolutions along the ground. And this manœuvre was repeated over and over again, until Baby Puppy became quite dizzy. My men were astonished that Takkar had been tamed so quickly, and that I could play with him as safely as with the puppy."

Hedin played with the dog each evening after that, and soon Takkar became his faithful protector, day and night. But, Hedin goes on, "He developed a violent hatred of all that was Tibetan. He would not suffer a Tibetan to approach the tents. His attacks were dart-like. I had to pay a number of silver rupees to peaceable nomads for the torn clothes and bloody wounds which he caused. He… would not let a soul approach my tent. And when we feared curious neighbours, we had but to tie Takkar before the flap, to assure me perfect peace."

Hedin expressed a great indebtedness to Takkar "for the successful outcome of the sixth crossing of the Transhimalaya; and consequently I bear his memory in warm regard." That mountain crossing took Hedin south out of western Tibet down the Satlej river into northern India. The scenery filled him with both awe and nostalgia. "My journey along the Satlej and across its deep-lying tributaries was one of the most interesting I ever undertook in Asia, because we crossed the Himalayas crosswise.

Words fail to describe the landscapes of overwhelming grandeur which our eyes encountered everywhere. To see them once, is to possess a lifelong memory of the high mountain-peaks, with their dazzling snow-fields, and the steep, rocky walls which enclose the valley of Satlej; and one even imagines hearing the mighty roar of the foaming river..."

The date was August 1908, and in his journal Hedin tells us that he had not seen another European in four years. As he and his companions descended out of the mountains and the days got warmer, the faithful Takkar's plight became miserable. For awhile, the big dog accompanied the party along the mountain trail alongside the river. And then—"From Poo, we descended to ever lower levels. It grew warmer, day by day. Takkar suffered agonies with his thick black coat of hair. With tongue hanging out, and dripping he ran from shade to shade; and he lay down, outstretched, in every brook, to cool himself. Half a year ago, he had come to us, while Tibetan winter storms hurled the drift-snow round our tents. As far as the Shipki-la [pass] he had breathed the fresh, cold air of his home-country, and had seen the last yaks. We had now brought to a land of infernal heat. He pondered and cogitated. He was conscious of loosening bonds. We had taken him from the nomad-herds by force; and now, again treacherously, we were luring him down to a country, the heat of which he could not endure. He felt more and more like a stranger among us. He was frequently out of sight the whole day; but in the cool of the evening he would come to our camping-place. He felt lonely and forsaken, and noticed that we left him heartlessly. One evening he failed to turn up. We never saw him again. He had doubtless gone back to Tibet, to the poor nomads and the biting snow-storms..."

## Colonel Younghusband and Big Dogs at the Opening of Tibet

By the beginning of the twentieth century, adventurers, spies, naturalists, treasure hunters and dog-watchers, alike, had breached Tibet all the way to Lhasa, the 'Forbidden City.' Their endearment to big dogs was demonstrated by sending a number of them as gifts to European royalty and as companions to other members of the privileged classes, although it is unclear which of the big dogs that reached Europe at this time were from Tibet proper and which were acquired in the Himalayan foothills. Nonetheless, all made deep impressions on dog-lovers in Europe.

The event that finally opened Tibet to the outside was the infamous Tibet Expeditionary Force led by Colonel Francis Edward Younghusband in 1903-04. This British military 'mission' (so-called) effectively changed the way the West looked at and dealt with Tibet, and how the Tibetan government coped with the British and other foreign governments. What led to the Younghusband Mission were British worries about increasing Russian influence in Central Asia, dating well back into the 1800s. At the turn of the twentieth century exaggerated concerns were raised about Russian rapprochement in Tibet, threatening British India.

Colonel Younghusband's march on Lhasa was one of the final dramatic acts in what came to be known as the 'Great Game,' an antagonistic rivalry involving counter-espionage and intrigue between Great Britain, Tsarist Russia and China in Central Asia. The Russians called it, euphemistically, the 'Tournament of Shadows,' a contest of national wills and strategic interests fueled by imperialistic ambition and one-upmanship between the then 'great powers' of the world, each of which sought economic and diplomatic supremacy over Central Asia. The Great Game began in 1813 and concluded with the Anglo-Russian Convention of 1907. During those years, it fueled a number of clandestine incursions by explorer-adventurers and spies, and became grist for historians and novelists. Rudyard Kipling's 1901 novel *Kim* provides one well-read, romantic account of the game, told from the perspective of a British colonial waif.[32]

After a lot of diplomatic sparring, Lord Curzon, the Viceroy of India, sanctioned Colonel Younghusband to take a military force of Indian and British troops (including a Nepalese Gurkha marching band) to Lhasa. Younghusband's objective was to impress the Dalai Lama's government into favoring the British in its international alignments. The military incursion lasted well into 1904, during which Younghusband's army routed and slaughtered hundreds of poorly armed Tibetan defenders—and shot dead some ferocious dogs. By the time Younghusband reached Lhasa, however, the Dalai Lama had fled to Outer Mongolia, and the colonel had to negotiate with priests of the Tibetan Buddhist hierarchy.

The Younghusband Mission succeeded in opening Tibet to British diplomats and, after some time, to European mountaineers. It also opened up Tibet as a source of big dogs imported into Europe. The journalists who accompanied Younghusband posted descriptions of the fighting; and they, as well as a few military officers, also wrote about the large and dangerous dogs they encountered on the road to Lhasa. When entering Tibetan courtyards, the soldiers quickly learned to have their pistols drawn and cocked in self defense against some very aggressive guard dogs. Some of the troops acquired pets of various kinds, including two wild asses. "But mostly it was dogs," says Charles Allen in *Duel in the Snow: The True Story of the Younghusband Mission to Lhasa*. Allen quotes a soldier, Mark Synge, who described "a touching sight... to see great bearded men sometimes leading, but as often as not carrying, on the march dainty little lapdogs... One or two Tibetan mastiffs—more like huge Welsh collies than mastiffs—also accompanied us."[33]

The most famous big dog brought out of Tibet after the Younghusband mission was 'Bhotean,' sent back to England by Major W. Dougall. Bhotean was highly regarded when exhibited at a dog show in England in 1906. According to Juliette Cunliffe, "This dog was considered a fine specimen of his breed, the best then ever seen in Great Britain."[34]

Cunliffe goes on to describe the high price Dougall got for selling Bhotean, and Dougall's intense interest in the breed. In addition, she adds, "Amongst many interesting observations made, he said that Bhotean was 'the long, low type, on very short legs, with great bone, and enormously powerful.' He talked of the breed being able to stand any amount of cold, but they could not endure damp. Their owners, he said, set the same value on them as an Arab did on his horse. They were used as guards and protectors only, and were in no sense sheep dogs. They often wore great leather collars with roughly beaten spikes on them, so that, in the event of a leopard or panther attack, they were protected from a fatal grip on the throat. When the herds were stationary for any time, the natives hobbled the dogs, by tying their forelegs together, crossed."[35]

After Tibetan mastiffs began to be seen in England, British interest in the breed grew. Rev. H.W. Bush informed the British public in 1908 about these dogs in an article in *The Kennel Encyclopedia*. "Since the return of the Tibet Expeditionary Force to India a couple of years ago," he wrote, "much that was previously known about the breeds of dogs in that land has been confirmed, and the existence of others, not so well-known, has been revealed. Up to the time of the Expedition, Tibet was a closed land, and all information to be obtained about the dogs was what the Tibetans visiting India either could or would give. The existence of the larger dog of Tibet, the so-called mastiff, has been known to the 'outer' world for a great number of years, and from time to time specimens, both good and bad, have found their way into India whence some have been taken to Europe and England. It is a well-established fact that the Tibetans have not as a rule parted with their best, and small blame to them for that."[36]

"About the smaller Tibetan breeds, the Lhasa terrier and the Tibetan spaniel," he went on, "they have been easier to get, and thrive both in India and England very well indeed. The same cannot be said for the

*Plate 2.6. The famous Tibetan mastiff named 'Bhotean', imported into England from Tibet in early 1900s.*

mastiff; he cannot stand the plains of India even in winter for any length of time, while diseases, especially distemper, prove very fatal!"

Thus, colonial officials (diplomats and military personnel) in British India (and in Tibet since Younghusband's foray) were the first to import Tibetan mastiffs to the British Isles, just as Count Széchenyí and others had introduced them to continental Europe. The next to feel a strong attraction to the big dogs were British mountaineers.

## Climbers and Canines on Everest

By the early twentieth century British mountaineers had begun to penetrate the Himalayas. Buried in a few of the official expedition records and books are mentions of big dogs encountered on the way to the peaks. But first, we must appreciate how Himalayan mountaineering, which brought so many new visitors to Tibet, got started. And we need to understand some of the attributes of the peaks that are visible for long distances all across the south of Tibet and occupy such a predominant place in the Tibetan psyche. There is a bloody great profusion of them—over a thousand above 6000 meters (19,685 ft) and fourteen over 8000 meters (26,246 ft), nine of which are within Nepal or straddle the Nepal/Tibet border. The highest is a great snow-blown pyramid of metamorphosed black gneiss and marine limestone piercing the sky along the northeastern Nepal/southern Tibet border. Its supreme height was first ascertained in 1852 (though not officially announced until 1856).

To the Tibetans this peak is known as Chomolungma, 'Goddess Mother of the World.' (The Chinese call it 'Qomolangma feng.') Chomolungma can be clearly seen from the Tibetan side rising high above all other peaks on the southern horizon. It is especially well seen from Tingri (or Dingri), a small town on the main highway between Lhasa and western Tibet. It is from Tingri that most modern expeditions from the Tibetan side of the peak begin their final overland approach to the base of the peak.

The Nepalese call it Sagarmatha, which literally means 'Head of the Sky.' The name is of recent derivation, however, given by a Nepalese historian named Babu Ram Acharya in 1956.

To the rest of the world it is known as Mount Everest.

During the 1850s, map-makers from the Great Trigonometric Survey of India had become convinced, from afar and after several false assumptions, that the peak first known to cartographers as 'b' and later as 'XV' was, in fact, the highest in the world. At the time, Andrew Waugh was Surveyor-General of India. After many years of painstaking triangulation and progressively more refined calculations by the Survey's field staff, Andrew Waugh (later Sir Andrew) confirmed the peak's great height. In 1856, Waugh announced it to the world by its English name, after his predecessor, the distinguished George Everest (later Sir George).[37]

The name was immediately protested, including by Everest himself.

## 2. THE NOBLEST OF THE PARTY 39

Even the noted naturalist and diplomat, Brian H. Hodgson, former British Resident in Kathmandu, weighed in on the debate. He pointed out that the mountain already had at least one, if not two Nepalese names. But when they were subsequently shown to be wrong (Sagarmatha was not one of them), "The name 'Everest' therefore was given to this king of mountains, and it has appeared in the English maps ever since," wrote A.L. Waddell, summing it up in 1906. One problem behind the controversy was that the surveyors could hardly get close, as both Nepal and Tibet were closed to outsiders, though that rule was broken many times. After the naming of Mount Everest was settled and Tibet and, later, Nepal were relatively more open to outsiders, the passion to climb her began to be actively expressed. It was a passion that has inspired countless mountaineers ever since.[38]

Mt. Everest was certainly not the most beautiful nor most difficult to climb, but it was the highest and, therefore, the most sought after. The earliest measurement put it at 29,002 ft (8840 m). Later, it was recalibrated at 29,028 ft. In 1999, during the Everest Millennium Expedition, the US National Aeronautics and Space Administration (NASA) with the National Geographic Society revised it upward again to its current official elevation of 29,035 ft (8850 m).[39]

In 1921, Francis Younghusband (knighted by then, and known forever after as Sir Francis) described Mount Everest with life-like characteristics, based on his knowledge and observation of it (from a distance) during his sojourn to Tibet earlier in the century: "Mount Everest for its size is a singularly shy and retiring mountain," he begins. "It hides itself away behind other mountains. On the north side, in Tibet, it does indeed stand up proudly and alone, a true monarch among mountains. But it stands in a very sparsely inhabited part of Tibet, and very few people ever go to Tibet. From the Indian side only its tip appears amongst a mighty array of peaks which being nearer look higher."[40]

In 1905, following the British mission to Lhasa, Younghusband sent two of his officers, Colonels C.G. Rawling and C.H.D. Ryder, on a reconnaissance to Gartok in western Tibet. On the way, they approached within sixty miles of Mt. Everest, but had no time to get closer. Rawling was especially intrigued and thought it might be feasible to climb by the North Ridge. Although other English climbers had already contemplated assaulting Everest, Rawling was the first to get close enough to consider a likely route. He was not the first to directly discuss climbing it, however. That distinction goes to Charles Bruce, a British soldier in the Gurkha regiments who proposed climbing it in 1893 to Francis Younghusband who was, then, still a fresh young officer. Although theirs was the first serious consideration, the notion was not taken forward until after Rawling's closer observations.[41]

And, the dogs? They're coming. (I can hear their barking...)

When he returned to England from Tibet, Rawling pondered climbing

Everest with other British climbers, no doubt over cigars and brandy evenings at the Royal Geographical Society in London. Their enthusiasm was boundless.[42] Three of them, A.L. Mumm, Tom Longstaff and Charlie Bruce, all Alpine Club members, raised the notion with the British authorities and began a determined but gentlemanly campaign to secure official government permission for an expedition via the north side through Tibet. (Nepal was still out of bounds.) Their plan was championed by the Viceroy of India, but neither his considerable influence nor his superlative credentials as 'The Most Honourable George Nathaniel Curzon, 1st Marquess Curzon of Kedleston' was enough to pull it off. Instead, their request was turned down flat by John Morley, Britain's Secretary of State for India.

Morley was an ambitious bureaucrat, determined to flaunt his authority. He was known at the time by cabinet colleagues, behind his back, as "Aunt Priscilla," and to others as a "petulant spinster." He was later remembered as "an elderly, austere, dry-as-dust little man." In rejecting the request for permission to allow a British expedition to Everest, Morley characterized the plan as "furtive" and as a potential "embarrassment" to Britain in the political climate of the time.

It was not until after World War I (and Morley was out of the way) that the first official British climb of Mt. Everest was sanctioned (though, sadly, the enthusiastic Colonel Rawling had died in the war). After a reconnaissance in 1921, the first assault on the peak was attempted, unsuccessfully, in 1922. Several more attempts were made by the ambitious Brits during the '20s and '30s, but none was successful, ostensibly "on account of bad weather, ill health or poor organization."[43]

It was the 1924 British assault on Everest that has special significance in the history of mountaineering. On June 8, shortly after noon, George Mallory and Andrew 'Sandy' Irvine were last seen high up on the slopes of Everest, heading towards the summit. It is unknown if they actually made it, however, and it was not until 1953 that Mt. Everest was finally 'conquered,' from the Nepalese side, by Tenzing Norgay and Edmund Hillary. That's when Hillary (later Sir Edmund) made his famous pronouncement: "Well, we knocked the bastard off."

In 1999, Mallory's frozen body was found high on Everest, but not Irvine's. Nor was their camera found, which might settle the question once and for all.[44]

The dogs now join the story, for on at least two early expeditions, big Tibetan dogs were duly entered into the mountaineering literature. The most remarkable were 'Tendrup' and 'Police-ie.'

### 'Tendrup'
The authors of several books chronicling pre-1940s British mountaineering expeditions mentioned big dogs, but none more eloquently than Hugh Ruttledge in his books *Everest 1933* and *Everest: The Unfinished Adventure*

## 2. THE NOBLEST OF THE PARTY

(of 1936). The latter of these two expeditions, for example, included a black mastiff picked up from the village of Tangu on the march towards the mountain across southern Tibet. Ruttledge describes him and some of his antics in three anecdotes that reveal genuine respect and affection.[45]

"... Smythe's slumbers were disturbed," he writes, "by the barking of 'Tendrup' (the bear), a young Tibetan mastiff which had been bought by Oliver at Tangu. Tendrup possessed a strong character and a sense of humour all his own. He readily accepted Oliver's instructions to refrain from barking at night, though all his instincts and training pointed to this as part of his duty. But woe betide any Tibetan who was found prowling round our dump. When short of real malefactors, Tendrup kept himself fit, and his teeth in good condition, by mimic warfare with us."

"... Tendrup was in his best form... when the drivers were tying the loads on to their animals. We watched as he stalked through the lines, seeking opportunity and audience. Both happening to coincide, he silently approached from behind a Tibetan who was bending over a box—a potent target. It was a perfect piece of slapstick comedy: No savage, slashing attack, just a gentle pinch at the right place; the Tibetan leaped into the air, and Tendrup walked off with his tail well up, laughing all over his face."

After the expedition was over, the climbers and Tendrup trekked out across Tibet and down through Sikkim to Kalimpong in Bengal. There,

*Plate 2.7. 'Tendrup' sitting in front of Hugh Ruttledge and other climbers on the British Everest Expedition of 1936.*

they found a home for the dog. "... At Kalimpong Tendrup, that original and amusing character, found a good home with Raja Topgye Dorje, into whose house he walked with an air of finality which there was no mistaking. Clearly he foresaw that life with Oliver in the hot valleys of Jandola would not suit his constitution, while here were good quarters and an excellent climate. That dog is a practical philosopher."

### 'Police-ie'

A lengthier and more eloquent account is found in the earlier expedition book, Ruttledge's *Everest 1933*, in a section written up by E.O. Shebbeare. The story of the Tibetan mastiff bitch engaged for guard duty by that expedition is certainly one of the classic tales about the big dogs of Tibet and the Himalayas. She was named 'Police-ie' and is described early in the book as "one-eyed and suspicious." She was brought by Nursang, the expedition sirdar, to guard the stores.[46]

Shebbeare tells the full story of Police-ie in the book's chapter about the expedition's transport arrangements. "In Tibet," he begins, "no convoy is complete without its watch-dog, whose duty is to bite all strangers approaching within a hundred yards of the baggage under its charge. Nursang had not failed in this respect, and had provided himself with the usual Tibetan mastiff, an attractive bitch about the size of a sheep-dog, but of more stocky build, with a thick, dark-brown coat and buff points, which he had christened 'Police-ie' in allusion to the duties she was appointed to perform. The name looks clumsy in print but was quite serviceable in practice, and she certainly lived up to it... [B]y her code all strangers were guilty until they had proved themselves innocent—a point which the victim could discuss, if he chose, while one of our medical officers was dressing the wound. She made no mistake, however, as to identity, and any man of our party of sixteen Europeans and seventy porters, once known, was never forgotten, or at any rate never bitten. How she managed to distinguish some of our men, who often wore Tibetan dress, from the local population has always been a mystery to me."

There was trouble at the start. Shebbeare describes the scene when Police-ie first met the climbing party: "The store dump at Kampa Dzong was about two hundred yards from the main camp, and therefore in the neutral zone as far as Police-ie was concerned. As I walked across to see Nursang, I noticed the dog in the offing and, from earlier experiences with Tibetan mastiffs, insisted on a formal introduction, but forgot to warn George Wood-Johnson, who came across a few minutes later, and Raymond Green was soon busy with iodine and cotton-wool. The rest of the party lost no time in making the acquaintance of our guardian, and I do not think any other member of the expedition was ever attacked."

"She was a friendly beast [who] liked to be made a fuss of, but would never enter a tent, nor did she show the slightest interest in any sahib's food. I suppose she lived on barley meal and an occasional mutton bone, but I never saw her eating. Her bed was the bleakest and most exposed part of the camp, preferably on snow or ice; she scorned the lee of a tent even in a snow-storm, and all that was visible on such occasions was a nose sticking out of a small rounded drift. After we had reached the Base Camp I expected that she would remain there in the comparative luxury it afforded, but I had under-estimated her hardihood. Completely independent, she would follow any party up the glacier, usually at a distance of a hundred yards or so; and, though friendly to all, she owned allegiance to no one, not even Nursang. She made several trips up and down the glacier camps, coming or going at her own discretion. The highest point she reached was the foot of the rope ladder below the North Col; had it not been for this obstacle I believe she would have helped to establish Camp VI."

Police-ie was a remarkable dog and a great companion for the climbers. About what happened on the day she disappeared near a glacier camp, Shebbeare writes only a few words: "Poor Police-ie, no one knows what became of her or when she came to grief, but she had been seen to treat crevasses with contempt that they did not deserve, and we fear that this was her undoing…"

## Pilgrims' Progress

In recent years, stories describing first encounters with the big dogs have tapered off in proportion to the shift away from adventures of exploration and mountaineering. There are still a few mid- to late-twentieth century accounts to relate, however, especially from pilgrims in the Himalayas and Mongolia. I use the term 'pilgrim' here in two ways—as the generic wayfarer, wander, traveler or stranger, and as someone who travels far on a religious quest.

### *Kawaguchi's advice*
At the turn of the twentieth century, a Japanese pilgrim, a Zen monk named Ekai Kawaguchi, crossed the Nepal Himalaya, alone, and secretly entered Tibet. The story of how he dealt with the dogs he encountered is told by Scott Berry in *A Stranger in Tibet* (1990). "His initial greeting was one he would soon become accustomed to… the hostility of a pack of Tibetan mastiffs." On their home ground, Berry says, "they are some of the world's best watchdogs" and Kawaguchi quickly learned that he stood little chance of getting near a house or tent without an invitation. Fortunately, somewhere along his route through Nepal, he had learned the accepted way of dealing with them—a way that "put the traveler and the dogs in a state of armed truce."

Kawaguchi describes it in *Three Years in Tibet* (1909). One night, as he approached a nomad encampment, he was met by "a pack of five or six ferocious-looking dogs... with long shaggy fur and very cruel looks. I had before then been told that when attacked by dogs of this kind I must not strike them, but that I should only ward them off, quietly waving a stick in front of their muzzles, and on this occasion I religiously followed that instruction, and found to my entire satisfaction that the dogs did not try to snap at me..."

Later in his travels, however, his luck changed. He was robbed by brigands and wandered about for three or four days, cold, hungry and suffering the painful symptoms of acute snow blindness. Finally, he was taken in by a group of Tibetan pilgrims who fed him and made him comfortable. The next morning, as his hosts were packing to leave, he met the camp dogs. "I finished my tea and went out of doors, while they were busily engaged in packing their effects," he writes. He walked to the end of the row of tents "when seven or eight ferocious Tibetan dogs attacked me, barking loudly. Handicapped as I was with the pain in my eyes, I could not deal with these dogs so deftly as at other times. At first, I kept my eyes open and brandished my two sticks, driving back the animals, which attacked me from all sides. But once I was obliged to close my eyes, and immediately a dog behind me seizes one of my sticks. The next moment another dog fastened his teeth on my right leg, and threw me down.

"I uttered a feeble cry for help, which brought several men on the scene, and they drove away the dogs with stones. But the blood flowed out abundantly from the wound, which I held fast with my hand, and I lay motionless until an aged dame brought me some medicine, which she said was a marvelous cure for such wounds. I dressed the wound with the medicine and bandaged it..." After that, the pilgrim and the nomads parted company and Kawaguchi went his lonely way.[47]

## *Matthiessen's mistake*

It is unfortunate that seven decades on, when Peter Matthiessen came to the Himalayas with the zoologist George Schaller in the 1970s on a research pilgrimage in search of the elusive snow leopard, they he did not know or heed Kawaguchi's advice about "quietly waving a stick in front of their muzzles." Early in their trek Matthiessen and Schaller encountered some ferocious dogs in a Tibetan refugee village on the Nepal side of the Tibet border. Their response was to lash out at the dogs, which put them in serious danger. It had been raining and as they entered the muddy lanes of a mountain village they were met by "snarling, straining animals on both sides." One of them broke or slipped its chain and attacked from behind.

Matthiessen was selected for the attack. "Luckily," he writes, "I heard it coming, and swung around upon it with my heavy stick," which only infuriated the dog. "Searching in vain for a heavy rock, I did my best to crack its skull, while the dog lunged back and forth at the tip of my stick in horrid fury. Meanwhile, [Schaller] had located a heavy split of wood; he hurled it at the dog with all his force. The brute dodged, then sprang after it, sinking its teeth deep into the wood. Finally, it was driven off by a Tibetan, who until now had watched calmly from the doorway of his hut to see how I might fare... I never walked without my stave again. If I had not cut it in the hour before (after eight days of getting by without one), I might have been hurt badly, and I marvel to this day at the precise timing."[48]

### Hedley's observations

Another pilgrim of sorts, a British missionary on a religious quest in Asia, has more to say about meeting the big dogs. In the course of an assignment in rural China, John Hedley set out on horseback to assess prospects for the Methodist church in Mongolia. In his book, *Tramps in Dark Mongolia* (1910), he describes how to deal with them. "Then, thinking of Mongolia," he begins, "you must take into account the dogs. They refuse to be overlooked or forgotten. Big and bony brutes, long-haired and shaggy, loud-voiced and vicious, they are to be feared and avoided. Never does a traveler approach a settlement unwelcome or unnoticed. Vociferously they rush out ever so far to meet you, barking and yelping at your heels, and even at times laying hold of the horse's legs in their excitement."

He goes on to advise horseback travelers to stay mounted when approaching a settlement, until someone comes out to drive the dogs off. "You are not charmed with them," he says, "but you soon learn to respect them." Travelers on foot were advised to carry two sticks, "one in the left hand to keep the dogs at bay, a shorter and stronger one in the right hand to drop on Towser's nose if he venture too near." The moment one enters the courtyard of the dogs' master, however, the clamor ceases "and they once again resign themselves to the wakeful dozing from which your coming has roused them. They bark at you and show their teeth no more, so long as you remain inside the yard or house. They will even condescend to make friends with you, and stand before you with eager eyes and wagging tails if you will fling them a crust or a bone. But put your foot outside the gate again, and the feud begins between you and doggie. War is once more declared, they are again your enemies..."[49]

### Snellgrove's peregrinations and a pup named 'Nying-kar'

So much for how to deal with the fierce and feisty. Another pilgrim who trekked along the Nepal Himalayas in the early 1950s has a much more pleasant story to tell, about the Tibetan mastiff pup he adopted on the long trail. David Snellgrove, a renowned British Tibetologist and Buddhist scholar, traveled for seven months across northern Nepal with his assistant Parwa Pasang Sherpa, their cook Takki Babu and their camp helper Lopsang. Their sojourn took them on foot from the high western

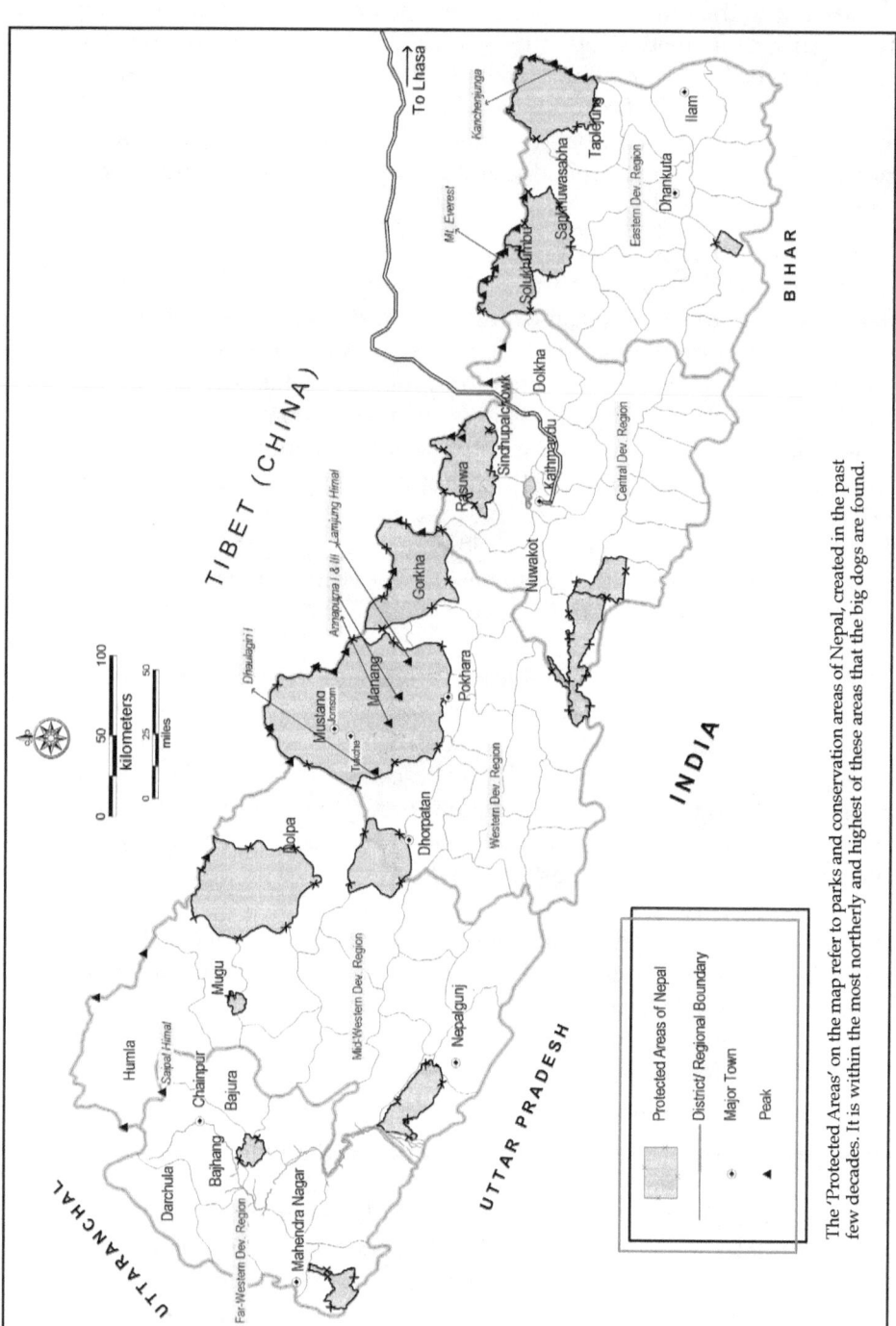

Map 2. Nepal (map by Ramesh Shrestha).

The 'Protected Areas' on the map refer to parks and conservation areas of Nepal, created in the past few decades. It is within the most northerly and highest of these areas that the big dogs are found.

## 2. The Noblest of the Party

Nepal district of Dolpa eastward across northern Mustang, Manang, Gorkha and Dhading Districts, ending at the northern edge of the Kathmandu Valley. (See Map 2, p.46) The story of their travels paralleling the Tibetan border, and of a dog named 'Nying-kar' has the makings of a modern epic.

*Himalayan Pilgrimage* (1961), "tells of a journey made through the Tibetan regions of western Nepal in 1956," Snellgrove begins. "Its title was unwittingly suggested by the people amongst whom we travelled, for used as they are to making pilgrimages themselves, they assumed that we too must be pilgrims. Such journeys are usually undertaken for religious merit, and although we were in pursuit of knowledge and experience, which is not quite the same thing, I would be sad to think that we had deceived our hosts in any way. The Tibetan word for pilgrimage means literally 'going around places' (*gnas-skor*) which describes our activities succinctly enough. To what extent the religious element is present in our case, the good reader must be left to judge for himself..." In the Preface he declares that his primary aim is "to give an overall impression of the peoples, amongst whom we were travelling."

"Our general plan," he goes on, "was conditioned by the weather, for we would have to visit the districts of interest to us on the southern side of the main Himalayan range between March and May before the monsoon gained force. Then we could conveniently cross over to the northern side and continue to travel there until September, for the high mountains would ward off the stormclouds and it would be warm enough to cross the highest passes, of which I estimated fifteen between 17,000 and 20,000 feet."

They started low, at Nepalganj on Nepal's southwest border with India. From there they ascended the Bheri river valley into the high, culturally Tibetan northern border district of Dolpa on the north (Tibetan) side of Dhaulagiri Himal. Then they trekked slowly eastward paying particular attention to the local Tibetan people, their culture and religion. On the way they visited Mustang and Thak Khola, then crossed the Thorung La Pass (17,770 ft, or 5416 m) into the district of Manang (locally called Nyeshang in Snellgrove's account) behind the Annapurna massif.

To avoid the monsoon they turned north and crossed another high pass over Kangla Himal into the ever more remote and much drier valley of Nar and Phu, two high villages near the Tibetan border. (This region has more recently and somewhat romantically been called 'The Lost Valley.') To ascend Kang La Pass (17,408 ft, or 5306 m), they climbed from Pisang to the village of Bangba (also known as Ngawal; 12,008 ft, or 3660 m). They had found many of the villagers of Manang to be (in those days) abusive and unfriendly. But those of Bangba and in Nar and Phu were generally more pleasant.

They stayed in Bangba two nights, time enough to explore various shrines, temples and monasteries round about. Throughout the trip,

Snellgrove kept accurate track of all religious images, frescoes and other artifacts seen, as part of his study of Tibetan religion.

When they returned to their camp the first evening, they discovered that someone had stolen their torch (flashlight). But, "The theft was repaid... in an unexpected and most happy manner"—by the appearance of a hungry black pup.

"He belonged to the fierce breed of the Tibetan mastiff, but was a mere puppy of six months or so," Snellgrove writes. "He settled outside the door while we were cooking our evening meal; at first we took little notice of him, but when one of the inquisitive villagers, who were hanging around, drove him away with a kick, he attracted my attention by a pathetic little yelp. We fed him that evening and the next morning he was still at the door, having slept outside in the rain. I fed him again, and when we left he followed at a respectful distance. When we stopped far up on the wet, misty mountain-side to eat a lunch of cold stewed meat and potatoes, he came and shared it with us, and I decided that unless an owner claimed him, this dog should be ours. This was not going to be easy, for food was scarce and it meant giving him food that the rest of the party could all too easily have eaten themselves. The others realized this from the start and when out of humour tended to resent his presence..."

Over the Kang La they "plunged down happily on the steep scree to the stream that would lead us to Nar...," commenting on the presence of blue delphiniums growing between the rocks alongside the trail. Nar consisted of "some thirty houses, most of them arranged along the sides of a wide track. Entering the village, we had to run a gauntlet of staring eyes and barking dogs. Our own dog made no response to this noisy welcome and we passed through to the upper end of the village where there was a pleasant site for camping on the edge of the grass-covered mountain-side..."

A few days later they trekked on north to Phu ("Upper Nar" in his account). *Phu* in the local dialect means 'end of the valley,' and it was a long trail with steep downs and ups, past ruins of ancient fortresses and monuments dating to medieval Tibetan times. When Snellgrove's party finally arrived at the cliff-top citadel of Phu they made camp, then set off to visit the nearby Tashi Lakhang temple. Inside, up a ladder on a second floor, several local men and women were engaged in preparing tea, to drink, and *torma* effigies for an evening ceremony honoring the 'Tranquil and Fierce' Buddhist divinities. As the salt butter tea was being served, "In hope of food our black pup clambered up the ladder to join us."

Their hosts showed little surprise at the sudden appearance of the strangers, and the dog, at the top of the latter. Everywhere they went, Snellgrove observes, the unexpected arrival of strange travelers, including an Englishman speaking fluent Tibetan, raised little concern or overt curiosity. The locals considered them to be like any other pilgrims on one or another of the many trail through the Himalayas. "These people were spontaneously friendly," he writes, "and even if we had been oblivious to the changed landscape outside, we should have known at once that we were in a very different land from Nye-shang..."

Eventually they returned down valley to Nar, and back across the high pass to Pisang, in Manang proper. "The black dog," he admits, "was now accepted as a member of the party and food was always somehow provided for him. In our present straitened circumstances it usually consisted of a bowl of buckwheat broth fortified with a lump of butter." And, by now the pup had been formally named 'Nying-kar,' meaning 'white heart' for the little spot of white on its chest.

Snellgrove's plan was to continue out of Manang, over another high pass, the Larkya La, eastward into northern Gorkha District. As they prepared to move on "There remained only the problem of the dog," he writes, "for we had heard that his owner was a man of Bangba who was away on a journey to Kathmandu. It would therefore be unwise to induce him to follow us, and I was persuaded to give my consent to the following plan." Because Snellgrove wanted to keep Nying-kar, they worked out a subterfuge to make it *look* like they had tried hard to leave it behind. "We were to feed him unseen by any villager that morning and otherwise to ignore him altogether. When he began to follow us, he was to be driven off with sticks and stones, but without actually being hit. Pasang and Takki Babu were sure that he would persist in following us, but the villagers would see that it was entirely against our will."

As expected, and oblivious to the abuse they threw at him, Nying-kar stuck close to the travelers, following them out of Pisang and down the trail in plain view of all. He was theirs now, for sure. The following week, however, as they were preparing to cross the Larkya La, it seemed as if both Lopsang the camp helper and Nying-kar the camp follower had second thoughts about continuing. The first to leave was Lopsang.

"To our surprise Lopsang said he was tired and would come no further with us. We were sorry to part with him... Our party was thus reduced to myself, Pasang, Takki Babu and the black dog." Over the next few days, however, Nying-kar also showed reluctance to continue on. While ascending Larkya La, Snellgrove writes, "Nying-kar would not come. Twice Takki Babu went back for him and twice he ran back to the [last] houses to keep company with the other dogs that were there. We could not force him to follow and so continued rather sadly without him." Then, the dog reappeared, "and to reward his faithfulness we shared our meagre lunch with him forthwith..."

Thereafter, the dog stayed with them—sort of. Several weeks passed as they trekked eastward and by the end of September Snellgrove's party was descending out of the high Himalayas on their final approach to Kathmandu valley. Some of the trail was along a river gorge across the face of a cliff, then up and down along a wooded track. It was there that the dog got into trouble. At one point they came to "a splendid waterfall, which plunged into a clear rock basin just beside the route, so we stopped to have a bathe. On the other side of the track there was a drop of about thirty feet down to rocks and shrubs. While bathing in the pool, we suddenly heard cries from Nying-kar, who had fallen down the precipice. He must have checked his fall, for he suffered nothing worse than a sprained back-leg, but even this made progress slow and difficult for him. He was very unhappy and clearly blamed us for his sufferings, since he refused to accompany us and would only follow at the heels of the last of the porters."

Now, as the dog hung back, Snellgrove's frustration begins to show. "If one of us kept to the rear, he would lie down and refuse to move. Thus we could do nothing but leave him to follow. He limped in late at our lunch-time halt and even later when we stopped for the night. The next day we tried to encourage him, but with no better results, and could only trust that he would continue to follow. By noon we reached Arughat and great was our excitement, for here we would join the main route through central Nepal, which connects the district headquarters of Pokhara, Gurkha [sic: Gorkha] and Nawakot with Kathmandu; it seemed we were already at our journey's end."

Takki Babu cooked a meal there, and although the party stayed for more than two hours, Nying-kar did not appear. They moved on without him, though on the following day Pasang went back to the trail, such was their attachment to the dog. The main eastbound track across the midhills now took them up the Ankhu river valley. The elevation was lower there, and they suffered from the heat. "It was irksome and oppressive as no other part of our journey had been," he writes. One wonders how the mountain dog must have suffered.

The next day Takki Babu bought eggs and rice in a village and cooked lunch. Pasang, who was still looking for the dog, had not caught up to them, so "we left a share for Pasang in the hope that he would soon follow with the dog. Pasang eventually caught up to them at a village called Katunje. "He arrived late in the evening leading the unhappy Nying-kar, whom I had scarcely expected to see again. By good fortune he had been found by two soldiers from the check-post at Jagat, who were on their way down to Gurkha. They recognized him and he consented to be led on a rope, an act of servitude to which he had never submitted before in spite of our well-meant efforts." At Katunje Snellgrove engaged new porters, including a small boy who was willing to carry the dog.

On October 4th, in the seventh month of his Himalayan pilgrimage, Snellgrove got his first glimpse of Kathmandu valley. "We could just see its western end and distinguish the little Newar town of Kirtipur. Pasang hastened ahead, so that he could get a jeep and bring it to meet us at

## 2. The Noblest of the Party

Balaju, where the motorable track begins. Meanwhile Takki Babu and I came along with the porters." When the trekkers arrived at Balaju, "The jeep stood waiting with Nying-kar by its side, and as we came to take our places in it, I was pervaded with a sense of happy fulfillment..."

That's where Snellgrove's *Himalayan Pilgrimage* ends. But, what happened to the dog?

Readers had to wait nearly four decades before Snellgrove revealed Nying-kar's fate. In 2000 he published *Asian Commitment*, an extensive retrospective volume chronicling his long life of research and adventure. And there, in a synopsis of his 1956 trip, he summarizes the affair with the dog. After they left the Manang village of Pisang, he writes, "A black dog had been following us for the last few days, a typical Tibetan black mastiff... We did our best to dissuade him..., but he attached himself firmly to our party, and accompanied us all the way back to Kathmandu. Pasang took him later to Solu-Khumbu."[50] Apparently, Nying-kar lived out his days in Solu-Khumbu, in the shadow of Mount Everest, the land of Sherpas and Tibetans, amidst yaks and herdsmen, and in the company of other big dogs of his breed.

I first read David Snellgrove's *Himalayan Pilgrimage* and faced my first Tibetan mastiff almost simultaneously, in 1964 while I was a Peace Corps Volunteer in Nepal. During a month's leave my friend Bruce Morrison and two other companions and I set off to trek and climb mountains in Mustang and Manang Districts near the Tibet border. Towards the end of our adventure we followed Snellgrove's route through Pisang village. Two weeks prior to that, in a yak pasture on a ridge in Thak Khola, high above the Kali Gandaki river, in full and striking view of the majestic peaks of Dhaulagiri and Annapurna, I encountered my first Tibetan mastiff. That mutually frightful and unplanned confrontation was the realization of a long held dream and the beginning of my own studies of these noble dogs.

# 3

# First Encounters

The British writer, Robert Macfarlane, author of the inspiring and philosophical *Mountains of the Mind* (2003), tells us that "To know a landscape properly, you must go into it in person." That's fine for mountains, but what about mountain dogs? I believe that we can read all we want about them, but (to paraphrase Macfarlane) to know great dogs properly, we must greet them on their own terms and meet them on their own ground, in the place where they come from. After reading stories about encounters with the big dogs by explorers and adventurers and pilgrims on the Roof of the World, I set out as a lad of twenty-three to meet them *in situ* as we say, in place, *their* place. I was fortunate to have been posted by the American Peace Corps to the Nepal Himalayas. I had a clear view of Himal Chuli and the Annapurnas and other great peaks a few miles north of my little village. They inspired me, and I knew that I would find big dogs amidst those peaks in the high valleys and alpine pastures. So, in the fall of 1964 three friends and I trekked off to climb and experience a true 'Himalayan high' and, at the same time (for me, at least), to find the original landrace dogs of the mountains.[1]

The first encounter occurred on a windy October morning high in the most magnificent mountains on earth. We were near the Tibet border on our annual leave, a week's walk north of our Peace Corps village. I was young and filled with admiration for the raw nature of the Himalayas, of icy peaks jutting into a cobalt sky, of rough-hewn shepherds of the high meadows, and of their yaks and their livestock guardian dogs—at least what I knew of all this from books. Coming face to face with the dogs of the mountains, however, aroused a jumble of emotions: shock, fear, surprise and, ultimately, respect and admiration. I recall it as clearly now as if it just occurred. It is a memory that has hovered in the back of my mind ever since, as a reminder, and a caution.

### The Alubari Mastiff

My friend Bruce and I call the dog that we met that day "The Alubari mastiff" because we met him at a place in the mountains called Alubari, the 'Potato Field.'

The view from the top of the ridge that crisp and blustery October morning took our breath away. We had just negotiated a dangerous cliff trail up out of Thak Khola, the local name for the high valley of the upper Kali Gandaki River in Mustang District. The last thousand feet to the viewpoint seemed endless, but once there, at close to 14,000 feet above sea level, out of breath and hurting from the climb, we stood and marveled at the view.

I shrugged off my Kelty pack and dropped it at my feet, then pulled the draw strings of my windbreaker hood tight to keep the biting wind off my sweaty head, all the while gawking at the scenery. From the ridge top we were not sure which was *more* astounding, the view *straight down* onto the flat roofs and yak caravanserais of Tukché village 4,000 feet below us or up another 10,000 feet plus above us to the snow peaks rising starkly against the blue sky.

"Amazing!...," is all I could say. "Simply amazing." We slowly turned in a 360 degree circle to take it all in. We'd read everything we could about these mountains and had planned this trek for almost a year. We had departed Tukché village at dawn, after a breakfast of salt butter tea, flat breads called *chapatti* (like a tortilla) and spicy omelet from the spotless kitchen of Chapal Singh and his gracious wife, our hosts of the Thakali ethnic group. We had spent a week walking that far from the mid-hills, then stayed the better part of another week in Tukché preparing for the highest part of our adventure.

Chapal Singh's house was a large caravanserai, a compound with a central courtyard suitable for keeping livestock (yaks, ponies and Tibetan mastiffs) and hosting traders. It was situated at the north end of Tukché, opposite the school yard and a Buddhist monastery; hence it was near the hub of much activity. The town appeared to us much like Maurice Herzog of the 1951 French Annapurna expedition had described it a little over a decade earlier—a "maze of alleys, and the houses, regular little fortresses, …mostly caravanserais where passing travelers can find lodging for the night." He then pointed out that "The majority of the 500 inhabitants are Buddhists, whose piety can be judged from the wall of prayer wheels, 50 yards long."[2]

Chapal Singh remembered the French and seemed to know everything else about the history, economy and culture of the town and of greater Thak Khola valley, Mustang District, and western Tibet a few miles away. When I asked in Nepali where we could see a Tibetan dog, a *Bhoté kukur*, he smiled. "Utha," up there, he said, pointing up the mountain towards the high yak pastures. He described them as large and imposing creatures that instilled fear in all predators and outsiders. We would know one when we saw one, he said. Indeed we would. Meanwhile, we enjoyed the antics of the yappy little Tibetan spaniel that accompanied his wife about the house on her chores.

Tukché, or 'Tukucha' on some old maps, grew up as a salt trading mart during the late 1800s and early 1900s when a clan of local ethnic Thakali entrepreneurs established themselves as astute long distance salt traders. During the height of the salt trade, Nepalese porters from the mid-hills carried rice in huge loads up to Tukché, where the Thakali *subba* (commissioners) bartered it for rock salt. The salt came on yak caravans down from the broad and inhospitable salt pans, the shallow lakes of the Changtang, Tibet's remote northwest region. Sometimes the Tibetan traders brought big dogs with them. The salt (and dog) trade continued until the 1950s, when it was almost totally halted following the Chinese occupation of Tibet.

After the week in Tukché buying supplies and hiring porters to carry our tents and food and to guide us up to our immediate goal, Dambush pass, we set out. We hoped to reconnoiter Tukché peak, north of the massive Dhaulagiri, for a future expedition. And maybe, while we were there, some of us could top its diminutive sidekick, Little Tukché.*

Almost directly after exiting Chapal Singh's caravanserai we began ascending the ridge, the first stage in our trek to the pass. Bruce and I set out first, while our friends Stuart During and Phil Brandt, accompanying the four porters we'd hired, would catch up to us later.

Now, standing on the top of the ridge far above Tukché, Bruce and I began putting names to peaks. South of us, etched boldly against sky, rose the magnificent and perpetually snow-capped Dhaulagiri, the 'White Peak,' the world's seventh highest. At the southeast was graceful Annapurna-I, tenth highest, named after the Hindu goddess of wealth and prosperity, the consort of Lord Vishnu. The three razor sharp Niligiris, the 'Blue Mountains,' were at our east. Closer still, southwest of us, was the cone-like Tukché Peak, so close that it seemed in the crystalline air that we could reach out and touch it. Tukché Peak is shaped like a great heap of Tibetan rock salt, apropos of the salt trade going on in the valley below.

Turning around, we looked north across the broad expanse of upper Mustang District to the edge of the Tibetan plateau. On the west side we admired the Luma Himal and, far off, a glimpse of Chungian Changma Himal and Mustang Himal along the border. In the northeast we could see the edge of Muktinath and Damodar Himal and a bit of Rijpuwa Danda. Few of these northernmost peaks of Mustang were over 20,000 feet, but were remarkable nonetheless.

Centuries ago upper Thak Khola and the whole feudal princedom of upper Mustang, known as Lo Manthang, were both part of a powerful old western Tibetan kingdom called Gugé. The southernmost outpost of Gugé was the Thakali village of Kobang (well inside modern Nepal), a few hours walk south of Tukché and within view of our ridge top aerie. Here we were

---

*Elevations (some are estimates): Alubari pasture c. 14,000 ft (4270 m); Annapurna-I 26,545 ft (8091 m); Dambush Pass (also called Dhampus or Thapa Pass) 17,225 ft (5,250 m); Dhaulagiri 26,794 ft (8167 m); Little Tukché Peak c. 18,500 ft (5640 m); Niligiri Peak (highest) 22,166 ft (7061 m); Tukché Peak 22,703 ft (6920 m); Tukché town 8,484 ft (2586 m).

standing on a ridge high in the sky, in a part of old Tibet, and although the modern political border was now situated a few miles north of us we were certainly still within *cultural* Tibet, in a land historically linked with the Tibetans. It is a land of snow peaks, glaciers and river valleys, of scattered Thakali and Tibetan villages and irrigated grain fields, of Buddhist and Bönpo monks and their monasteries, of trans-Himalayan traders carrying rock salt and animal hides and wool, of vast herds of lumbering yaks, goats and sheep, and (we had been told) of some remarkably big dogs of ferocious demeanor.

As we pondered all this, and took in the view, I wondered where the dogs were. "There's gotta be at least one big dog on this mountainside," I said. Indeed, there was.

## *Face to face with fury*

The Great Himalayas often seem larger than life, with the world highest peaks and some of the deepest valleys and, not least, big yaks and dogs in the high pastures. The first sight of the peaks has been known to leave outsiders speechless. Similarly, the first sight and sound of a Tibetan mastiff at these heights typically fills one with foreboding. Getting close to either one, peak or dog, is a formidable challenge. That morning, after our reverie at seeing the peaks so clearly, so close, Bruce and I came face to face with such a dog. *Too* close.

Turning away from the scenery, we looked across the windswept yak pasture we had entered. We saw no potatoes at Alubari, only grass, alpine scrub and a few yaks grazing. It was not they, however, that caught our attention. It was the sight, and now the sudden frenzied barking, of a very large dog.

Silhouetted against the skyline on the hill above us was a magnificent black-and-tan Tibetan mastiff wearing a heavy collar and bell, and securely staked out on a chain. He'd been napping, but was now wide awake to our presence. He was huge, wide-eyed, big-headed, with a fully feathered tail curled over his back. His hackles were up and his fury at our intrusion was immense. His bark was deep, sonorous, commanding.

"Forget Dhaulagiri!," I told Bruce, as I cinched tight the belt on my pack. "We've got *that* to deal with now," I said, pointing to the dog.

We approached the yak camp and the dog with apprehension. The closer we got the more agitated the beast became until, in a great rage, he threw his whole weight at us to the end of his tether.

This was our first introduction to a yak herder's protective big dog. Sure, we'd read about them in books and had both seen mastiff-type dogs in the towns and villages of the valley below, but none were as big, as close or as imposing as this one. I realized some years later, after studying these guardian dogs more carefully, that the Alubari mastiff's reaction to us was typical. First, he challenged us with loud barking, then lunged at us to the end of his tether. And then...?

Plate 3.1. *A typical* do-kyi *(tied watchdog) at Thak Khola, Nepal.*

The dog's tether suddenly snapped, and rocketing down the slope at us was a frothing mass of canine fury. The dog, it seems, was more shocked than we were. He was genuinely bewildered, and as soon as he regained his feet he stopped in his tracks, spun 180 degrees around and raced pell-mell up over the top of the ridge out of sight. Unexpectedly free in front of two strangely dressed foreigners was apparently too much for him. We were equally shocked that one of these fabled guard dogs would swap ends and disappear so quickly. He was gone. We never saw him again.

Off to the side, two herd boys laughed at us. They had watched the whole scene unfold, seemingly unconcerned about the disappeared dog. They knew he'd come back, sooner or later. When we asked the boys for directions to the pass, they politely pointed at the track going westward along the ridgeline up and out of the meadow.

### On to the Pass

After Alubari we saw no more dogs, but when a snowstorm blew in from the south that afternoon we faced something potentially more dangerous—a snowy whiteout. We survived that, and eventually made it to our destination, Dambush Pass. There, our Thakali porters had dumped their loads and retreated hastily back towards Tukché, leaving us to make camp. We stayed a week, taking day trips out and about. One afternoon we encountered a herd of feral yaks in what was called the 'Hidden Valley' on the French expedition map, north of Dhaulagiri. The beasts were so wild that we could not get close enough for pictures. The huge shaggy master of the herd, a large white male, snorted and tore vigorously at the ground with his hooves, signaling the rest of the semi-wild beasts to move away and for us to stay back.

We also discovered the tracks of a Yeti, the fabled Abominable Snowman of the Himalayas, in deep snow along the eastern and northern slopes of Tukché Peak. To our chagrin, however, when we followed them we realized that they were merely snow leopard tracks that had stretched out to Yeti size as they warmed up in the midday sun.

# 3. First Encounters

On the west side of the pass we came across the wreckage of a brightly painted red and yellow Pilatus Porter aircraft, a PC-6 model STOL (Short Takeoff and Landing), with the name 'Yeti' appropriately painted on its nose. It had crashed there four and a half years earlier, on May 5, 1960, while ferrying supplies to a Swiss Dhaulagiri Expedition. The pilots were lucky; they had stepped out of the wreck unscathed and walked safely down to Tukché. Eight days later, on May 13, without the further assistance of plane or pilots, the Swiss summited Dhaulagiri for the first time.[3]

Our main intent to reconnoiter Tukché Peak for a possible future attempt by some ambitious Peace Corps friends was thwarted, however, by too much snow and too little mountaineering equipment. We could not even get close. As a compromise, three of our team climbed a minor peak on the west side of the Hidden Valley.[4]

Looking back on it, the most excitement we got out of the trek to Dambush Pass was our startling encounter with the yak herders' dog and the spectacular mountain views. At one point while admiring the view from our camp at Dambush Pass we watched with astonishment as a cornice broke off the top of Tukché Peak and plummeted over 12,000 feet into the valley below. The power of Mother Nature was not lost on us.

By the end of the week we were out of time and food. We descended back along the same route we had come up, in the clear sunshine this time; no whiteout. Since we had no porters, and had lightened the load a little by eating our rations, we shouldered the heavy packs ourselves, and started back down the trail to Tukché. While crossing a steep scree slope I stumbled and fell and started sliding downward feet first. To stop, I dug my feet and hands into the shale and came to a stop grasping a hand full of stones. Turning them over, I discovered that one was a small black 165-million-year-old fossil sea shell. I still have it on my desk as a reminder and talisman of the trek. Strange how when I pick up that little fossil now, all these years later, it evokes so many memories of the trek, and a glimpse into Asia's geological prehistory. All those millions of years ago the shale found in these high mountains had lain at the bottom of the Tethys Sea that once separated the central Asian landmass (now Tibet) from Gondwanaland (now India). In light of all that antiquity, our brief sojourn for a week in 1964 amounted to less than a millisecond on the clock of time.

As we passed back through the Alubari yak camp on our way down to Tukché we saw neither yaks nor dog. No doubt they were out of sight and out of the wind around a corner where the grass was greener and more tasty. One of the herd boys met us and said that after breaking his tether and running off the dog did not return until nightfall. Were we that scary?

## Dark Nights and Dog Stories*

Over the years since that first encounter at Alubari, I met, acquired, bred and trained several Tibetan mastiffs, and exhibited my best dog, Kalu, in dog shows. Well before raising Kalu, however, I had several other memorable encounters with big mountain dogs. I remember one angry critter close-up and furious in a Sherpa yak herder's camp while on trek in the Mount Everest region. After passing through a dense oak and rhododendron forest, the trail opened out into a broad, green meadow full of yaks. The herder's camp was at the far end. As I stepped into the open, I saw the angry dog, and he saw me, and judging by what happened next I can say that he was one clever canine. Unable to reach me by lunging to the end of his chain, he spun around, backed out of his collar, and then attacked. I stood stock still at his charge. It was the yak herder who moved, making a spectacular flying tackle to catch the dog by the hind legs before it could tear me apart. If I had been recruiting line-backers for the Pittsburgh Stealers, I would have signed the fellow up on the spot (the dog, too, as team mascot).

In the early 1970s, while conducting research for my PhD dissertation, I spent part of one monsoon season in the high altitude Gurung shepherd camps of Lamjung Himal, in north central Nepal. Among other topics concerned with Gurung life and culture, I wanted a close hand look at the culture of transhumance pastoralism; I was especially interested in the role of livestock guardian dogs with the herds of sheep and goats. I documented the place of canines in the lives, culture and subsistence economy of the shepherds, working out of a rain-soaked tented camp in a high mountain pasture.

On many days Naresh, my Gurung research assistant, and I were engulfed in clouds. Ignoring the wet, we spent each day observing and interviewing shepherds (*gothalo*) in their camps called *goth* (pronounced 'goat,' with a breathy 't'). We learned early on to be extremely cautious when approaching a gothalo's camp. As soon as we were close enough to be heard we shouted, in Gurung: *"Nagi khad-Oh!"* ("Hold the dogs!"). That emphatic final *"-Oh!"* in Gurung has the same effect as shouting *"Listen up!"* in English.

Whenever we approached a shepherd camp the dogs set up a stir. Most were tethered (the classical Tibetan *do-kyi* or 'tied dog'), protecting the entryway to the shepherds' shelters. Usually someone came out to hold the dogs while we approached and entered. If nobody showed, we

---

*This account is adapted from my article 'Himalayan Highlands', *Summit* magazine (June 1973); see also 'Trekking high in the monsoon, and the shepherds of Lamjung Himal', *ECS Nepal* magazine (September 2008). See also my journal article, 'Gurung shepherds of the Lamjung Himalaya' (1974).

*Plate 3.2. Cream colored (white) Himalayan mountain dog in a shepherd camp of Nepal's Lamjung Himal.*

stayed back, for if we went closer alone it was not safe. Once greeted by a shepherd, however, we were inevitably invited inside, safely past the dog. The entryway into each hut is narrow, low, and lightless, and it takes a moment to adjust to the darkness inside. Most dogs are tied outside in the weather during the day, and those that remain untied come and go, or nap curled up by the door out of the rain. When departing, they usually ignore visitors whom they now recognize as friends of their owner.

A typical Gurung shepherd hut is made from large bamboo mats laid over arched saplings tied down securely to make a long, low watertight structure. There are many types of bamboo in the Himalayan forests, each with a variety of traditional uses. Among the best for roofing and fencing is a small variety called *ningalo*, hardly more than an inch in diameter. It grows profusely in the high forest (the *jangal* in Nepali and Hindi—from which English 'jungle' is derived). The local men cut, split and weave the ningalo into mats while it is still green and pliable.

Each hut is large enough to sleep two to four people (or more when needed). There's a fire pit for cooking, heat and light at one end, and a sleeping area and storage space for clothing, tools and equipment at the other. Among the shepherd's essential clothing are woolen blankets called *radi* and a sleeveless woolen cape called *boku*. The blankets and capes are made from sheep's wool sheared by the men and boys, spun by the village women, then shrunk in hot water and kneaded by the men with their feet to a felt-like density. The natural lanolin and tight weave make them warm and waterproof. Cooking utensils and wooden jugs for holding water comprise most of the rest of a gothalo's paraphernalia. (Nowadays, plastic jugs and buckets are ubiquitous, replacing traditional wooden and bamboo containers.) Some shepherds also own old flintlock guns and they are adept at making snares and traps for hunting birds and small game for food. Not least among a gothalo's supplies is an iron dog collar with spikes on it, which is placed around the animal's neck for protection from predators such as leopards and wolves, and sometimes from bears lower down in the dense forest.

One evening the shepherds from a camp a half hour's walk from our tent invited Naresh and me over for a meal. It promised to be a night of story-telling, singing and drinking strong rice or millet *raksi*, a Nepalese wine. We knew that our host had several fine guardian dogs and we hoped to get good information about them and their role in the sheep camps. At dusk our host arrived to guide us across the meadows. There were four of us: Naresh and me, our Sherpa camp master, and the shepherd guide. In case we encountered aggressive dogs, we each carried a long stave, and the Sherpa filled his pockets with stones. We had to cross the territory of other shepherds where the dogs were loose, and as we approached their territory they sounded the alarm by barking ferociously. Typical guardian dog behavior, I thought. A few years later it was confirmed when I read more about livestock guardian dogs in *Mother Earth News*. There Ed Andrews and Randy Kidd wrote that when disturbed "the dog... will first sound a barking challenge," and if that doesn't repel the intruder "the canine will 'rush' the stranger with tail upraised."[5]

Within minutes of starting out we were surrounded by five vicious canines. They darted in and out at us, snarling and snapping menacingly, creating a din that could be heard for miles. We shouted to a nearby goth for help, but nobody came out. We swung our staves and threw stones, and when one dog was turned away yelping in pain his mates attacked him and we made our escape. A few minutes later we reached our destination, shaken but unhurt.

The Hungarian Count Bela Széchenyí, who traveled to Tibet in 1879–80, describes a similar encounter. "One dark evening during an excursion," he writes, "I spotted the Monastery of Tun-Kurr in the distance. But no sooner did I move towards it when five of these fierce dogs came leaping at me out of nowhere surrounding me and trying to tear me apart. I swirled my spiky walking stick around to keep them at bay and I also drew my hunting knife. Unfortunately for once I did not carry my revolver. It took about ten minutes before I could fight these beasts off. My stick must have touched and hurt one of the dogs since it retreated howling. The other dogs followed me for a while and then retreated too. Later on I discovered that, during the darkness, I had walked by mistake into a yak herd which these dogs were guarding. Though I called out many times, none of the herdsmen came out of their tents. Apparently they were frightened of the dark."[6]

I know exactly how Count Széchenyí felt!

We spent a grand evening with the shepherds, eating, drinking, discussing dogs and listening to the old songs and sayings about life in the high pastures.

I took notes by candle light. Hours later, when it was time to return to our camp in the dark, we took a longer but safer route back along a hill above the pastures, to avoid the dogs. They barked loudly in our direction, from a distance, but none came up to find us.

The following day, sitting outside my tent in the early morning sun, I wrote up my notes: "By evening the sheep and goats have been herded back from the pastures to camp and the big dogs are turned out to guard them. These shepherds call their dogs *bugyu-nagi*, which in the Gurung language literally means 'highland-dog'... They are not herding dogs in the sense of steering or rounding up their charges (although they are sometimes seen to help), but they are guard dogs. Their sole purpose is protection. The dog's attention lies beyond the herd to the danger of bears in the forest [on the route up to the high meadows], leopards... [and] strangers... At night, when the dogs are the most active and alert, if a leopard, wolf, *buwaso* (the Asiatic wild dog, or dhole [*Cuon alpinus*]), or human intruder is detected, the darkness comes alive with frenzied barking and snarling, and shepherds shouting."

A Gurung song extols their dogs' virtues. One stanza sums it up, which I translated as: "Thrice in the night, our dogs make watch rounds of the *kharka*s (meadows)."

## Helambu Pup*

During the winter of 1966 I acquired my first yak herders' pup. I was trekking in the high pastures of Helambu, a mountainous region near the Tibet border north of Kathmandu. One afternoon I was invited into a Sherpa yak herder's black woolen tent and offered a cup of salt butter tea. When my eyes became accustomed to the dark inside of the tent I saw a black-and-tan pup asleep in the corner. After some bargaining I bought him for a few rupees, and when I left I tucked him into my shoulder bag to carry back to my apartment in Kathmandu. I named him 'Amjo Gipu,' Sherpa for 'Long Floppy Ears.'

Some months later I trekked back to Helambu with Amjo at my side. Here's what happened.

Amjo became fully alert when, stopped on the hill above the herder's camp, he detected the familiar smells and sounds of his birthplace. While he was growing up in my home, I had trained him to walk obediently at heel beside me on a leather service lead so that both of my hands were free to hold my walking stave and use my camera. Excited now, he tugged at the lead, begging to be let loose. When I unhooked him, he raced round and round me, barking loudly. Then he bolted down the hill

---

*Part of the story of 'Amjo', my Helambu dog, was published in the article 'Himalayan highlands', *Summit* magazine (June 1973), slightly revised and elaborated here.

and across the meadow towards the yak herder's wife who had come out of the tent to see what the fuss was. He recognized her, and in a flash was leaping with joy around her legs. Then he turned and charged back up the meadow for another race around me. He did this twice as I walked towards the tent, laughing at his every move.

While I sat inside the tent sipping traditional salt-butter tea, Amjo remained off-lead and for awhile I lost track of him. He was last seen nosing around outside, no doubt reliving some residual memories from puppyhood. After copious cups of tea—one of the essentials of Sherpa hospitality—I went out to find him. Amjo was nowhere to be seen. It was late in the day and, while I had been in the tent, the yak herd and a flock of sheep had returned from grazing to the nighttime safety of the camp.

Where was Amjo? The herder and I searched and eventually found him curled up sound asleep amidst the sheep. You cannot imagine a more contented looking dog. It was where he belonged.

On that trip I bought two large gong-like bells, the kind that the herders place around the neck of big yaks to keep track of them in the pasture. Back in Kathmandu, I hung the bells on the wall of my small flat. Occasionally, when I jangled them, Amjo would leap up out of a nap and race around the room, knocking over lamps and vases, and barking loudly in sheer ecstatic joy. I did it no more than twice, given the havoc he created.

Amjo was a strong, lithe dog, more of a Himalayan mountain dog than the heftier, hence more 'ideal' type of Tibetan mastiff. He was, nonetheless, a fine companion and watch dog. The Helambu Sherpas call this type of dog a *shikari*, 'hunter.' They are sometimes used to stalk deer in addition to guarding camps and livestock. In Kathmandu, Amjo faithfully guarded my home compound against strangers, stray dogs and occasional cows that wandered in to graze in the vegetable garden. I trained him to come, sit and heel, and to walk with me on the service lead. I also taught him to scale the six-foot brick wall around the yard, though I regretted it later when he used that trick to get out of the compound and fight street dogs. He was never hurt in those fights, however; any blood on him after a fight was always from the other dogs.

I often trekked with Amjo along the northern areas where Nepal borders Tibet. He loved it. After two years, however, I had to give him away, as I was about to leave for home in the USA where I had no place to properly keep a big dog. His new masters were a Swiss couple who worked with Tibetan refugees at Dhorpatan, a high altitude village near the Tibetan border in Nepal's northwestern mountains. During my last week in Kathmandu, about two months after I had given Amjo away, I met an American couple, Gail and Bob Bates of American mountaineering fame, recently back from a trek through the Dhorpatan valley. We were all dinner guests of another famous American mountaineer, Willi Unsoeld,

*Plate 3.3. The author with Amjo, a Himalayan mountain dog on trek in Nepal.*

who lived in Kathmandu with his wife Jolene and their four children. That evening over dinner, the Unsoelds and I were told an intriguing dog story.

"Usually," Gail began, "our Sherpa guide trekked with us each day, keeping us on the right path. He always carried a long stave and a pocket full of rocks to throw at any aggressive dogs we met. On the morning we entered the Dhorpatan valley, however, the Sherpa was far behind somewhere with the porters. Suddenly, we saw a pack of big dogs running out of the village and up the trail towards us. A big, dark mastiff was in the lead, barking furiously."

Then Bob took up the story. "We wondered what to do. There was no safe place, no Sherpa to protect us, and no rocks to throw. And we were in the direct line of attack," he said.

Then, to their astonishment, "The big dog in the lead leaped up, kissed each of us on the face," Gail put in, "then raced round and round our legs, barking. Apparently all he wanted was an affectionate hug. He led us all the way into the Swiss camp."

After a pause at the end of the story, I confessed that the big dog was my Amjo. "It's his typical behavior," I said, "and he especially loves Americans."

Amjo was eventually taken to Switzerland where he lived out his final years, but not before his Swiss owners spent a few months working on a project somewhere in central India. It's well known that most of India is not a good environment for big Himalayan dogs. It's too hot, and the locals treat them badly out of an innate but not unreasonable fear of large dogs. Nonetheless, Amjo survived all that while performing quite admirably his watchdog 'duty.'

One day, while driving through the Indian countryside, the Swiss project jeep broke down. "To find a mechanic," they told me later, "we had to walk a few kilometers back to the last town. The jeep was loaded with our personal baggage, so we hesitated to leave it; but no one wanted to wait there alone, either. But, we didn't worry," they said. "We put Amjo in charge, sitting on top of the baggage before we set out together to find help."

By the time they returned a few hours later a large crowd of villagers had gathered around, but not too near, the jeep. As the couple approached, they saw the attraction. There sat Amjo inside the back of the jeep on top of the duffel, barking whenever someone ventured near. "Nobody dared," they said. Amjo was too big, too protective, and too scary for anyone to come close. The baggage was very safe.

## Dogs at Kesang Guerrilla Camp*

During the 1960s and early 1980s I trekked again northwest of Pokhara in the mid-hills up into Mustang District on the border of Tibet. At one time in its history, Mustang was an integral part of western Tibet. The history of Mustang as part of Tibet has been written by modern scholars, and was recognized as early as 1793 by William Kirkpatrick, the first Englishman to visit the kingdom of Nepal. Kirkpatrick wrote in his 1811 book that "Moostang is a place of some note in upper Tibet...," by which he meant high western Tibet. In time, however, Mustang became a vassal principality under the suzerainty of the House of Gorkha and the Raja of Nepal and is now one of Nepal's most northerly districts.[7]

On one trip to Mustang a companion and I were especially keen on finding big dogs. On our way through the midhills, we encountered many dogs in villages and pastures that fulfilled the basic functions of watchdogs and livestock guardians, but few came up to standard as 'true' Tibetan mastiffs. They might better have been called Himalayan mountain dogs, though still exhibiting significant Tibetan mastiff ancestry.

The Dalai Lama once described Mustang as "one of the few places in the Himalayan region that has been able to retain its traditional Tibetan culture unmolested," and it was there that we found excellent dogs. The best of them were at Kesang Camp.

Kesang Camp was located in a remote high mountain valley. After the Dalai Lama had fled Tibet in 1959, Tibetan Khampa guerrilla fighters

---

*The story of Kesang Camp is an updated version of my article, 'The mysterious mastiffs of Tibet: Finding the dogs of Kesang Camp', *Dog World* (USA) magazine (1984).

established a safe haven there, from which they proceeded to fight a war of resistance against the Chinese occupation of their homeland. The local Khampa chief was Wangdu, a renowned resistance fighter. Kesang Camp was located a few miles south of the Tibet border close to the district town of Jomsom. In Tibetan, the resistance group was known as *Chushi Gangdrug*, an ancient name for the eastern Tibet region of Kham from which most of the partisans came. When they first came and set up Kesang Camp, Wangdu and his rag-tag fighters brought several big mastiffs directly out of Tibet with them to serve as watchdogs.[8]

From 1960, Wangdu's resistance force, supplied and supported by the American CIA, spent over a decade staging raids across the border against the Chinese troops in southern Tibet. Under political pressure from China, however, outside support eventually ceased and in 1974 the Tibetans were forced to flee Kesang Camp just ahead of the Royal Nepal Army. In doing so, they left the dogs behind. Some days later, as the Khampas were attempting to reach safety in India, Wangdu was killed in a Nepal Army ambush at Tinker La, an 18,000 foot pass in far west Nepal.[9]

Kesang Camp was then taken over by the Nepal Army and converted into a mountain warfare training base. The dogs, including one that the Nepalese named 'Wangdi' (after Wangdu), remained to serve their new masters.

I made two visits to Kesang in those years, once surreptitiously (out of a naïve curiosity and without permission) in the mid-1960s at the height of the Tibetan resistance movement, and again in the early 1980s (openly, by invitation) in company of the Nepal Army camp commander after it was established as a mountain warfare training base. When the news of my first visit reached the ears of the Tibetans' CIA handlers, I was abruptly called to the American Embassy in Kathmandu and issued a stern reprimand by the embassy's political officer. Two decades later, my second visit merited only a wink and a nod. It was during the second visit that my trekking companion and I found another Wangdi, a descendant of the first canine patriarch of Kesang Camp, and several other big dogs.

## *Magnificent mastiffs!*
The legend of their size and ferocity is told in every village in the region. 'You want to see Tibetan mastiffs?', villagers would say, laughing. 'Go to Kesang—*if you can get in!*'

On my first visit to Kesang I had traveled with two other Americans, an ethnomusicologist named Terry, and an adventurous high school boy named Regon. When we arrived at the district town of Jomsom (Dzongsam in Tibetan), we were stopped from going further north by a strictly enforced ban against foreign trekkers in the border area. Tibetan guerrilla action made this Tibetan frontier region dangerous. Even the Nepalese border authorities were hesitant to venture very far out of town.

Jomsom is a dry, wind-blown place with flat-roofed Tibetan-style houses and large caravanserais for traders and their pack animals (horses, yaks and yak-cow hybrids called *chowri*). It is only a few miles north of Tukché, the old salt traders' village. Its inhabitants are a mixture of ethnic Thakali and Tibetans. It is also a town of yapping dogs, including Tibetan 'spaniels,' popular as house pets. This is one of the few places where this breed is found in the Himalayas, probably brought back from Tibet proper decades ago by traders. Lack of scientific breeding or any form of control, however, lowers the overall quality of these dogs. As a result of mixed breeding the town has many dogs of a type that a friend of mine calls 'Spastiff'—a mixture of Tibetan 'spaniel' and Tibetan 'mastiff.'

In addition to my two human companions, my seven-month old Himalayan mountain dog Amjo trekked with us. Amjo was black-and-tan, a common color. Even at a young age Amjo was large, and his presence on trek with us created both admiration and concern from fellow travelers. Most Nepalese have a great fear of the big Tibetan dogs, which they call "Bhoté kukur."

There were Tibetan mastiffs north of Jomsom we were told, at the palace of the Raja (King) of Mustang (Lo Manthang), and also east, up in the mountains at the Khampas' guerrilla camp in Kesang meadow. I had heard of the impressive red mastiffs belonging to the Mustang Raja, but the presence of big dogs at Kesang was news to me. The Raja's walled fortress town was off-limits to us and a long walk north if we could have gone. It was well within the Nepal government northern border area restricted zone where no foreigners were allowed. But Kesang Camp, high in the mountains nearer to Jomsom, was only a half day's walk beyond the small village of Thini. Legally, Kesang was also off limits, but when no one was looking we sneaked out of town and into the hills to go see it.

At 4 o'clock one morning, Terry, Regon and I left Amjo tied in front of our tent at the back of one of the town's large yak caravanserais, scaled the compound wall in the pre-dawn light, then slipped quietly and unseen out of the town. The track took us uphill past the Thakali village of Thini and into the mountains. Kesang Camp was up there, directly beneath the awesome, icy north face of Tilichho Peak (23,405 ft, or 7133 m) the northern most summit of the Niligiri Himalayan massif. It was a beautiful morning with clear blue sky overhead. It was also quite cold. The sun did not come up over the mountains until we were well along. For the first few hours we saw nobody, though we had an uncanny feeling that we were being watched.

We arrived within sight of Kesang about noon, and were met and surrounded by eight or ten burly Khampas. None spoke Nepali or English, but they did the inevitable—took us to their leader. He was a determined young man who by outward appearances looked more like an illiterate, rough-hewn yak herder's son than a guerrilla army commander. We knew by his demeanor that this must be the fabled Wangdu, the resistance leader, though he did not formally introduce himself. His reaction to our arrival was abrupt and commanding. He did not expect us, nor want us there. We looked nothing like the CIA operatives he was used to meeting.

"What in hell are you doing here?" he sternly demanded to know, in colloquial English. (He had learned English while being trained in guerrilla warfare at Camp Hale, Colorado, a secret CIA base in the American Rockies that we had heard rumors about.)

Terry was the first to reply, saying that he was a musician, an ethnomusicologist studying Himalayan folk music. He was searching for a rare Buddhist musical manuscript among Tibetan monks, he said, and wanted to meet the lamas in the camp's *gomba* (temple), which we could see on the far side of the camp.

"We have no gomba!" Wangdu declared. But from where we stood on a knoll at the northern edge of the camp we could plainly see the tell-tale red temple building.

For his part in this dialogue, Regon, explained that he collected butterflies as a hobby and that he was looking for some on this trip. At this, our interrogator was taken aback. Tibetan Buddhists have trouble taking any life, even an insect's. (One wonders how they fought a war!) The Khampa leader promptly informed us what we had already guessed, that there were "no butterflies" up here either. But of course, there are many varieties of butterflies even at this high altitude, though at the time we did not argue the point with him.

My own excuse was dogs. "I want to see the famous Tibetan mastiffs of Kesang Camp," I said.

You know the answer: "We have no dogs!" Wangdu declared without hesitation.

Just then, a commotion in the camp below us set off a clamorous uproar from several of the camp's—"we have no"—Tibetan mastiffs. The Khampas did have dogs, of course; they were chained out of sight behind the buildings. The commotion began when a Tibetan man in combat attire emerged from a hut along with large Alsatian at heel. As man and dog crossed the open compound, they demonstrated a well-rehearsed attack-dog training routine (to impress us?) then disappeared into another building. This put the unseen mastiffs into an uproar, and it reminded us that we were the sort of intruders those big dogs were prepared by instinct and temperament to repel. From the sounds of it, they could do that quite well.

My thoughts were interrupted by our host's abrupt repeat pronouncement: "No gomba! No butterflies! No dogs! You've no business here!"

This was neither the time nor place for sightseeing. This man and his companions had only one thing in mind, waging guerrilla war; and though we were not their enemy, we were a nuisance to be rid of as quickly as possible.

"You must leave," he said firmly.

We agreed, and turned to go. Behind us the dogs' barking rang out clearly across the mountain sanctuary. As we trekked down out of the mountains we were escorted part way by a two stern Khampa partisans. We were also watched from the ridge tops by men armed with binoculars and guns. We could see the sunlight glinting off of them as we descended the trail away from Kesang Camp.

When we arrived back in Jomsom late in the afternoon, Amjo met us with great excitement. So did the district police, who ordered us to leave the region immediately. We were informed that we had broken the law by going to Kesang Camp and that we were no longer welcome. We departed Jomsom that evening, and within two weeks we were out of the mountains and safely back in Kathmandu, somewhat wiser but still curious.

At Kesang Camp we had found the big Tibetan dogs that I sought. Yes, and we had heard their distinctive deep challenge bark. Loud. Deep. Dark. Authoritative. But closer inspection of the Kesang dogs was out of the question. It was disappointing. My goal was unfulfilled. I had not actually *seen* the dogs. Over the next months and years they became a fading memory.

### Finding Wangdi, the canine patriarch

During the 1970s, the saga of the Tibetan guerrillas hiding out in the Himalayas of Nepal suffered an abrupt about face. These tough mountain men had earlier been well supplied with American arms to keep up their resistance movement against the Chinese. But the American CIA's clandestine support to them suddenly collapsed following Henry Kissinger's secret trip to Beijing in July 1971. His trip eventually led to the opening of Communist China to the West. The CIA was abruptly ordered to stand down, and the Khampas were virtually abandoned. As a result, the Khampas had to flee the resistance movement and their Himalayan hideout. Some of them attempted to go west and then south into India, leaving their beloved guardian dogs behind. When the Royal Nepal Army moved into Kesang Camp, the officers and men took care of the dogs as best they could. Under the army, Kesang continued to remain off-limits to strangers.

In the early 1980s I returned once again to Mustang District, accompanied by a friend named Lance. Along the way (most of a week's walk) we looked for good dogs, but saw few that measured up to 'true' Tibetan mastiff. Then I remembered the dogs of Kesang Camp and told Lance the story of my earlier encounter. We wondered what had become of them.

One evening at a trekker's lodge in Jomsom we were joined at our table by a Nepal Army officer. (He'd probably heard me speaking Nepali with the staff, and knew that we were not typical tourists.) He introduced

## 3. First Encounters

himself as the commander of Kesang Camp. We chatted awhile, and as we sipped locally made apple brandy, a specialty of the inn, he told us that he had trained in the United States in tank warfare. Besides the fact that the Nepal Army owned no tanks, we wondered how they could operate in these rugged mountains. (They don't.)

Before long our conversation turned to Tibetan dogs, and I asked him about those in his camp.

"Yes, they're still there," he said, and assured us that they were just as large, protective, respected and admirable as ever. When he saw how interested we were, he surprised me by inviting us up to see the dogs. As camp commander, he said, he could bend the rule against foreigners going to Kesang.

A few days later Lance and I joined him on the trail up to the camp.

The sky that winter morning was exceptionally clear. The Himalayan peaks around us were icy white, cold, forbidding, and seemed very close. The north wall of Niligiri massif and Tilichho Peak hung ethereally over us as we trekked upward into the mountain sanctuary. The great white bulk of Dhaulagiri peak filled the horizon at our south, its familiar snow plume streaking off the summit like a flag. We were clearly on the Tibetan side of the Himalayas, although still within the political bounds of Nepal.

We saw only one other person on the trail that morning, an old man with a basketful of radishes for the Kesang Camp kitchen. A little brown Tibetan spaniel trotted at his side, keeping him company on an otherwise empty track.

Our host rode a small white Tibetan horse, which he reined in and rested frequently on the switchback trail, allowing us to keep pace afoot. From a small portable radio he carried we listened alternately to Kenny Rogers singing ballads, static from a Voice of America broadcast, and Kiev jazz from Radio Moscow. (It reminded us how close we were to the steppes of Soviet Central Asia.)

On the trail in this otherwise vast mountain landscape we were mere tiny specks. The scrub-strewn brown canyons and arroyos were empty, arid, rocky and deeply eroded. A brisk chilled wind blew down on us off the peaks, and Himalayan Griffons soared gracefully on the thermals overhead.

In time we entered the sparsely forested Kesang mountain sanctuary, sweaty and tired from the steep climb. At the camp's perimeter we were smartly saluted and passed into the compound by neatly uniformed Nepal Army sentries.

The silence with which we were greeted was uncanny. Where were the dogs, I wondered. We expected to hear their frenzied barking challenge, or to be 'rushed' as unwelcome strangers. But nothing so dramatic or frightening happened. Rather, we saw four or five large dogs napping in the sun, at rest after their all-night vigil around the camp.

One large black dog was chained. Others were loose. The commander ordered several men to hold them as we passed. "They are not to be trusted with strangers," he said, none too reassuringly.

Each dog raised its head and watched attentively, fully alert as we approached. None, however, seemed overly concerned at our presence. They surely recognized our host, his horse and the bridle bells, but detected no alarm in his manner nor in ours.

Tibetan mastiffs are rarely aggressive unless seriously provoked or badly abused and mistreated by prolonged confinement or constantly chained, for example. Their behavior at Kesang, we noted, was typical. While we assumed that strange intruders would be summarily repulsed, as guests we were allowed in without any trouble, especially after being gently introduced by the dogs' master. After a brief examination, as visitors we were considered safe by the dogs. So much for the myth of the innately ferocious and uncontrollable Tibetan mastiff.

Before we could examine the camp dogs more closely, however, we were invited to the officers' quarters to rest from the trek. It was a chance to catch our breath in the rarefied air (11,500 ft). It was then that I realized we were in the courtyard of the very temple I had seen years before. No longer a *gomba*, a Tibetan Buddhist shrine for meditation and prayer, it was now the Officer's Mess.

We were joined by several other officers and as we sat in the courtyard in the warm sunshine drinking Star Beer from liter bottles we chatted about sports, mountaineering, hunting and, ultimately, dogs. I asked them about the protective temperament of the big Kesang dogs. One officer replied that they are basically "night dogs". That is the time when they roam the perimeter of the camp repelling any two-legged strangers that might approach, and all four-legged marauders such as jackals and leopards. The officers also told us how the members of a German expedition had stumbled unexpectedly into the camp a few years earlier, after losing their way down out of the mountains. The dogs raised the alarm and in the uproar the climbers beat a hasty retreat to a safer route.

The officers asked us about how to improve both the feeding and the breeding of their beloved dogs. We did our best to answer their questions.

From what we saw of their size and general conditions, the dogs seemed to be getting enough to eat—mostly leftovers from spicy rice, meat and vegetable curries. But quantity does not make up for quality. Under Nepal Army care the dogs may have had plenty, but were probably not getting the high protein from meat and milk byproducts, nor the wheat and barley-based diet that their former Tibetan masters had fed them. That traditional diet, which the ancestors of these dogs had eaten for hundreds of generations, is not the same as what is available from the plates of their new masters from the lowlands. The present generation of dogs appeared to us to have suffered as a result of the change in diet. Nutritional problems may account for the somewhat smaller than expected size of some of the dogs. Poor nutrition may also account for other problems, like the slightly splayed forelegs of one of the camp's largest dogs. But, as neither of us was a veterinarian or nutrition specialist, we could only speculate.

When we heard about the camps' breeding program we knew where the most serious problems were. The officers told us that they regularly gave away the best of each litter and that they allowed the dogs to breed indiscriminately.

"That's a mistake," I said. "If you really want to improve breeding to produce better dogs, you'll have to keep the best bitches and permit them to breed only with the best, most well-conformed males."

They agreed, and the conversation turned to some of the faults and the favorable points in the dogs' conformation. We spoke of such characteristics as color, size, voice, and temperament or disposition. The camp officers knew generally what to look for, but had applied that knowledge haphazardly to breeding practices.

After talking, Lance and I went out to inspect and photograph the dogs. Several officers held the dogs while I measured each at the withers, and Lance took notes. Under the circumstances, however, we were unable to determine their weight.

We were most impressed with two large black males, one named Lata (it literally means 'dumb' or 'mute,' from *lato* in Nepali) and another named Kali ('Blacky,' from *kalo* meaning 'black'). They were heavily built dogs and strong-boned, each with a massive head and shoulder, a thick mane, a well-formed, broad muzzle with properly pendulous upper lips at the flews. The large head, solid blocky muzzle, heavy mane, broad, strong fore body and deep brisket are diagnostic of Tibetan mastiffs. So is the deep, sonorous bark. But, we never heard these dogs bark (though the dogs' barks that I had heard during my earlier visit perfectly fit the description.)

Kali and Lata were superb specimens. Their coats were thick and healthy to the look and the touch. We had no trouble handling them, after properly cautious introductions. Once they were roused from napping, they stood proud, noble and alert, and untroubled by our attentions.

Lata was chained, but not aggressive. He was coal black, with a white spot on his chest—the 'white heart' so much desired on black dogs by the Tibetans. He stood 26 inches at the withers, about average according to international standards of the Fédération Cynologique Internationale (but slightly on the small side by the standards written later by American Tibetan mastiff clubs). This strong five-year-old was father to Kali, his look-alike.

Kali was 12 months old, not fully mature. He was only twenty-three inches at the withers, but the rest of his looks more than made up for

*Plate 3.4. Lata, the Tibetan mastiff of the Khampas, in Kesang Camp on the Nepal-Tibet border.*

his youthful shortness. He was solid black and well-proportioned with a thick coat, mane and full, bushy tail. Kali had free run of the camp, but seemed to prefer hanging around the Officer's Mess where he napped in the sun in front of the great wooden door.

Next we met the two bitches of the camp, four-year-old Kanchi ('Little Girl' or 'Youngest Sister') and one-year-old Julie. Both were quite small and did not impress us. They each stood only 21 inches at the shoulders. They were thin, and Julie, in particular, was ill-proportioned. Their generally low protein diet of leftovers allowed for less than full potential in growth. Kanchi had good coloring, black with tan eye spots, and tan on her muzzle, chest, legs and vent. Julie was yellowish-grey, with light-colored facial markings and light spectacles or 'mask' that reminded me of a Malamute's face.

The fifth dog, Tiger, was led to us on a rope leash. He was considered dangerous, but displayed no aggression towards us. His battle scars attested to some ferocity in the past, however. He had clearly fought with other dogs and probably with jackals in the forest. We were even told that he had mated with a wild jackal—the stuff of legends.

Tiger was a poor example of a Tibetan mastiff. His coloring was not good. His neck and back were light brown, accented by an even lighter undercoat and markings. His undersides were yellow-grey, and his muzzle was grey-white and marked like that of Julie, his daughter. Only Tiger's size impressed us; he was 28½ inches at the withers, large by any standard. Nonetheless, we politely suggested that he be removed from the breeding pool of Kesang dogs. (Tiger may, in fact, have been descended from the Alsatian I had seen here twenty years earlier.)

The officers mentioned two other dogs of lower quality, but we did not see them. There was one more dog at Kesang, the patriarch of the camp, old Wangdi. We would see him, they said, after lunch.

Lunch was served in what had been the God Room of the former Buddhist temple, now the Army Officers' Mess. We sat with the officers at a large table in the center of a room filled with military memorabilia, medals and trophies of the regiment. Mountaineering equipment covered one wall—ice-axes, pitons, crampons, ropes and other expedition gear. Silver cups, plaques and other awards from various competitions filled a case at one side that had once housed Tibetan sacred books. Pictures covered another wall, with the King and Queen of Nepal in silver frames prominently at the center. A large cast iron woodstove stood at the front of the room, a necessity on cold winter evenings.

We ate in the traditional Nepalese style—with our fingers, and with little conversation. We were served great quantities of boiled rice, curried vegetables and copious helpings of spiced meat in a thick sauce (wild goat that one of the soldiers had shot). The bones and leftovers all went to the dogs, particularly to Kali who was called into the yard.

After eating, we gathered in the sunny courtyard once again to take our leave and bid farewell to our generous hosts. It was afternoon, and we had to start down for Jomsom now if we were to arrive back before dark.

As Lance and I were escorted back across the camp compound, we came across old Wangdi asleep on the flagstone path. In his prime, this big dog had been the best in the camp. He was black, now going to grey, with tan points. He was named after the last Tibetan guerrilla commander of Kesang Camp. Legendary Wangdu-the-man was killed in September 1974 by the Royal Nepal Army as he tried to flee Nepal following collapse of the Tibetan resistance movement. He was ambushed on Tinker-Lipu La, a high pass in far west Nepal near Saipal Himal. Wangdi-the-dog, however, was still alive, but old and lame, his formerly lustrous coat matted and ragged. A cyst-like growth protruded from his rump. The men tried to rouse this tired old dog, but he was uncooperative. We left him reclining in the sun.

Wangdi was probably whelped not long after the first big dogs were brought from Tibet to guard the Camp. The descendants of those first dogs still guarded the mountain stronghold; but though they retained their dignity, their circumstances had changed dramatically. They looked to us like old nobility from a distant past—no longer enjoying the same respect they had commanded nor the prestige to which they were previously accustomed in old Tibet.

As we walked out of camp the dogs watched, but were as silent as when we had arrived.

## Notes from Mustang and Dolpa
## Tibetan Mastiffs on the Nepal-Tibet Border
By Alton C. Byers*

The Mustang and Dolpa Districts of northwestern Nepal are largely arid, treeless extensions of the Tibetan plateau, sheltered from the summer monsoon rains by the Annapurna and Dhaulagiri massifs to the south. (See Maps 3 and 4, pp.75-76) Rainfall averages less than 300 mm/year [12 inches], and sod grasses and steppe shrubs cover the rugged, high altitude landscapes. The regions are sparsely settled by Tibetan and Tibetan-like peoples (the Bhotia) who practice Tibetan Buddhism and Bön religions. Barley, wheat, potatoes and buckwheat are grown in oasis-like irrigated fields, which generally surround the clustered, flat-roofed houses of rammed earth construction. In addition to agriculture, the inhabitants have traditionally been involved in the salt, wool and cereal grain trade between Tibet and southern Nepal, and also engage in raising yaks, yak hybrids and sheep and goats. Life is hard in these "last strongholds of Tibetan culture in Nepal," but to visitors from outside, these places possess a timeless mystique that is not soon forgotten.[10]

If you mention Mustang or Dolpa to Nepalese hill and mountain folk, it nearly always elicits excited descriptions of the *'thulo kukur'* or *'Bhoté kukur'* (Nepali for 'big dog' and 'Tibetan dog,' respectively) that inhabit these regions. The dogs protect villages at night, guard and herd the yak and sheep, assist their masters in hunting *tom* (Tibetan bear), *lau* (or *lawa*, musk deer) and *na* (blue sheep), chase away bandits, and have been used in warfare, possibly for thousands of years. They are most often described as "fierce and dangerous" animals that "will attack any stranger coming near." This is the Tibetan mastiff—a powerful and practically fearless breed of dog, highly prized by the people of northern Nepal and cautiously respected by all others.[11]

The mastiffs of Mustang and Dolpa are large dogs, with males often reaching weights of 44-55 kg (100-120 lbs). They are characteristically square-headed with massive jaws, and have long, thick hair, a distinctive black-and-tan (sometimes all black, or golden) color, with a bushy, curling tail. Their presence clearly marks the transition from Hindu Nepal to the Buddhist realms of northern Nepal and Tibet, as one treks northward through the mountains along ancient trade routes. One sees smaller, apparent cross-breeds at the transition points in lower elevation towns and villages, and larger, more purebred mastiffs further north and higher in elevation. Because of their relatively greater importance in the

---

*This account of dogs in Dolpa (with two maps) was prepared by Alton C. Byers specifically for this book, based on research conducted during the 1980s. Alton Byers is Director of Research and Education at The Mountain Institute of Washington, DC.

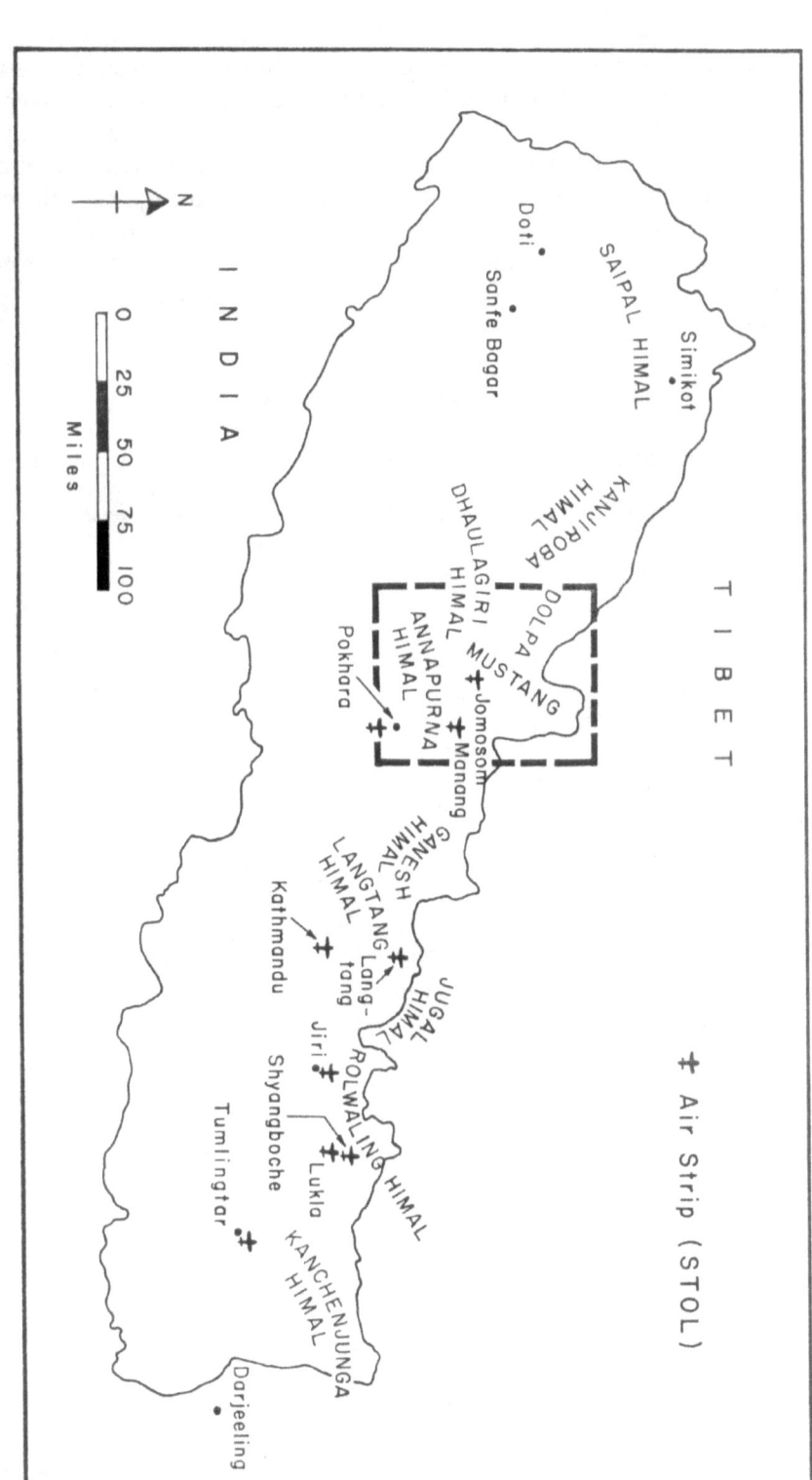

Map 3. Nepal Himalaya.

pastoral economy, mastiffs are treated with much greater deference by their owners here than are other common village dogs of the lower hills.

In these more remote Nepal-Tibet border regions of the Himalaya, the primary roles of the mastiff are to protect villages, private compounds and monasteries against thieves, and animal herds against thieves and predatory wild animals. The Tibetologist Robert Ekvall, who analyzed the role of dogs in Tibetan nomadic society in the 1930s, later wrote that the security created by the dogs, "is not only a defense raised against attack and thievery from outside the community, but forestalls pilfering and casual or surreptitious borrowing within that community... [B]ecause of the watchfulness of the dogs, nothing may be casually picked up or appropriated, and the community itself is spared much bickering and suspicion between its members because possessions have gone astray."[12]

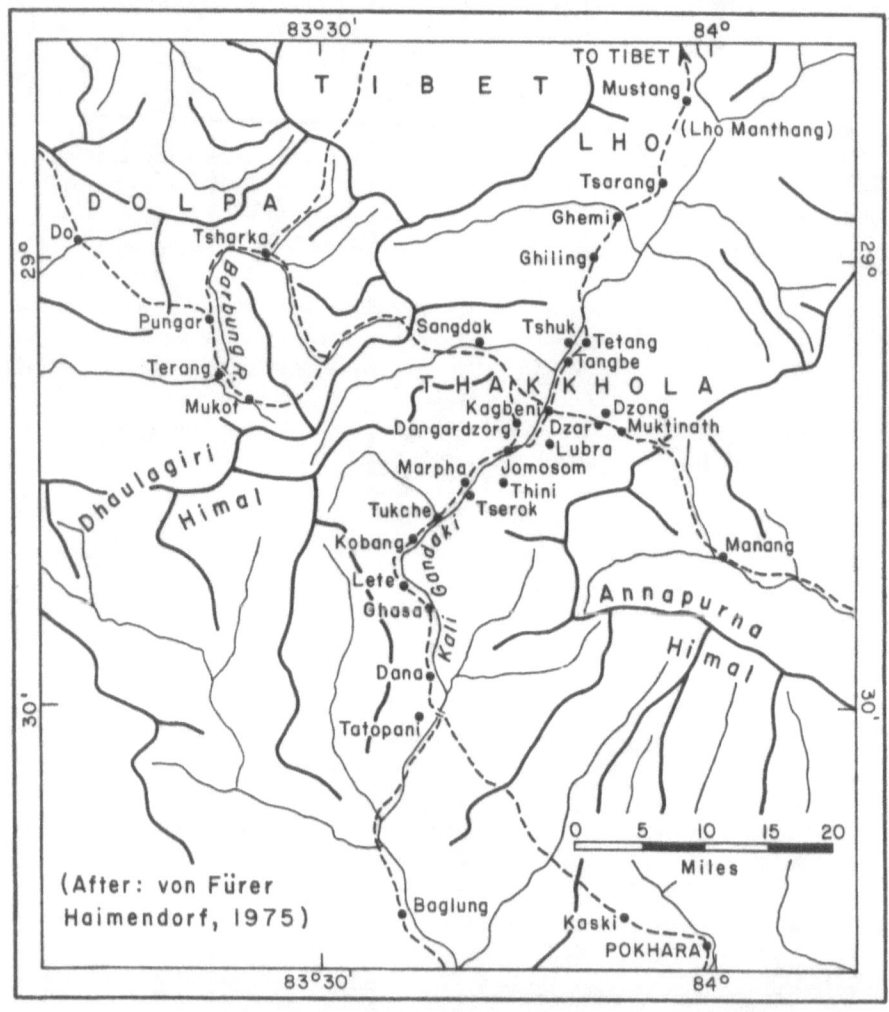

Map 4. Eastern Dolpa and Mustang Districts, West Nepal.

## 3. First Encounters

In villages, the dogs are chained during the day, which tends to encourage viciousness, and let free to roam the alleys by night, which induces fear in all who venture out after dark. They are a particularly strong deterrent against intruders in the walled villages of upper Mustang, from Kagbeni north to Lo-Manthang, and in Tibet, where escape routes are practically non-existent and all doors are firmly bolted at nightfall. Tibetan mastiffs (and related dogs of mixed breed) also discourage any unnecessary wanderings in the more dispersed villages of lower Mustang in the Thak Khola region, from Kagbeni south to Ghasa (see map). For example, in the 1980s while I was working on a conservation project in the administrative center of Mustang District at Jomsom in Thak Khola (the valley of Thak), many a jolly late-night session eating *momo*s (traditional Tibetan meat-filled dumplings) and drinking *raksi* (local wine) with my Nepalese colleagues in the local Thakali lodges was cut short by the knowledge that the mastiffs kept in the nearby Nepal Army camp would be unchained at 10:00 pm. The dogs effectively enforced this informal night curfew, as by 9:45 we were all safely home.

Even during the day, the mastiff guard dogs are a source of concern to most travelers. The naturalist and travel writer, Peter Matthiessen, in describing his search for the illusive snow leopard in Dolpa in the 1970s, found that the dogs are considered to be "so fierce that Tibetan travelers carry a charm portraying a savage dog fettered in chain; the chain is clasped by the mystical 'thunderbolt,' or *dorje*, and an inscription reads, 'the mouth of the blue dog is bound beforehand'." (See p.26.) Matthiessen then mentions a rather typical account of his and naturalist companion George Schaller's efforts at thwarting one such "blue dog" attack, and for the rest of their journey through Dolpa he writes: "I never walked without my stave again." (See p.45.)

The fortress-like architecture of many villages in Mustang indicates the high level of inter-kingdom warfare that occurred in the medieval past in this part of the Tibetan borderland, and mastiffs have played substantial roles in past and present battles. The French adventurer Michel Peissel, writing of his travels in *Mustang: The Forbidden Kingdom* in the 1960s, states that the Tibetan Khampas (the militant guerrillas who fought the Chinese during the 1960s and early 1970s), "trained these fierce Tibetan Mastiffs to fight in battle with them. The dogs... were taught how to jump on a rider and knock him from his horse. I believed this, as the dogs we had begun to encounter... were all of tremendous proportions." It is very likely that combat training pre-dates the Tibetan-Chinese skirmishes, and that mastiffs most certainly performed sentinel duty for the roving armies of the Mustang Raja ('King').[13]

Robert Ekvall described quite clearly what happens to unwary horsemen who are attacked and brought down by the big dogs. One of the dog's natural inclinations is to hamstring the intruding horse, which then typically panics, bolts and bucks wildly. The unwary riders, in two of Ekvall's accounts, fell into the midst of the dogs and never got to their feet. He also describes certain "clearly defined rights" as to how guard dogs may or may not be treated, including what weapons may legitimately be used against them and which are banned. Because of the dogs' great importance, particularly among nomads, weapons of immediate destruction, such as spears and guns, are prohibited—"unless, indeed, one is prepared to become embroiled with the owners."[14]

Mastiffs have also traditionally been used as guard dogs for the many yak and sheep herds of Dolpa and Mustang. Yaks are usually grazed in the high pastures over 3675 m (12,000 ft) above the villages. Wolves, bear, lynx, jackal and both spotted and snow leopards inhabit many of these remote regions, and the mastiffs play critical roles in the protection and herding of yaks, and as companions to the isolated yak herders.[15]

### Chokgya Lama's dogs

Much of my information concerning mastiffs in Dolpa was obtained from Chokgyel Lama ('Chokgya'), a former yak herder from a village near Tarakot who later moved to the Tashiling Tibetan refugee community at Pokhara, in Nepal's middle-hills. Over the years, Chokgya and I trekked and worked together throughout northern Nepal, and he was a storehouse of practical information regarding life in the mountains. He was born in Dolpa around 1940 and until he was 20 years old he tended the family yaks, ploughed his brother's fields, cut firewood and accompanied his brothers on trading excursions to Jumla in far western Nepal, and to Tibet. His appreciation and respect for the hard-working mastiffs frequently surfaced in accounts of his early years in Dolpa. For example, Chokgya told me that from the time he was twelve he stayed in the *goth* (pasture) herding yaks where his only companions were two large dogs. "In Dolpa," Chokgya said, "it is very difficult to see other people because the region is so remote, and people so few. Once I was in the pastures for four or five months and never saw another man. During this time the dogs were very good friends and were especially useful animals, especially for protecting yaks. I would feed them tsampa (ground, roasted barley) and water curd (buttermilk), and from time-to-time would feed them meat."

*Tsampa*, butter and *churpi* (hard cheese) are the staples of most Tibetan people, and yak herders will carry up supplies of tsampa, cheese and spices to the pastures. They are also entitled to eat all they wish of the milk and milk-products from the herds they tend, in addition to the meat of a winter- or predator-killed yak- thus the "time-to-time" schedule of feeding meat to the dogs. For Chokgya, the best friend of the yak herder was a good dog. "I had two," he said, "one for leading the herd when we changed pastures, and one for guarding. The first would always be at the head of the herd, and the second would stay behind to bring in the stray yaks and also to protect them."

Chokgya claims that the lead dogs actually guide the herd to new pastures. I have not seen this, but have witnessed the dogs participating in the rounding up of stray yaks during the hours of dusk when they are brought into the camp area for the night. The herdsman's yak-hair sling (*shakpo pumba*), however, seems to be more useful than the dogs for bringing in the yaks. The *gothalo* (herder) simply lobs a large stone in the general direction of the truant yak, which induces it to start towards the camp.[16] The dogs will then finish the job by running behind the yaks and herding them back to camp. "Sheep and yaks will listen to a dog," Chokgya continued. "When he barks or gives a warning they will come together. At night they know the dogs will protect them."

One of Chokgya's more intriguing stories concerning his mastiffs involves an encounter with a *mithé* ('jungle man,' another name for a yeti, or abominable snowman) when he was 15 years old. During the early hours of a foggy morning, the dogs began to bark and run around the camp, whereupon the yaks began to group together in a protective circle. The dogs appeared to be crazed with both fear and anger. Shortly after Chokgya heard the distinctive, high-pitched *"che-che-che"*—the whiney cry of the mithé—and he remembers a "very bad smell." A large man-like image did appear momentarily out of the fog, which Chokgya fired at with his old muzzle-loader. The dogs took off in pursuit of the creature, but returned a short while later, exhausted. Chokgya found a pool of "very black blood" nearby a stream where the creature was seen. Such stories are common throughout much of the Himalaya and remain a mystery in terms of what the mithé may or may not be.

"There are two types of dogs in Dolpa," Chokgya continued. "One is the small one, and the other the big one. The big one is not so good for hunting. The big dog is good for guarding yak and sheep." The hunting dog, the smaller one, locally called *sheuke* (Tibetan *sha-kyi*), is colored black and white. A good hunting dog costs a thousand rupees.

"In Dolpa we used to go hunting with the help of the dogs." Chokgya said. "The dogs would chase the lau (musk deer) and na (blue sheep) through the mountains, sometimes into traps, and sometimes until the deer were so tired that we could simply walk up and shoot them. There are many hunting dogs in Dolpa, male and female. The females are much faster and better than the male dogs. In Dolpa it is the poor people who mostly go hunting; the rich don't go because they have yak and sheep to eat."

The reader may wonder about the practice of hunting among these mountain dwelling Buddhists, whose religion prohibits killing animals. Survival in the high pastures, however, both in the Himalayan mountains and out on the Tibetan plateau (where field agriculture is virtually non-existent and vegetables are usually unavailable), requires the nomads to hunt game animals and to periodically butcher herd animals for food as well as for hides, horns, wool and other by-products that are converted into clothing, tools, weapons and used in barter. As Ekvall notes in his discussion of the contradiction between religious proscription and the predilection for hunting, "for the nomadic pastoralist... hunting is exciting pleasure."[17] More recently, however, especially in western Tibet north of the Nepal Himalaya, the anthropologists Goldstein and Beall have observed that hunting is common but a relatively non-essential component of the nomad's subsistence; at best it is considered to be an economic reserve. Among nomads and villagers alike, however, where hunting occurs success brings the hunter considerable prestige; and hunting is clearly a practice that predates the relatively recent inculcation of the Buddhist doctrine renouncing it. Ekvall describes the practice of hunting by Tibetan Buddhists as "a heritage from earlier times, when high-pasturage ones were uninhibited by the preachments of Buddhist saints like the eleventh-century poet-saint Mila [Milarepa], one of whose poems tells of converting a huntsman from the killing of a red doe. Very early legends tell in great detail of hunting, equally the sport of kings, commoners, and beggar outcasts."

As a youth, Chokgya was proud of his and his dogs' hunting abilities. It is difficult to say, however, exactly what he means by "small hunting dogs," unless they are a smaller, faster cross- or mixed-breed such as those found in the "transition zones" mentioned earlier. The full mastiff is not generally regarded as being a good hunting dog, although they are used for such on occasion. In comparison to the heavier, slower Tibetan mastiffs, some of the more lithe, faster hunting dogs used by the nomads may be considered a type of 'sighthound' similar to the Saluki, Greyhound, Borzoi and other hunting hounds of eastern Europe and central Asia. A sighthound is defined as a dog that runs or courses game by sight, rather than by scent. The few accounts (and equally few photographs) of Tibetan-Himalayan hunting dogs describe a comparatively shaggy dog, more like those of Afghanistan.[18]

In their detailed account of life of Tibetan nomads, Goldstein and Beall quote one hunter of na (blue sheep) who described his dependence on dogs for a successful hunt: The dogs "tilt the odds in our favor," he said. "Their job is to corner one of the na among the crags, and bark loudly to lead me to the spot. The best dogs will even try to run back a ways to make it easier for me to find them, all the time, of course, keeping their prey at bay."[19]

Chokgya and I spent many months together in the yak pastures of lower Mustang over a period of ten years (during the 1970s and 1980s), and had ample occasion to meet many Tibetan mastiffs. Some, as in the yak pasture above the village of Marpha in Thak Khola, were nothing less than vicious upon seeing intruders such as ourselves. Chokgya's skill with the braided yak-hair sling was greatly appreciated at these times. On another occasion, while climbing in the Annapurna Sanctuary in the early 1980s, I witnessed a young Sherpani porter from another expedition being severely bitten in the leg for straying too close to a herd. Some dogs we met were quite timid, and were kept more for sounding alarms than for attack purposes. By and large, they had the characteristic black-and-tan markings, although several golden mastiffs were seen, as well as much smaller dogs (the size of Tibetan terriers and Lhasa Apsos), probably kept more for company than for work. Regardless of their temperament, all were treated with a quiet affection and respect by their owners.

Tukché and Lete are villages in Thak Khola known for their superior mastiffs. There, a healthy male puppy, the pick of the litter, sold in the early 1980s for around 250 rupees, and considerably more to foreigners along this popular trekking route. (Rs. 250 was approximately $10, a considerable sum at that time; for some nomads, a prize Tibetan mastiff was considered to be equal in value to a full grown yak, or a good horse.) The walled town of Lo-Manthang, the former capital of Mustang, about three days walk north of Jomsom, is also famous for its mastiffs. The large, 'pure dogs' bred there, especially those belonging to the family of the former Raja of Mustang, are highly prized, and the people of Thak Khola, and even of Kathmandu, frequently make arrangements with traders to deliver puppies to them on their next visit.

As noted, the mastiffs' viciousness is probably more a result of how they are kept and of their conditioning. As with any breed of dog, ferocity comes from being tied up throughout the day while strangers pass by, or from living in remote places combined with their innate territorial instincts for which these dogs are so well known. Certain dogs kept only as pets, however, are quite affable and appear to be excellent (although somewhat aloof) additions to the large and busy households of the mountain people.

The big dogs of Mustang and Dolpa play integral roles in the economic survival of the local people. By comparison with other regions of Nepal, food, fuel, water and arable land are very scarce here, and roving bands of brigands are still somewhat common. Those resources that do exist must be zealously guarded and protected—functions performed well for countless centuries by Tibetan mastiffs.

# 4

## Saipal's Kalu

For most of my life in the Himalayas, almost since the first day I arrived as a Peace Corps Volunteer in 1963, dogs have influenced much of my leisure time, my outlook and thinking. I have lived and worked in Asia for over four decades pursuing anthropological research, development work and personal interests that have taken me across the mountains of Nepal into the high Tibetan borderlands, and on to Tibet, Bhutan, north India and Pakistan. Wherever I went I encountered big dogs—Tibetan mastiffs, shaggy-faced KyiApsos and Himalayan mountain dogs, in their natural settings as livestock guardians and sentinels in private compounds, and as pets and show dogs. I have raised several of them as companions, the most famous being Champion Saipal Baron of Emodus, aka 'Kalu.'

When I joined a large rural development project in Nepal in 1980, seventeen years into my Himalayan sojourn, my wife and two small children joined me to live in Kathmandu. We all wanted a companion and guard dog. In 1982 we were given a puppy named 'Saipal Baron' as a gift from Jay N. Singh of Kathmandu's Saipal Kennel. He was a fine dog with a proud pedigree traceable directly to southern Tibet, a noble black-and-tan Tibetan mastiff in full conformation with international breed standards.

At the same time that I joined the Nepal Kennel Club, I chose the kennel name of 'Emodus' and registered the pup as 'Saipal Baron of Emodus.' To the family, however, he was simply 'Kalu' ('Blacky'). Together we trained and showed him, and bred him. Today, his name is found in the pedigrees of some of the best Tibetan mastiffs breed lines in Europe and North America.

### What's in a Name?

Naming is something akin to a fine art in the international dog fancy. In describing where the kennel names like 'Saipal' and 'Emodus' come from, both of which have specific linguistic, geographical, historical and personal significance, I am reminded of an observation by the Himalayanists S.B. Burrard and H.H. Hayden. Over a century ago they wrote about how things are often named to reflect a place or period in history: "Languages are the basis of geographical names," they said, "and languages have their origin in history; and thus it is that geography and language and history are all parts of the whole." Similarly, how we name our pet dogs, breed lines and kennels, and sometimes our cars and our estates (if we are fortunate enough, that is, to own an estate) often reflects our knowledge and understanding— and occasionally our ignorance—of language, geography, history and culture.[1]

## Emodus Tibetan mastiffs

'Emodus' is a term used by ancient Greco-Roman geographers for what we know today as the Himalayas. Giving my breed line a name as obvious as 'Himalayan' kennel sounded too plain and predictable; the more exotic 'Emodus' had a ring of distinctiveness rooted in something ancient and enduring. One of the first to use the term Emodus (sometimes transliterated as Emódio, Hēmōdus, or Imaus) was the Sicilian historian Diodorus (c.80–20 BCE). In his *Library of History*, a collection of forty books on the history of the world from Creation to his own time, he describes "the mountain chain of Hēmōdus" (according to one translation) lying north of the river Indus. He based some of his second-hand observations on the even earlier writings of Megasthenes (third century BCE). A near contemporary of Diodorus, the Greek savant Strabo (c.63 BCE–24 CE) wrote a seventeen-volume *Geography*, an extraordinary compendium of peoples and countries of the (then) known world. It includes mention of Emodus in Book IV: 'On India'. The term also appears in Book V of *Natural History* written by Pliny the Elder (23–79 CE), no doubt borrowing from both Strabo and Diodorus. Thus, Emodus was well established in the writings of the Ancients.

In the sixteenth century CE it appears again in the great Portuguese epic poem, *Os Lusiadas* (*The Lusiads*, 1572), by Luis Vaz de Camões, where the poet describes the travels of Vasco de Gama, the explorer who opened up sea trade between Europe and India. Vaz de Camões mentions 'Emodia' (Emodus) in a part of the poem labeled *'Descrição da Índia'*:[2]

| | |
|---|---|
| Além do Indo jaz, e aquém do Gange, | Outside of Indus, inside Ganges, lies |
| Um terreno muito grande e assaz famoso, | a wide-spread country, famed enough of yore; |
| Que pela parte Austral o mar abrange, | northward the peaks of caved Emódus rise, |
| E para o Norte o Emódio cavernoso. | and southward Ocean doth confine the shore. |

Much later, in 1903, during the height of the Raj, when the British ruled colonial India, Henry Yule and A.C. Burnell listed Emodus among many terms that the Ancients used to refer to the Himalayas. In *Hobson-Jobson*, their unique and entertaining Anglo-English dictionary, they wrote: "HIMALY'A. n.p. ...common pronunciation of the name of the great range 'Whose snowy ridge the roving Tartar bounds,' properly *Himălăya*, 'the Abode of Snow'; also called *Himavat*, 'the Snowy'; *Himagiri* and *Himaśaila; Himādri, Himakūta*, &c., from various forms of which the ancients made *Imaus, Emōdus*, &c."[3]

## Saipal Tibetan mastiffs

Jay N. Singh's Saipal Tibetan mastiff kennel in Kathmandu was named after Saipal Himal, an impressive Himalayan peak rising above Bajhang District in Seti Zone of northwestern Nepal bordering western Tibet. Mr. Singh was raised in Bajhang, the son of the last Raja of Thalar, a former principality located alongside those of Bahjang and Salyan whose royal families were all related by marriage. (Part of old Thalar is now Khaptad National Park.) In his youth, Jay looked north from his mountain home to snow-capped Saipal peak (23,048 ft, or 7025 m) for inspiration. It was from somewhere beyond Saipal Himal that his father acquired the first of several Tibetan mastiffs that Jay remembers from childhood.[4]

One day, Jay recalls, his father, Raghab Narayan Singha (he spelled it with an 'a'), was bargaining with a trans-Himalayan trader. In those days, "My father always had many dogs—Tibetan mastiffs—in Thalar, and various other breeds [later] in Kathmandu, including German Shepherds, Lhasa Apsos, Great Danes, Japanese spaniels and Bull terriers. The mastiffs sometimes came with the ethnic Byansi shepherds of this region, who brought them down from western Tibet over the Tinker-Lipu Pass and through Darchula District. One time in the 1930s I watched him bargain for a fine Tibetan mastiff. The trader wanted two and a half rupees, but my father would only pay two. When the trader would not reduce the price, the deal was off. He was a hard bargainer."[5]

Plate 4.1. *Jay N. Singh with the bitch, Bounty, in Kathmandu, 1983.*

"Some years later, in 1958 or '59," Jay N. Singh goes on, "I acquired my own first Tibetan mastiff, a two-week-old male. I paid only five rupees for him. He had large bones and excellent markings, black-and-tan." Note that five rupees was equivalent to about 50¢ US, a considerable amount in those days when money was scarce and most transactions were by barter. Jay called his dog 'Bhoté,' meaning 'Tibetan.' Later, in the 1970s, long after Thalar ceased to exist as an independent principality and Jay N. Singh was living in Kathmandu, he started his Saipal breed line. In time, his Saipal dogs became well known among Tibetan mastiff aficionados in Europe and North America.

When I first met Jay N. Singh in 1980, 'Bhotoo' was his finest bitch. She was six years old. She had been acquired from Ang Tharkey Sherpa, whose dogs were directly descended from landrace mastiffs brought into the Khumbu, the Mount Everest region of northeast Nepal, directly from Tibet across the Nangpa La (a pass at 18,700 ft, or 5700 m). Singh bred Bhotoo to 'Tiger,' also of Tibetan ancestry. Their most famous pup, 'Tü-bo,' became the foundation dog of many European bred Tibetan mastiffs dating to the late 1970s. It was also about this time that Mr. Singh helped establish the Nepal Kennel Club, whose first patron was G.B. Shah, the last Raja of Salyan.

I was impressed with Bhotoo, and once described her as a large, graceful, steel-grey and black bitch that exudes 'class' and 'nobility.' "When she is around," I wrote, "all eyes and conversations naturally shift to her. It is not unlike the reaction when a strikingly beautiful woman

*Plate 4.2. Tü-bo in Germany, the half brother of Saipal Baron (Kalu).*

enters a crowded room and is greeted with an appreciative hush. Her nobility and class were inherited by her son, 'Baron,' now snoozing at my feet as I write this..."[6]

In 1981, Jay N. Singh bred Bhotoo again, to a dog named 'Sheroo' from Rasuwa District in the Ganesh Himalayan massif northwest of Kathmandu. Sheroo's forebears came from yak herders in the Kyirong ('Happy Valley') in southwestern Tibet. Saipal Baron was one of their outstanding pups. At the age of twelve weeks, he joined our household in Kathmandu, and after scoring well in dog shows in Nepal and India he became known officially as 'Champion Saipal Baron of Emodus.' Our family's life with Kalu was a cherished time.

## Life with Kalu

Kalu was a survivor. When he was six weeks old he contracted parvovirus, a disease that causes severe diarrhea and vomiting in puppies. In those days, the early 1980s, parvovirus was a little-known canine illness for which a medical cure was not yet known.[7] Jay N. Singh nursed him back to health by feeding him glucose water from an eye-dropper every hour, day and night for a week. This lucky pup recovered and was soon registered with the Nepal Kennel Club. Mr. Singh then gave Saipal Baron to us, as a companion for our daughter Liesl and our son Hans Dietrich.

Henry Ward Beecher once said that "The dog was created specially for children. He is the god of frolic." To young Liesl and Hans, Kalu was a remarkably frolicsome god, indeed. They played endlessly with him, and helped raise, train and prepare him for dog shows. Kalu, in turn, helped the children grow up as gentle, dog-loving human beings. In time we bred Kalu to Saipal Bounty (no relation), and some of their pups went on to become the foundation dogs for some of the best Tibetan mastiff breed lines in North America.

Training Kalu was a challenge. A Tibetan mastiff is first and foremost a guardian dog. *His* first inclination was to be our watchman. *Our* first inclination was to teach him rudimentary commands. We spent many hours drilling him on the basics: sit, stay, come, heel, and stand still for grooming. *We* thought we had succeeded, but Kalu knew otherwise, and proceeded to train *us* in the ways of a guard dog. Training sessions ceased abruptly, for example, whenever someone came knocking at our gate. Nothing deterred Kalu from investigating the intrusion and letting everyone know who was boss. It is instinctive and in the nature of the Tibetan mastiff to guard things—commercial property, a Buddhist monastery, a herd of yaks, or (in our case) a house and yard and children—always with one eye on the gate. In the long run we were happy for it. We felt secure when Kalu was on duty, day or night.

Raising Kalu's and Bounty's pups was a great experience for the children. They observed nature at work from conception to birth and upbringing.

*Plate 4.3. Kalu (Saipal Baron) with the author's son, Hans, 1983.*

*Plate 4.4. Kalu (Saipal Baron) with the author's daughter, Liesl, 1983.*

They were especially fascinated by the dogs' affectionate muzzle-nuzzles before mating (canine foreplay). We were all intrigued after whelping by the way both dam and sire shared in the pups' socialization. Liesl and Hans still speak with wonder at the close relationship they observed between mother Bounty and father Kalu before and after the birth of their first litter of seven pups.

The pups spent their first few weeks in our house, in a box in the back hallway. Kalu was quite aware of them from Day One. He'd sit outside the screen door about twenty-five feet from the box, eager to come in for a closer look. But Bounty would have none of it. If he slipped through the open door, she would warn him off with a low growl.

Bounty was the boss, and though Kalu was twice her size, he obediently backed away at her warning. For several days he sat outside the door, whining quietly and looking wistfully in for a glimpse of his progeny. Gradually, Bounty made allowances. After some days he could sit just inside the door, then closer, near the entrance to the back hallway. As the pups grew older, she let him to come to the edge of the box and look in while they slept. After a few weeks we put the frisky pups on an old blanket in a large three-sided box outside in the yard and covered it with a piece of plywood as a rain and sun shade. Between nursings, Bounty

*Plate 4.5. Kalu's pups in 1983.*

turned 'pup duty' over to Kalu, and while she went off to rest quietly by herself in a shady corner of the yard he played dad.

Kalu was a marvelous house guard, especially at night. Though he never bit anybody hurtfully, he made sure that visitors knew the limits. He'd occasionally nip someone with a sharp sting on the leg, to let it be known who was in charge; but he never drew blood. That nip left only a small bruise mark where his teeth stung flesh. Kalu never allowed strangers to approach the pups unattended, and wild predators dared not give them much more than a passing thought. When visitors came, he would stand by the box and eye them suspiciously. His pups could only be picked up and petted after one of us gave proper introductions and permission.

Kalu's bark was a deep sonorous *WOOF!*, one of which was enough to repel any intruder. One night I was awakened by his warning bark. "*Oh-oh*, trouble," I told my wife, and went out to the porch in time to see Kalu chase a large barn owl away from the puppy box. On another occasion after being awakened by a commotion in the yard I turned on a spotlight. There was proud Kalu with a piece of torn trouser in his mouth while a frightened young thief scrambled over the wall into the night empty-handed, but wiser, and in need of a new pair of pants.

In those days the Nepal Kennel Club held a national dog show in Kathmandu each year. We entered Kalu first as a pup and the next year as a one year old. Each time he won Best of Breed over his elder, slightly larger and equally well-bred brother. At the first show Kalu also took Best Himalayan Dog (the Wakeman Trophy), Reserve Best of Show Opposite Sex, and Best Pup. All that was against a small but impressive array of other fine-looking Tibetan mastiffs. The first year he also earned his first Challenge Certificate, or 'CC,' and the second year he won more top prizes and his second CC, awarded by a well-known all-breed judge from the Kennel Club of India. The third year I took Kalu to India and entered him in the Delhi Dog Show. There, against over two hundred other canines, he earned more honors.

Getting Kalu to the Delhi Dog Show, however, wasn't easy.

## The Delhi Dog Show*

One day before the show, Kalu and I took an evening flight from Kathmandu to New Delhi. At the Delhi airport, while waiting for his crate to appear on the luggage beltway, I was approached by a nervous airline representative and asked to come directly to the plane. The baggage handlers refused to go near Kalu's cage! I had to unload dog and crate.[8]

---

*This account is adapted with edits from my article 'Putting on the dog at the Delhi dog show', *Dog World* (USA; 1985).

Awhile later I attempted to check into a hotel near the showground. A friend in Nepal had assured me that the hotel I had chosen allowed pets. But when the night clerk saw the impressive big dog (which he incongruously called "doggy") on leash beside me in the foyer he said "No!," cancelled our reservations, and pointed us towards the door.

What does one do with a suitcase and cage under one arm, and a large dog on lead on the other, in a strange foreign city on a cold dark winter night? I tried to phone the emergency number that the secretary of the Delhi Kennel Club had given me. No answer. I also tried the home number of an American diplomat friend, thinking he might take us in. Again, no answer. I found out later that all New Delhi phone numbers had been changed, but that a new directory was not yet published.

Eventually, with some trepidation, the night clerk phoned the hotel manager at home. He was so enraged at being disturbed so late that his answer was an even louder *"No!"*

When I threatened to sleep in the lobby, the hotel staff finally condescended to make a few calls around the neighborhood until they found lodging for us at a small inn. So off we went to the House of Lords Guesthouse, a no-stars establishment with a reputation about which I was quite oblivious. When Kalu and I arrived there by taxi well after midnight, the sleepy night clerk informed me that we could stay for the price of a double room. After all, he reasoned, there were *two* of us! Once I got that sorted out and paid in advance for a single, he said quite firmly: "You must take 'doggy' outside for easing."

"For *'easing'*?" It took me a moment to realize that "easing" was the Anglo-Indian euphemism for pissing. I assured the clerk that Kalu was a remarkably clean critter and would let me know when he wanted to relieve himself. "Of course," I replied, "we [both 'doggy' and 'sahib,' for that's what he called me] would go out into the garden" at the appropriate time.

After Kalu and I had settled into our seedy, mosquito-filled room for the night (after the mandatory trip to the garden), there was a rap at the door. Kalu was instantly alert, looking steadily at the closed door, a low growl rumbling deep in his throat.

"*Kyā hai?*" I asked in Hindi. "What is it?"

A woman's voice came back at me inquiring if I wanted "a companion" for the night. *'House of Lords,'* indeed! Room service.

I was tempted to open the door to her in such a way that she would see only Kalu sitting there with his big pink tongue lolling out—surprise, surprise! But it was so late that both dog and lord wanted nothing more than some uninterrupted sleep. I told her (through the closed door) to go away. Kalu then curled up beside the bed, my faithful protector from wily women of the night and other intruders. We fell asleep while the ceiling fan whirred squeakily overhead, and the plumbing in the bathroom dripped annoyingly.

## 4. SAIPAL'S KALU

Over the next few days, Kalu made a great impression wherever we went in New Delhi. Big furry bear-like mountain dogs are uncommon on the Indian plains, and the sight of Kalu scared most passersby. The first morning I had trouble finding a taxi driver who would consent to take us to the dog show. Finally, a big, burly, bearded Sikh with an incongruous pink turban agreed. He drove us along Delhi's boulevards with fear and trepidation, all the while muttering to himself, hunched close over the steering wheel as far as he could be from the dog panting down his neck from the back seat of the tiny Ambassador sedan.

When we arrived at the dog show, the taxi-wallah doubled the metered fare. Hazard allowance.

The timing of the Delhi Dog Show could not have been better. It was held on a weekend in January well ahead of the intense heat of India's summer. Nonetheless, during the two days of the show the daytime temperature was 65° Fahrenheit (18° Celsius). That was mild for me but comparatively hot for Kalu in his heavy winter coat. Four months later we would have baked in Delhi's hundred degree-plus summer broil. These big dogs do not fare well in such temperatures.

The show venue was grand, across a moat from the Purana Qila, Delhi's historic 'Old Fort,' a sixteenth-century edifice whose crumbling walls stand as a reminder of India's glorious medieval past. The high walls of the fort frame one end of a two-mile vista across the city to the impressive Rashtrapati Bhawan, the Presidential Palace, formerly the colonial Viceroy's House. What a dramatic backdrop for the show! Between the show and the fort, the lagoon-like moat was alive with water birds, and Delhi's National Zoological Park was nearby. The zoo covers 214 acres and houses over 2,000 species of animals and birds. Its gardens are popular for family gatherings and it typically draws large crowds on weekends. Many picnickers came directly over from the zoo to see the dogs, swelling the show crowd.

On Saturday, the first day of the show, we watched the obedience trials. The highlight was an impressive demonstration of Alsatians and Dobermans trained and handled by an elite corps of men of the Indian Border Patrol. I hoped that Kalu might learn a thing or two by watching their skilled routine; but, instead, he lay quietly at my feet beside the ring, eyes closed, seemingly disinterested, yet cannily aware of our surroundings.

Entirely by chance that afternoon, I met an American friend who had come to see the show. He lived in Delhi, loved dogs, and invited me to stay at his house. I needed no persuasion. I checked out of the House of Lords without hesitation, rode across town with another fretful taxi driver to my friend's home where both the dog and I settled in for the evening. No more late night knocks!

Kalu was scheduled to show on Sunday afternoon. While waiting his turn in the ring, he enjoyed the attention of scores of interested dog fanciers from all over India. Among his admirers were three teenage girls of obvious refinement, dressed in bright colored Punjabi pants and matching blouses. They stared admiringly at Kalu as he lay snoozing contentedly on the grass beside me.

"Oh, Uncle!" they exclaimed. "What breed is he? Is he dangerous? Will he bite? May we pet him?" They hesitated to come close.

I answered in Hindi that he was a *"Bhotea kutta"* (Tibetan dog) and was very gentle. Their eyes widened. They had never seen such a critter, though they had certainly heard hair-raising stories about the aggressiveness ascribed to this breed in the popular press and dog lore of India. Then, as if to assure them, when they reached out gingerly to pet him, Kalu gave each girl an affectionate lick on the hand.

I explained many times to curious onlookers that weekend that the keys to Kalu's gentleness were his good breeding, careful upbringing, patient training and the fact that he was never chained up. Aggressive behavior can be instilled in any dog by constant tying, confinement or abuse. Many of the big dogs in the shepherd and nomad camps of the high Himalayas and Tibet are tied up during the day, poorly fed and often mistreated. As a consequence they have a reputation for ill-tempered ferocity.

One comment I heard at the dog show summed up much of the popular belief about Tibetan mastiffs: "I saw your dog," a gentleman remarked. "It even appears you've been able to *train* him!" he said in a tone of disbelief.

I replied that training the dog was one thing, but that *training the dog's owner* was the more formidable challenge.

Training a Tibetan mastiff takes dedication and time. It is not difficult, so long as distractions such as strangers entering his territory do not divert attention away from the lessons. These dogs are first and foremost guardians and sentinels, and one cannot severely curtail that important instinct without undermining the dog's innate inclinations.

We also met twelve-year-old Shekar Rao, a boy from the north Indian industrial city of Jamshedpur. Shekar had come with his father to show the family's yellow Labrador retrievers. The boy immediately fell in love with Kalu and for two days they were special friends. Some who saw the boy playing gently and freely with Kalu were intrigued and surprised that a dog of Kalu's size and (to them) fierce looks, with his imposing head and thick mane (hallmarks of the breed), could be so friendly toward a total stranger.

The Sunday show was judged by the Vice President of the Kennel Club of India. Thirty-nine breeds were represented, ranging in size from tiny Pomeranians and Pugs to Saint Bernards and Great Danes. Eight breeds were of Asian origin, including four Tibetan: Lhasa Apso, Tibetan spaniel, Tibetan terrier, and Tibetan mastiff. There were also Pekingese, Salukis, Afghan hounds and one very fine Rampur hound. The latter was the only exclusively indigenous Indian breed shown, a magnificent bitch named

'Bathsheba' from the north Indian state of Uttar Pradesh. Sleek, fast Rampur hounds were traditionally used for hunting wild boar and deer, but they are barely known beyond the subcontinent. The indigenous Asian dogs, I feel, deserve greater attention than they got at that dog show.[9]

India has many fine native dogs, mostly hounds such as the Sindh, Vaghari, Maratha or Pashmi, Poligar and Rampur, as well as the Rajapalayam dog and the Chippiparai dog, all with exotic names, origins and functions. But, it must be in our nature to find "the grass greener on the other side of the fence," as the saying goes. Hence, European breeds tend to receive greater attention among the elite Indian socialites who can afford the luxury of raising exotic dogs. A journalist writing in the Indian press explained why: "During the colonial period," he said, "educated Indians, in their efforts to ape the rulers, preferred foreign dogs and many indigenous breeds petered out due to neglect. Symptomatic of this attitude is the reference to all Indian dogs as 'Pariahs.' Collies and spaniels came to dominate the scene and lowly Indian breeds were forgotten."

Among the colonials, the reverse attraction was true. Many of the English in India became enamored with some of the local exotics, especially Tibetan and Himalayan breeds. It was they who first exported big Tibetan dogs to England during the nineteenth century, and then founded several UK-based oriental breed clubs.

By late Sunday afternoon, all the dogs had been shown, the various winners announced and awards presented. No surprise: European breeds took top honors. First, Second, Third and Fourth Best of Show went to a German shepherd, Irish setter, Doberman pinscher and French bulldog, respectively.

And Kalu? He made a brief showing in the finalists' ring of a dozen dogs selected by the judge, but was sent off when he snapped at a rather prissy Pug. Nonetheless, he was the only canine featured that weekend in the *Times of India*, where his photo took up more column space than one of Prime Minister Indira Gandhi in the same issue. And a few months later Kalu was also pictured with several other dogs in *Swagat*, the Indian Airlines magazine.

On Monday morning, Kalu and I returned home to Kathmandu. Liesl and Hans were there to meet us, excited to see Kalu's impressive wins: Best of Breed for Tibetan mastiffs, Best of Breed in the Special Limit and Open Class, and another CC. The judge was quite complimentary of Kalu's high quality and conformation, and had awarded him accordingly. It was that third Challenge Certificate and his new designation as a "Champion" dog that made us the most proud. Thereafter, he was recognized by the kennel clubs of India, Nepal and Great Britain, and by the Fédération Cynologique Internationale, as "Ch. Saipal Baron of Emodus." While that was his formal name, he remained forever Kalu to us.

Kalu lived a long life and died at the old age (for a dog) of twelve. His passing was a sad event for our family. We saved his ashes and several years later Hans, Liesl, Kareen and I solemnly buried them in a sacred forest high in the Himalayas near the Tibet border.

## Under the Great Wall of Pisang[10]

In 2004, my daughter, Liesl, published a little book about the lore and legends of Manang, a high Himalayan district near the Tibet border. In it, she briefly describes a unique place associated with death and burial in Himalayan society and, not least, with big dogs and their association with the Afterlife. She writes that for the ethnic Gurung people of the mountains, the Great Wall of Pisang "is an important place to visit during one's life and an important location on the final adventure—death... [It] is the entrance to the land of the dead..."[11] Part of my career was spent studying the culture of the mountain-dwelling Gurungs, so I thought it fitting that Kalu's ashes also be buried under the sacred mountain.

Dogs have a special place in the lives and culture of Tibetan and high Himalayan people. During the early 1970s, while living in a Gurung village doing my PhD dissertation research, I noted that dogs have importance in human life and in death. I heard many tales about them.

Gurungs celebrate a number of human life crisis events, beginning with birth, then the naming ceremony, first haircut (for boys), coming of age, marriage and so forth, ending in burial or cremation and, finally, the ritual act of sending the soul (*plah*) off into the afterlife to the Land of the Dead. The ceremony for sending the soul off is called *pwé laba* in Gurung and *arghun* in Nepali, and it is usually held within weeks, at most a few months, after death. It is one of the most elaborate and critically important life crisis rituals of all. Seeing the dead off properly towards the afterlife has tremendous significance to the living. At the death of an elder, for example, his arghun highlights the transitions that society goes through (especially family and clan) in adjusting to his absence. It is when social roles are exchanged and renewed; when, for example, the new head of a household or a clan is formally acknowledged by others to replace the deceased. It is also a time when close kin and in-laws gather to celebrate the passing of a relative and, at the same time, when marriage arrangements for the young are worked out.

An arghun is an elaborate affair, sometimes lasting several days. The rites performed by a shaman whose traditions date back to practices so old that they pre-date Tibetan Buddhism. While researching this particular life crisis event, I heard the story of the very first arghun in Gurung mythology. It was conducted not for a man or woman, but *for a beloved dog*. I realized then how significant the dog is in Gurung society. That first dog led the way, though all such celebrations today are for people.

I was also told about the sacred place in Manang District, a few days walk north of my research village, where many Gurungs take the ashes of their deceased elders to be buried. The site is deep in a dark evergreen forest high in the mountains at approximately 10,000 ft (3000 m) elevation. The funeral ground is above the west (right) bank of the upper Marsiangdi river (called the *Mha-ri-syo* in Gurung), directly opposite a sheer mountain wall of solid gneiss stretching upwards over 6,000 feet into the sky. That mountainside, the so-called Great Wall of Pisang (after nearby Pisang village), has several local names: Oblé (or Wablé) in Gurung, Nghurung Khang among neighboring Manang people, and Swarga Danda or Paungda Danda in Nepali, each of which means essentially the same thing, 'Mountain of Heaven.' The funeral ground in the forest is at a Myju Deurali, the site of an ancient village and a place of worship. From there, the souls of the dead leave the earth and ascend up the sheer mountain wall, the Oblé Dome, toward Nirvana, or Heaven.

Anthropologist Judith Pettigrew has described the physical approach to Oblé during a 1992 visit: "We cross into Manang district by cutting through the mountains, up and over Yekre where the souls of those who are dying 'prowl' in the hours and moments before death. In Manang our route takes on an added dimension. As well as being the trail of the shamanic soul journey, the ancestors' migration route and the salt trade route, it becomes other peoples' trails—an important tourist route, the trail down which the Buddhism spreading Tibetan refugees travelled, a route dotted with small Tibetan and Tamu [Gurung] villages. Tamu gods, shamans and ancestors thus share their landscape with foreign hikers, Buddhist lamas, Tibetans, and Brahmans from Kathmandu imported to run the local bureaucracy."[12]

The mountainside that rises directly above the river, from the east (left) bank, is a steep escarpment that appears to have been thrust vertically towards the sky in some ancient tectonic cataclysm untold millions of years ago. Looking out from the funeral ground, the eye naturally follows the contour of the wall up and up to several rock outcrops silhouetted against the sky at the top. Gurungs will tell you that those skyline outcrops seen from a distance look to them like a stone building (a house or a temple) with a man and big dog standing alongside. They call this outcrop the *Sa Yi Gompa*, 'Temple of the Dead.' According to local belief, it's the shaman's job to guide the soul on its journey to the afterlife from the funeral grounds in the forest, across the river, and up the mountain to the top, past the house, the man and the dog, then onward into the heavens.

During the ritual, the shaman offers a gift of grain to the spirits of the four directions and eight regions, requesting them to open the way for the departed soul to reach the Sky World of the Dead. Arriving there, the soul is greeted by previously deceased ancestors. It is believed that the shaman momentarily transforms himself into a vulture and once the soul is released at the top, he flies back down the Great Wall to complete the celebration of death among the mourners on the ground. People say that the sight of the huge Lämmergeier vultures that often soar on the thermals over the top of the Great Wall confirm this belief.

The proper ritual begins with the living relatives placing the ashes or a piece of bone of the dead (or both) in a small *chorten* (a monument of stones) in the forest, within which a tree seedling is planted. This grave marker is typically festooned with strips of colored cloth, flowers, fruit and incense sticks. Juniper branches are burned and the name of the dead is shouted loudly in the direction of the Great Wall, echoing up and down the mountainside. If the tree seedling grows well, the soul certainly enjoys a peaceful afterlife. If it grows poorly or is stunted, the soul is in torment and, if the sapling dies, the soul stays back to trouble the living. When that happens, a special exorcism is performed to placate the soul so that it can complete the upward journey into the sky and leave the living in peace. In any case, the soul is hailed and is finally ushered on into the afterlife by the man and Tibetan mastiff-looking dog that stand silhouetted against the sky the top of the mountain.

Given the significance of the dog at the first Gurung funeral, and my own research into Gurung life and culture, what more fitting place to leave Kalu's ashes than in the sacred forest beneath the Great Wall. In early October of 2006, my wife, son, daughter and son-in-law stopped in the forest while trekking in Manang to remember our canine friend's life with us two decades earlier. After placing Kalu's ashes and a few personal mementos in the ground under a big tree, we stood solemnly, remembering his important place in the life of our family. There was no shaman with us, but we hoped our simple ritual would suffice. Several American and Nepalese trekking companions joined us in silence to commemorate Kalu's farewell. We wished him a good journey across the river and up the mountainside to the man and dog guarding the gate to heaven at the top. His spirit has never returned to bother us, so our short ritual must have been satisfactory. After awhile, we saw a Lammergeier riding the thermals near the top of the mountain...

Kalu had had a good life with us and, over the years, his offspring and theirs have become the prized possessions of other dog-lovers in Nepal and North America. One of them was a young pup named Bhalu ('Bear'), whose life and death is another tale worth telling.

## Bhalu's Story
By R. Bruce Morrison*

Bhalu had never been away from his mother or littermates until he made the long flight from Nepal to Canada. The breeder in Nepal believed that a puppy should be kept with its mother for as long as possible. When the pup arrived in Edmonton, I picked him up at the airport and took him directly to my small condominium. Thinking he might need some exercise after being cooped up for so long, we walked around an adjacent field so that he could explore and mark his territory to his heart's content. When we returned to the condo, I kept a close eye on him lest he decide to urinate inside. But he showed no inclination; instead, after thoroughly exploring the premises, he sat down and let loose a forlorn howl. It was heart breaking. I realized that he had never been away from his litter until then! His sense of loss and loneliness did not go away easily. I did my best to comfort him and it seemed to work until I penned him in the kitchen for the night.

I was just falling asleep when he commenced to howl again. Because I was living in a condo complex I could not ignore the noise he was making. I went downstairs and tried to comfort him. The only way he would quiet down, and then only reluctantly, was if I slept on the kitchen floor beside him. But it was obvious that I was a poor substitute for his lost parent and siblings. I spent several uncomfortable nights on the floor silently cursing his breeder who believed in keeping pups with their mother as long as possible!

### *Confronting coyotes*
A short time later I moved to a farmhouse in the country. As soon as we arrived I let Bhalu out of the car to wander around while I unloaded the baggage, all the while watching him out of the corner of my eye. The old farmhouse was located on patch of grass partitioned off from

---

* Bhalu's Story (originally subtitled 'A Tibetan mastiff's life, death and afterlife') was written by Bruce Morrison for this book. Bruce is an anthropologist who served with me in the Peace Corps in Nepal in the early 1960s. In 1984, when Bruce was teaching at a Canadian university, I gave him a Tibetan mastiff pup registered with the Nepal Kennel Club as 'Emodus Boris KC' (from the second litter of Kalu and Bounty). Bruce nicknamed him 'Bhalu', meaning 'Bear', and flew him home to northern Alberta. When he sent me the story, Bruce wrote: "Calling up memories of Bhalu is difficult because his death was so tragic. After many years, the memory is still sharply painful." He also described Bhalu's first attempts at adapting to his new environment: "The pup made the trip in good order, although he howled like a demented soul the first two nights—but most puppies do that anyway. My children were absolutely delighted with him... He was one of the family and a considerable source of pride..." An important message in this true story is the strength of a dog's instinctive behavior as a guard and companion.

a grain field. As we drove into the yard I saw a coyote in the middle of the field, beyond the fence, hunting mice. He stopped and stared at us with great interest, then returned to his task. Seeing coyotes in fields is not an unusual sight in this part of the world. In the past I would enjoy watching them. But they are worrisome if you own a dog, particularly a puppy. Coyotes regularly lure unwary dogs to their deaths, and most of the farmers around have told me stories of their family dogs being killed by coyotes.

When Bhalu got out of the car, he looked around and immediately spotted the coyote, about one hundred yards away. He stared at it for a couple minutes and then, with his eyes fixed on the wild critter, walked slowly and resolutely towards it. In a way it reminded me of John Wayne walking down the street on his way to a gunfight, but the little fellow was only three months old. There was no display of aggressive bravado, just intense concentration and purpose.

When he got to the wire fence that demarcated the rental property from the field, Bhalu stopped, stared at the coyote for a few minutes, then sat down. His gaze only wavered once when he turned to me as if to say: "What now, boss?" I kept a close eye on him while unloading the car, to be sure he did not commit the mistake of making a closer investigation, something the coyote would surely have welcomed with mouth-watering anticipation. But Bhalu did not move; he simply stationed himself behind the fence, and sat alert and watchful. The coyote seemed truly puzzled by this behavior. He stared at the pup, cocking his head first to one side and then to another, as if trying to get a better perspective on the situation. At one point he trotted off a ways, then stopped, turned and stared intently back at the pup. Then he strode stealthily closer to investigate, but stopped before he had gone very far. The two canids, the naïve but wary pup and the canny adult coyote, sat facing each other in mutual fascination, before the coyote continued his journey across the field. At the time I did not realize that Bhalu's approach to the coyote typified his approach to most intruders. He walked toward them with a silent, intense, unblinking stare!

Later, when fully grown, Bhalu used the same technique quite effectively against unwanted guests, such as the carloads of proselytizing Jehovah Witnesses who frequently called on the rural folk in our neighborhood. It was only when he had decided that the intruder was clearly a threat, however, that he resorted to more aggressive behavior. When strangers arrived, he first barked to announce their arrival. Then he would get up from where he was lying on the top step of the front porch and do his gunfighter's walk toward the driveway. The direct eye contact along with the slow measured stride seemed to discourage people from getting out of their cars. Unless I came out to call him off, most people invariably turned around and left, whereupon Bhalu meandered back up the

## 4. Saipal's Kalu

walkway to resume his indolent position as the household lookout.

We started obedience training soon after I moved to the farm. At first I focused on teaching him to come when I called him. In a rural area his safety depended it. I focused mostly on the command "come" because I feared looking up some day and seeing him running across a field after a coyote, a deer or a cow. The sight of a farm dog chasing a coyote was often the last my neighbors ever saw of their pet. Coyotes were very successful at luring unsuspecting dogs into an ambush. One coyote, sometimes a female in heat, would entice the farm dog into pursuit only to discover that it had friends waiting in the bushes. By that time there was little hope of escape. I wanted to be able to run out and stop my dog in mid stride with the command: "Bhalu, come!"

It was also important to teach him not to chase either deer or domestic livestock. It is an unwritten law that dogs that chase livestock or big game are shot, period! Just to cover my bets I went around to my closest neighbors and told them I was training the dog not to run livestock, but that if he should stray from the straight and narrow to please phone me and I would not only bring him home but also pay for any damages. To the best of my knowledge, Bhalu's only departure from that rule occurred with my own horses. Oh yes—he also liked to chase my cats, but that is really part of the horse story.

### *Confronting horses*

When Bhalu was about a year and a half, I married and moved to another, larger farm, which I had bought. We owned seven or eight horses at the time. I went through my usual procedure to wean him away from chasing horses—or I thought I had. Actually, it was the horses that convinced Bhalu of the folly of his ways. One day as I watched out the window I saw the horses begin to buck and play in a display of high spirits. At the time, Bhalu was snoozing in the sun on the front step. As they began to cut up, Bhalu raised his head and watched them with what I thought was calm disinterest. When the commotion started they were near our corral, about fifty yards from the house. Suddenly, they spun around, lifted their tails and began galloping toward a hill at the other end of the field. As they ran, they kicked up their heels, snorted and generally let the world know what a wonderful time they were having. This was too much for a young fun loving dog. He was off and through the fence in a flash, in hot pursuit. By running flat out, or perhaps because the horses saw him coming and slowed up a little, he was able to close with them as they reached the top of the hill.

Just when he thought he had them, they turned with almost military precision, half to the right and half to the left. Suddenly he was facing eight horses thundering towards him, ears flattened to their heads, teeth bared! He was astonished, to say the least. But it did not take him long to figure out that he was in deep trouble. In a flash he swapped ends and raced for the safety of the corral fence. As he ran the horses moved up onto his flanks, nipping at his rump, and each time they connected he ran all the faster, to escape. For a moment I feared for his life! Then I realized that the horses could easily have caught and killed him, but they did not want to. They were simply having a wonderful time teaching a puppy to mind his manners among his betters.

When he reached the fence he shot through without hesitation, no doubt leaving some fur on it behind him. As soon as he was safe, he spun around and sat down, facing the horses and giving them his most ferocious glare. Fear had been replaced by indignation! The horses, still enjoying themselves, pranced up and down along the fence swinging their heads, inviting him to come and play some more. From then on he only approached the horses in the field if I was with him. The leash, as well as the command to 'heel,' was totally unnecessary.

He must have learned something in that little foray, other than to avoid horses in open fields. He soon adapted what he learned to his nemeses, the household cats. I owned several house cats that resolutely rejected Bhalu's puppyish invitations to play. He liked to bounce around in front of them dodging one way and another inviting them to play. Their low growls and sharp hisses of warning did little to deter him. It took a couple of good solid stinging smacks on his nose to convince him that they were not about to soften up and join the fun. The cats enjoyed hunting outside and often wandered through the high pasture grass quite a ways from home. Although they managed to evade the neighborhood coyotes, they often fell afoul of Bhalu, who would lie in wait for them. When he appeared from nowhere they would sprint for home if they could not find a convenient tree for refuge. That was just what he wanted, for then he gleefully raced after them. Like the horses, he positioned himself on their flank, where he could nip at them from time to time. The sight of their spurring on to even greater efforts following one of his nips seemed to cheer him on. When they reached the safe haven of the house, however, the cats usually disappeared into some dark corner muttering to themselves.

### *Obedience training*
Bhalu was selective about whom he would obey. He liked most people and loved children, with whom he could play for hours. But he ignored most commands they gave him. Similarly, he selectively ignored the commands of adults, depending upon who they were.

He seemed to think of the two of us as a team and of most others as part-time players. After I married, my wife noticed that he largely ignored her commands, unless it pleased him to do so. If she gave him a command, and we were all three in the room, he would look over at me as if to say, "Well, should I do it or not?" If I repeated her command he obeyed without hesitation. Deferring to me did little to ingratiate him to her.

# 4. Saipal's Kalu

When Bhalu was about a year old I began to work more seriously on his obedience training. Things like walking on a lead, "heel," "sit" and "stay" were worked into every day activities. I did not give him any special guard dog training because the breeder I got him from assured me that it was unnecessary. The very best Tibetan mastiff, he said, possessed a natural ability to discern who or what was potential danger or not. Based upon what I saw of Bhalu's sire in Nepal, I believed it. I always wondered whether this was evidence of some sort of telepathic or empathetic ability, or strictly centuries-old instinctive behavior.

While he was an intelligent and willing student, Bhalu never exhibited the fawning need to please, as some breeds do. He reminded me of someone who was going along with whatever his friend wanted him to do, with a more or less detached air of amusement. All pretensions to sincere involvement dropped, however, when it came to my wife, Joyce.

When her frustration with Bhalu's reluctance to take her seriously became a source friction between them, I suggested that she take him to obedience school, which I thought might induce him to take her more seriously. Although she was a little concerned at what a potentially ferocious Tibetan mastiff might do in a roomful of other dogs, she agreed to attend the course. They arrived together on the first day to a gym-sized room full of dogs and owners. The instructor told them to form a circle with their dogs. She nervously found her place in the circle and told Bhalu to sit, but he was staring with great interest at an Irish setter bitch across from him. Joyce interrupted his reverie by pulling back on the collar and pushing her hand down on the top of his head at the same time she gave him the command 'sit,' in her most authoritative voice. To her embarrassment, his response was to collapse on the floor and roll over, four legs in the air, tongue hanging out and wiggling his rump so that she would scratch his belly. He decisively employed that trick through the rest of that day's session, to the amusement of other owners and her great annoyance. Bhalu seemed to think that Joyce had brought him to one hell of a party. He just wanted to meet all the other dogs and romp!

When I came home from the office late that evening, I found my wife sitting in her favorite chair, a glass of scotch in each hand, looking very annoyed. She said that one scotch was for me, but it looked to me as if she could have benefited from both of them! I ask how the class had gone, and she glared at me as if I were personally responsible for my dog's embarrassing behavior. Bhalu, for his part, was asleep on the floor at her feet, obviously feeling no guilt at all. While they both finished the class and Joyce got a certificate, it was clear that Bhalu was not only disinclined to change his relationship with her, but he suffered absolutely no remorse. He always liked Joyce, but he never seemed to believe that she had any real authority over him, although from time to time, he pretended to let her think so.

## Bhalu's presence

When he was an adult I did not chain him to keep him around home. He was not inclined to wander as a hunting dog might. I had also been warned that he might become mean if chained. He liked to sleep during the day and patrol a perimeter he established around our house at night. I was glad that we lived in the country when, at various intervals, he sounded his deep booming bark as wild creatures passed near the house. The first winter I discovered that the coyotes took him seriously. There was a fifty-yard circle around our house within which there were absolutely no coyote tracks. After he died, the snow revealed the coyotes making exploratory forays within that circle. It took a couple weeks of careful probing, but eventually they were not only walking around under our windows but also venturing up on the front porch which had always been his redoubt. No one else in the neighborhood had dogs that invoked such respect from the coyotes.

Bhalu died in convulsions, over a period of a week, following a few days spent at the local vet's while we went away one weekend. There were no other local kennels in which to house him. He had been there before and knew all the staff, so we did not worry about him. The morning after I brought him home, I found him in convulsions on the living room floor. In a few moments he wanted out and I thought he wanted to relieve himself so I let him out. Meanwhile I went upstairs to wash up and shave. When I came back down he was not waiting as usual to be let back in, nor was he in the immediate yard. With my heart in my throat I grabbed my coat and jumped into my four-wheel drive car to search the property. As I drove along the road bordering our place, I found him walking unsteadily along the edge of the gravel road about a mile from home. Where he was going, and why, only he knew. He seemed dazed and did not respond to my call. So I gently picked him up, put him in the car and drove home, and soon after that took him back to the vet. The vet examined him, but was unable to determine the problem. We left him there, under the watchful care of the vet and his staff, for a week. I visited him every day, sitting by his kennel with my hand on his head. I would like to think that my presence in some way reassured him, as one friend to another—but who knows? Whenever I came, he could not get up to show his pleasure at seeing me, as he usually did, but he knew I was there. Lying stretched out on the kennel floor he would raise his head a little, slowly wag his tail and look at me

Plate 4.6. *Bhalu, son of Saipal Baron (Kalu) with Elliott Johnson in Canada, 1986.*

intently with soft luminous eyes. When he died, I did not take his death well. I realize now, as I write this, how much I still deeply miss him.

I have had the pleasure of owning a number of good dogs in my life, but Bhalu was unique. What I remember most about him is that he was always his own dog, so to speak. He lived with me as a friend, and showed unabashed affection to my friends and family. He willingly cooperated by observing the few rules I laid down. He guarded our house with sensitivity, not out of a blind instinct that compelled him to attack trespassers. For some reason I never felt that I really owned him; rather, that he lived with me. There was something about his quiet dignity, a kind of nobility, that commanded respect. This may sound strange when talking about Tibetan mastiffs, but he seemed profoundly gentle and good-natured. (His father had a similar, laid-back temperament.) It was difficult to rile him, but when something did upset him he suddenly transformed himself into the ferocious mastiff of legend.

I remember once, when he was about a year old, a friend brought his Pit Bull of about the same age, to visit. Bhalu immediately wanted to romp. He picked up a stick and held it up for the Pit Bull to grab the other end. Then the two ran around the yard tugging at the stick. Bhalu was the picture of good spirits and fun. Suddenly, the Pit Bull dropped the stick and lunged at Bhalu's neck. For a split second Bhalu seemed dumbfounded and then it was as if something clicked and his heritage came alive. For a few moments, there was a fierce rough and tumble fight. We broke it up at the Pit Bull owner's insistence—because he did not want the Pit Bull to kill my dog, he said. I thanked him for his concern, but could not help wondering aloud: if my dog was in such danger why was all the blood coming from the other dog?

Like his father in the Himalayas, Bhalu seemed to be able to read people. He liked most visitors, particularly if I introduced them as friends. But sometimes, after studying the situation, he would block the stranger from coming up to the house. Those who chose to ignore his fixed eye-to-eye stare and advanced further were treated to a low rumbling growl, which very few chose to ignore. While he exercised discretion and seemed to give some people and animals some latitude, once he made up his mind that they were a threat, he brooked no nonsense.

After his death I had a number of rather bizarre experiences. The first happened one night not long after he died. When Bhalu was a pup and wanted outside at dawn, he would walk silently into our bedroom and stare intently at me until I woke up and let him out. One night after he died I awoke feeling that he was there, again, staring at me—and, sure enough, he was. I realized in a few moments that I was dreaming. Yet, it was unlike any other dream I had ever had, in part because it was so vivid. As I looked at him in astonishment I realized that his image glowed, radiating vitality and good health. I was thunderstruck, and when I got over my surprise and was fully awake, I wondered what it meant. I had no sooner ask myself that when I felt that it was to let me know that he was alive and well. Grief flooded over me, but also a kind of simple joy.

Several weeks later I had another dream. In it, I was sitting on the floor and in my lap was a Tibetan mastiff puppy happily chewing my fingers. I remember wondering what in the world this was all about? Just then the puppy looked up at me and said, by thought (and this is not a Disney talking dog story): "I'll come back when you need me." Then he went back to wiggling and nuzzling my fingers. I awoke and wondered if I was concocting the dream as a way to deal with my grief. If I was, it was certainly vivid, and somehow very real. It seemed, in a strange way, as if I had received a message from Bhalu. Stories of communications from the dead are highly suspect in our society, all the more so if they purport to come from family pets. I make no claims about these experiences; I just recount them for whatever they are worth to the reader.

A friend called one day to say that she had taken some pictures of Bhalu on her last visit and would I like copies? She was going to get them developed and would bring them around to the house. When she came she was very apologetic because she said that of all the pictures she had taken only one came out, and it was weird. Apparently she had taken the picture facing into the sun. There stood Bhalu, head up, looking off into the distance, a classic pose, radiating light. That picture hangs on the wall in my study, a reminder of his very special presence.

*Plate 4.7. The strange halo of Bhalu, son of Saipal Baron (Kalu) in Canada, 1986.*

The final strange episode came some months later. To help out some friends who were going away, we had agreed to look after a rambunctious Saint Bernard puppy. Since the dog lacked even the most basic training, I decided to build a kennel so I could be sure he was safe when I was away from home. One day while I was stapling fence wire to one of the posts, I looked up and saw a grey transparent image of a Tibetan mastiff strolling nonchalantly towards me. His eyes, in typical fashion, were fixed directly on me. I could not believe what I was seeing and felt a shiver of apprehension. (At this point let me say that, as an anthropologist, I am not in the habit of seeing ghostly apparitions of dogs, or of anything else for that matter. Nor have I ever had the sort of vivid dreams mentioned earlier after the loss of a pet.) As the dog advanced towards me, I knew that I was seeing something in my mind, but somehow real. While these thoughts bounced around in my head, I tried to make sense of what I was seeing or, perhaps, imagining. When the dog got up to me he rubbed against me for a moment and then wandered on his way, disappearing behind me. I knew by his carriage and manner that it was Bhalu, especially when he rubbed against me. When he was alive and I was working he liked to come up, affectionately make contact for a moment, then return to whatever else he was doing. On this occasion it was sort of like a relaxed 'How are you, boss?' sort of visit. When he had gone on I sat down in the sunshine, somewhat shaken, and though I could not see him, I still felt his presence.

I still grieve for Bhalu sometimes, but I have had no other strange or bizarre experiences. It does not take much, however, to conjure up a lot of affectionate warmth when I think of him.

# 5

# The Shaggy KyiApso

Some years ago, two decades before I finished this book, Ann Rohrer and Cathy Flamholtz wrote a short introduction to what they called the "Apso-Mastiff" and "Apso-Do-kyi," a type of dog that today is best known as KyiApso. "Tibetan fanciers," they said, "continue to be intrigued and fascinated by accounts of the rare Apso Do-kyi. Some have sought to import these exotic dogs to the United States. Surely, in years to come, we will learn more about the type, personality and, perhaps, even the origins of the breed. Until then, the famed 'Apso-Mastiff' will remain a mystery."[1] By now we know a great deal about this dog, some of it innovative, some of it contentious, and all of it remarkable and fascinating to dog fanciers.

## Introduction

In some of the nineteenth- and early twentieth-century writings about dogs from Central Asia, Tibet and the Himalaya countries there are occasional descriptions of big "shaggy" dogs. Today, we call them KyiApso from Tibetan *kyi-apso*, meaning 'bearded dog'; i.e., a dog with relatively long hair on its muzzle. Since the breed name KyiApso is well established (as one word, with capitals K and A), I will use the term here to refer to all of these dogs, much the same as the English term 'Tibetan mastiff' generally refers, outside of Tibet, to several types of big Tibetan-Himalayan livestock guardian dog. The Tibetan word *kyi* (sometimes spelled *khyi*) means dog and *apso* refers to the long hair found all over the dog's body, but especially on the head and muzzle. It also qualifies in common parlance as a type of *do-kyi*, meaning a dog that is tied, or a door dog, in Tibetan.

Irma Bailey wrote this about what she called the "Do-Kyi-Apso" in 1937: "The late Dalai Lama, himself, kept many dogs, among them one described as 'Do-Kyi-Apso.'" She understood the term *apso* to be a corruption of *rapso*, which she interpreted to mean "goat-like." She thought that the apso/rapso she saw at the Norbulingka, the Dalai Lama's summer palace, "may have been a cross between a Tibetan mastiff and the dog known in Tibet as the large Apso."[2]

Jigme Wangchuk Taring goes further into the etymology of the term and tells us that apso is derived from *ara* meaning 'mustache' and *sog-sog* meaning 'hairy,' and that "Pure breed *apsos* are only found in Tibet and come in two sizes, large and small. The large mastiff *apsos* are bred for watch dogs. These mastiff *apsos* are usually big, ferocious and filthy

Do-Kyi-Apso means a "Long-haired, tied dog." The breed is rare. This photograph was taken by the husband of the author. The dog was the property of His Holiness the late Dalai Lama

*Plate 5.1. The Dalai Lama's Do-Kyi-Apso (KyiApso); photo from Mrs. Irma Bailey (1937).*

as it is difficult to look after their thick coat and particularly their face and mouth after a meal. ...[They] are mostly kept by nomads and country houses as watch dogs."[3]

The generic 'kyi-apso' refers to *any* long-haired Tibetan dog—large or small, mastiff, terrier or Lhasa Apso, even a shaggy mutt. The ones that have intrigued big Tibetan dog aficionados in the West since the 1970s, however, are the large ones, some as big as their not-so-shaggy, unbearded Tibetan mastiff brothers and cousins. Some fanciers believe the KyiApso to be a separate breed, while others consider the Tibetan mastiff and Tibetan KyiApso to be the same breed—the Tibetan *do-kyi*.

### Tibetan KyiApso, a chronology
1886    "large shaggy white dog"—Andrew Wilson (1886)
1890s   "Great Dog of Tibet"—G.A. Graham (source unknown, but see 1931, below, and the 2001 article by J. den Hoed)
1903    "wolf-dog"—Henry Yule (1903)
1931    "Tibet sheepdog or wolfdog"—Graham, in Phyllis Gardner (1931)
1933    "Great Dog of Tibet"—W. Mut, as noted in Jan den Hoed (2001)
1937    "Do-Kyi-Apso"..., "tied long-haired dog"... , "large apso"—Irma Bailey (1937)
1983    "Tibetan Khyi Apso"... , "Kinnauri kutta"—Christian Ehrich (1983, personal communication)
1980    "long haired Tibetan Mastiff"—Dan Taylor (1980?)

| 1981 | "shepherd dogs of the Kinnaur"..., "Kinnauri kutta"—Hedy Nouc (1981) |
| 1988 | "Bearded Tibetan mastiff"..., "Kinnauri kutta"..., "Humli kukur"—D. Messerschmidt (1988) |
| 1988 | "Apso Mastiff"—Jay N. Singh (1988, personal communication) |
| 1989 | "Apso Do-Kyi"..., "Apso Mastiff"—Anne Rohrer & Cathy Flamholtz (1989) |
| 1992 | "Tibetan KyiApso"—Dan Taylor (1992) |
| 1996 | "Tibetan KyiApso"—TKC (Tibetan KyiApso Club) (1995) |

There is a belief among some fanciers that these dogs are found almost exclusively in the far west of Tibet, in the vicinity of sacred Mt. Kailash and Lake Manasarovar. But I have seen many of them in southern Tibet along the route from Tingri to Lhasa. They have also been reported from eastern Tibet and Bhutan, and are common in parts of the north Indian and Nepal Himalayas.[4]

Among Western KyiApso fanciers it is the dog's size and shaggy appearance that are its most distinguishing and attractive characteristics. What is important in Tibet, however, is not what it is called, nor how it looks, nor some mythical place of origin, but the dog's *function*, what it instinctively *does*: guard livestock or property. Of course, conformation (dog's shape or structure) often influences what it can do. Conversely, some KyiApso fanciers, in considering that 'form follows function' will say that different shapes and functions may be 'correct' so long as the dog can function.

Compared with the more well-known and popular Tibetan mastiff and other Tibetan breeds (terrier, spaniel and Lhasa Apso), relatively few Tibetan KyiApsos have reached North America or Europe. Among the first to North America were imported in the late 1970s, or perhaps earlier, but none of those early dogs seem to have left pups that continued to be bred. One KyiApso breeder I know met a veterinary ophthalmologist from Alaska in 1997, who recognized the breed and said that she had once worked with two dogs "just like it" some years earlier. She remembered that one of the dogs "had to be returned to Tibet" because the owner could not handle him. Unfortunately, the actual date for this sighting is unknown. My source also reports that another woman told her about seeing a "shaggy, bearded dogs from Tibet" at a breeder's kennel in the Upper Peninsula of Michigan during the 1970s. The woman "couldn't remember what they were called, but she remembered the distinctive walk the dogs had." (Some KyiApsos, especially large males, have a distinctive 'Mae West' roll to their hips when they walk slowly and leisurely, a style of walking that is easily recognized and remembered.) Thus, she concludes, one or a few such dogs may have already been in America prior to the arrival of the Tibetan KyiApso Club's first dogs, imported by Daniel Taylor and Melvyn Goldstein in the 1970s, described below.[5]

# 5. THE SHAGGY KYIAPSO

One of the earliest accounts of a large shaggy Tibetan dog in the literature is from Captain George Augustus Graham in the 1890s. Captain Graham is best known among dog fanciers as the founder of the Irish Wolfhound Club of Great Britain, in 1885. At one point, he used one such dog as an outcross while creating the Irish Wolfhound. Phyllis Gardner, in *The Irish Wolfhound: A Short Historical Sketch*, describes that particular dog (including a quotation from Graham). "The one used by Graham," she says, "was a specimen of a different breed from [the Tibet mastiff], the somewhat rare Tibet sheepdog or wolfdog, resembling a large English or Russian sheepdog; he has been described [by Graham] as 'standing 36 inches at the shoulder, and covered with a profuse coat of lion-coloured hair. His limbs are straight as gun-barrels, and he is exceedingly active and playful. Like all big creatures, he has a perfect temper, and is docile, affectionate, and intelligent.' His tail was carried hanging down, sometimes with the tip curled up, in a manner that would not be at all wrong in an Irish wolfhound. He was obtained from a London livestock provider's shop, and had no antecedents. I believe Queen Victoria once had a dog of this type."[6]

Hilary Jupp of the Irish Wolfhound Club completes the story: "The Tibetan, called Wolf, was mated to Lindsay's Tara (by Bhoroo ex Ch. Sheelah) and produced in the litter (whelped 16th June, 1892) two bitches, which are the ancestors of all modern wolfhounds..."[7]

The English traveler, Andrew Wilson, also wrote about a dog that, from his brief description, sounds like a KyiApso. He encountered it in western Tibet, directly north of the Indian Himalaya, in 1873. While encamped at a village called Nako, Wilson adopted a "pretty large shaggy white dog" that accompanied him for many months across Tibet. "We called it Nako, or the Nako-wallah," his story begins, "after the place of its birth; and

*Plate 5.2. Captain Graham's 'Great Dog of Tibet', from a newspaper clipping (Jan den Hoed, 2001).*

never did poor animal show such attachment to its native village. It could only be managed for some days by a long stick which was fastened to its collar, as it did not do to let it come into close contact with us because of its teeth. In this vile durance, and even after it had got accustomed to us, and could be led by a chain, it was continually sighing, whining, howling, growling, and looking piteously in the direction in which it supposed its birthplace to be. Even when we were hundreds of miles away from Nako, it no sooner found its chain loose than it immediately turned on its footsteps and made along the path we had just traversed, being apparently under the impression that it was only a day's journey from its beloved village. It had the utmost dread of running water, and had to be carried or forced across all bridges and fords."[8]

He goes on to say that "No dog, of whatever size, could stand against it in fight, for [it] had peculiar tactics of its own, which took its opponents completely by surprise. When it saw another dog, and was unchained, it immediately rushed straight at the other dog, butted it over and seized it by the throat or some equally tender place before the enemy could gather itself together. Yet Nako became a most affectionate animal, and was an admirable watch[dog]. It never uttered a sound at night when any stranger came near it, but quietly pinned him by the calf of the leg, and held on there in silence until some one it could trust came to the relief. The Nako-wallah was a most curious mixture of simplicity, ferocity, and affectionateness."

Wilson's story concludes when he left Nako "with a lady at Peshawar, to whose little girls he took at once, in a gentle and playful manner; but when I said 'Good-bye, Nako,' he divined at once that I was going to desert him; he leaped on his chain and howled and wailed. I should not at all wonder if a good many dogs were to be met with in heaven, while as many human beings were made to reappear as pariahs on the plains of India."

Other than these few accounts, other references to the KyiApso in Tibet are rare. The following two stories come from the Nepal and Indian Himalayas.

## "They have pleased me extraordinarily..."*

In December 1982, I saw my first shaggy KyiApso look-alike in Mustang District, the ethnographically and geographically Tibetan region inside the political bounds of northern Nepal. It was a red pup, proudly shown to me by a Nepalese government official who had obtained it from the Raja of Mustang. The Mustang royals, though vassals to the former kings

---

*This account is based on my article 'The bearded Tibetan Mastiff of the trans-Himalaya', augmented by more recent notes. The original article appeared in *Dog World* (USA) (1988).

of Nepal, have had long-standing cultural and economic ties with the traders and nomads of western Tibet. Upper Mustang (known locally as Lo Manthang) was once a part of the ancient western Tibetan kingdom of Gu-gé (or Zhang-Zhung), which included Mount Kailash and Lake Manasarovar in its territory. I hasten to add, however, that *political boundaries have little to do with a dog's origin and identity.*

A year later, while traveling in the north Indian mountain state of Himachal Pradesh, I came across another KyiApso, the so-called Kinnauri *kutta* ('kutta' is Hindi for 'dog,' and Kinnaur is a district in the Indian Himalayas). And in 1986, in northwestern Nepal in the district of Bajhang adjacent to Humla and western Tibet, I met a dog that I called the bearded Tibetan Mastiff, or Humli *kukur* ('dog' in Nepali).

Most recently, in 2007, I encountered a large handsome all-black male KyiApso in upper Manang valley of Nepal, east of Mustang District on the northern side of the Annapurna massif. He was one of a kind, but the locals whom I quizzed knew nothing about his origins other than he might have strayed in from another village up valley. For most of one day he lay in the sun, aloof but not unfriendly, passively eyeing all who walked the trail past our lodgings. The next day he followed our trekking party several hours to another village down valley where several similarly shaggy but smaller dogs drove him off with their aggressive territoriality.

Back in 1983, a Kinnauri kutta was shown at an informal shepherds' dog show that I attended high in the Dhauladhar mountains of Himachal Pradesh (see Map 1, p.iv) in May 1983. The 'dog show' was organized by Christian Ehrich who directed a rural development project in the district, funded by the German government. Ehrich also had what he called a Tibetan mastiff. He was keen on encouraging the Dhauladhar shepherds to take greater care in breeding good dogs, to perpetuate the best of the Himalayan breed. Ehrich hosted two high mountain dog shows, in 1981 and 1983, to which he invited the shepherds of the region along with several Tibetan dog fanciers from India, Nepal, Europe and America. The shows were advertised locally on placards in the Hindi language, posted where Gaddi shepherds, the traditional Dhauladhar mountain sheep and goat herders, would see them. As an incentive, Ehrich put up valuable prizes of hybrid Merino breeding rams for the Best of Show and Reserve Best of Show, and dog collars and flashlights to the owners of other select dogs.

To get to the high pasture for the Dhauladhar dog show most of the shepherds had to trek over the mountains from pastures in other valleys. In 1983, I was one of Ehrich's invited foreign guests. From the Himachal Pradesh hill town of Palampur (the project headquarters), we drove into the mountains by jeep, then walked for most of a day to a high pasture called Palachik. We camped in local forester's hut. One of us, Frau

Plate 5.3. Kinnauri kutta (bearded Tibetan mastiff) of the north Indian Himalaya, with a man from Kinnaur.

Hedy Nouc of Germany, had also attended the previous 1981 dog show where 52 dogs were shown. Writing about the first day of that year's show, she described how the shepherds and their animals "seemed to come out of every corner. I will never forget the sight," she said. "Sheep herds mixed with Indian, and above all Tibetan, goats. The shepherds had colorful head gear, and with each herd was 1 to 3 mastiffs. It took hours for all to arrive on the narrow trampled paths." The sun shone brightly that morning, and the whole event had a festive air about it, according to her description.[9]

In 1983, however, the sky was dark and it rained steadily most of the day. Only when the show was over did the sky clear. Even in the rain, however, the sight of all those sheep, goats, dogs, and distinctively dressed shepherds was no less enchanting. The 1983 dog show was also attended by a Punjabi Sikh and avid big dog fancier, Group Captain Amarjeet Singh Grewal.

Four types of Himalayan dogs were shown both years. The Gaddi kutta and Bara Benghali kutta were the two most common. Gaddi is the name of the shepherds' ethnic group, and Bara Benghal is the name of a Dhauladhar mountain pasture. A few Lahauli kutta type were also shown, from the district of Lahaul on the India/Tibet border. The fourth type was the Kinnauri kutta, from Kinnaur, a remote district in northern Himachal Pradesh. The Gaddi, Bara Benghali and Lahauli dogs could be best classified as Himalayan mountain dogs, or Himalayan sheepdogs, though they appear to be genetically close to the more 'ideal' Tibetan mastiffs. The Kinnauri dog was a KyiApso, long-haired and shaggy, with a hairy muzzle. Given their proximity and traditional trade relations, the ancestors of all these dogs were undoubtedly related to the big dogs of western Tibet.

Kinnaur is a district in the northeast corner of India's mountainous state of Himachal Pradesh, adjacent to the north Indian regions of Tibetan culture: Zanskar and Lahaul. This region is dominated by the high Dhauladhar and Zanskar mountain ranges, part of the Greater Himalayas along the border with western Tibet. The local people speak

# 5. The Shaggy Kyi Apso

several languages related to Tibetan. There are numerous ancient Tibetan Buddhist monasteries in the region and the indigenous people practice Lamaism (Buddhism), as do their Tibetan cousins across the international border. The people of Kinnaur were once known to the British colonials as 'Kinners.' In the mythology of their neighbors they were thought to be half human and half god. Trade and travel through this high mountain region has long been accommodated by the old Hindustan-Tibet track which follows the Sutlej River and crosses the Shipki La pass (15,000ft, or 4572m).

While good dogs were shown at each of the informal Dhauladhar dog shows, it was the Kinnauri dogs that caught our eye. Christian Ehrich wrote this to me about them: "The *Kinnauri kutta*... does not come very near the ideal TM [Tibetan mastiff] type, but contains elements of one or more other dog breeds from the higher Himalayas. A distinct wiry hair of medium length, and a bearded snout with certain Apso (long hair) characteristics hint at a distinct relationship to the legendary Tibetan Khyi Apso, or 'long-haired dog.' All colors occur in this breed, although spotted animals are rare. A dark slate-grey seems to be quite frequent in the breed. This dog is kept as much as a shepherd dog as all TM type strains, and crosses between the TM and the Kinnauri dogs are not infrequently found with the shepherds."[10] (Ehrich could just as well have called them variations of one breed.)

After seeing the Kinnauri kuttas at the 1981 dog show, Hedy Nouc wrote that it was "a breed of dog that until now I didn't know existed. They have pleased me extraordinarily, these herd dogs of the Kinnaur area... At first sight, they seemed somewhat like a Tibetan terrier with... shorter hair. The height at the withers is... between 50 and 60 cm [roughly

*Plate 5.4. High pasture 'dog show' with shepherds and dogs in the rain, in the Dhauladhar Himalayas of north India.*

20-24 in.]. They have all the color variations of the Tibetan Terrier, and many essential similarities. These Kinnaur Kuttas (*kutta* = dog, *kutti* = bitch) serve the same function as Tibetan Mastiffs; however, they appear to me to be more sociable or docile. ...[All] twelve Kinnaur Kuttas... were astonishingly the same, and they had the proper color variations. Black and tan, bright gold or coal black. We couldn't decide which was more beautiful. Spontaneously, I put up a cash prize and looked for the most beautiful pair."[11]

A few years later, it was my turn to be pleased extraordinarily. It happened when I encountered a beautiful cream-colored KyiApso on a trail in remote far western Nepal.

## A Bearded Muzzle in My Face

In March 1986 I met the Humli kukur face-to-face. At the time, I was a consultant to a community forestry development project in Nepal under auspices of the United Nations Food and Agriculture Organization. My work took me to the remote Bajhang District, adjacent to the even more isolated high mountain region of Humla, part of Nepal's northwesternmost Jumla District. "Humla-Jumla," as the locals call this region, borders Tibet at the north and India at the west. Ever since I first heard of it in the early 1960s, Humla-Jumla has meant, for me, a place of rugged isolation and adventure, of trans-Himalayan/Tibetan traders and their big dogs.

One of the overland routes through the hills and over the mountains to Humla from the south passes through Bajhang District. Part of this foot track is narrow and dangerous where it skirts the roaring Seti River gorge. This difficult track serves as a main trade route linking north India through northwestern Nepal over Saipal Himal into western Tibet.

We began our trip by flying 225 miles from Kathmandu west to Nepalganj, then another 100 miles north over Nepal's rugged western mid-hills and down steeply into a narrow canyon to a remote gravel airfield at Chainpur, Bajhang's small district town. To get a better fix on our location vis-à-vis Tibet, note that Bajhang is 600 miles due west of Lhasa, on the same latitude as Cairo, Egypt and north Florida (both slightly shy of 30° N.). Bhajang is only 30 miles south of the western Tibet border and, from there, it is only 40 miles farther north to the shores of Lake Manasarovar. Mount Kailash is another 55 miles north of that. This is all Tibetan mastiff and KyiApso country.

After making courtesy calls to the District Forest Office in Chainpur, my companions and I set out to walk down the Seti *khola* (river), visiting forest development sites and staying in tiny hamlets along the way. We were a party of seven, including a forest guide and two porters. As the trip anthropologist, my job was to observe and document the social and cultural aspects of village forest and woodlot development. My companions included a British forester with a degree from the Oxford School of Forestry, a young VSO

## 5. THE SHAGGY KYIAPSO

forester (the UK's Voluntary Service Overseas), and a senior Nepalese forest officer. We wanted to see the types of forests, how they were managed and utilized, and so forth. Our individual interests overlapped and intertwined in complementary ways.

Seti khola means 'white river,' named after the grey-white glacial sediments that it carries southward out of the high mountains. Bajhang's Seti river gorge is as spectacular and rugged as any in the Himalayas. The river and the trail beside it look like two narrow ribbons winding along in parallel through a steep and austere countryside. At every turn we faced new vistas of a lonely, sparsely settled and forested landscape. Our trek took us past isolated hamlets clinging to the high hillsides, looking from a distance like tiny postage stamps pasted precariously onto a panoramic landscape painting. Over the course of the next few days, we passed through three districts with the rather poetic roll-off-the-tongue names of Bajhang, Baitadi and Dadeldhura. Much of our time was spent walking the river track, hour upon hour, up, down, over and around the cliffs along the gorge and across narrowly terraced farmers' fields that cascaded like giant stair steps down from the ridge tops to the river.

The fields were ripe with winter wheat, a staple crop in this locale. Most houses were constructed of rough cut stone, plastered over with ochre mud and topped with thatched roofs. To our north were the snow peaks of Saipal Himal, after which Jay N. Singh, Nepal's foremost Tibetan dog expert, named his kennel. For part of the trek, we passed through the tiny former principality of Thalar, where Singh was born and raised (see Chapter 4).

The few villagers we met were mostly conservative Hindus of the (so-called) "higher" castes. As they seemed reluctant to provide strangers such as we with nighttime accommodations, we camped in the open, or in cow barns and primitive porters' sheds. We did our own cooking, meals of boiled rice garnished with what little meat, potatoes, green vegetables or beans we could find to buy; that is, whatever the locals would deign to sell us. Their strict caste rules about commensality with strangers limited their generosity. Both they and their mongrel dogs were suspicious and distrustful of us, and inhospitable to foreigners in general.

Each day, the young British VSO and the Nepalese forest officer kept us apprised of the forest types we were walking through, helped us with translations of the local dialect (quite distinct from what is spoken elsewhere in the mountains), and explained the intricacies of the local resource management systems that we observed. While walking the many hours each day between stops and our work, we each pursued a favorite trekking pastime: bird watching.

The Oxford forester carried binoculars and took great pleasure identifying most of the colorful birds we saw. Tiny Nepal is renowned for its exotic bird life, and boasts over 800 species—more than are found in all of North America. Himalayan ornithology is a challenge.

And I amused everyone as the group's unofficial dog watcher. You must understand that dogs are not highly respected in Nepal's low elevation conservative Hindu villages. They get more respect among the largely high altitude Buddhist livestock keepers. On our relatively low trail and in the hamlets we saw mostly small brindle or black pariahs, or 'pi-dogs,' basically scavengers of no special interest. Therefore, my curiosity about local canine lore puzzled my Nepalese companions and the local folks. If I had told them that I was writing a book about dogs, they would have laughed at such a silly notion. But I had a goal, a hope, that Humla-Jumla's trans-Himalayan traders who regularly plied this route and whom we expected to meet on the track might have some good mountain dogs with them.

There is little to draw outsiders to this part of Nepal. It is a remote food-deficit region. There are few tea stalls and no guesthouses for outsiders to stay, and it is not featured on any trekking agency itinerary. Though it is a sparsely settled region, we were rarely alone. On the second day out we began to meet the traders with their pack animals—hundreds of sheep and tough little Tibetan goats. Each animal was laden with one or two saddle bags weighing as much as 15 kg (33 lbs) each, transporting rice from the towns in the lowlands along the border with India to hungry villagers in isolated villages upriver.

The men and boys from Humla who tended the flocks were a hardy lot from the high Himalayas and Tibetan border area a few miles north of us. These mountain folk, typically of short stature by our standards, spend much of the dry winter season on the trail, conditioned to a Spartan life in the outdoors and to the exceptional rigors of their inter-mountain trade. The ones we met wore coarse woolen garments and smelled inoffensively of campfire smoke and livestock. They greeted us with curiosity and pleasure; their smiles were genuine and captivating. Each day we encountered five or six flocks, each with several men and boys and one or two dogs. Most of their dogs, however, were disappointingly small, scruffy critters that looked only vaguely like the sheepdogs of my dreams. They had few outstanding aesthetic qualities, and were not worthy of much attention. Nonetheless, they play a vital role in the local economy as livestock guardians against predators.

On the third day we found ourselves walking some of the worst and most dangerous trails of the entire trek. The cliffs of the Seti gorge rose above us for hundreds of feet along the north (right) bank. In places the trail was cut out of the solid rock face, dreadfully narrow at dizzying heights above the river.

That afternoon, while negotiating a particularly hazardous section across the face of a cliff, we rounded a corner and came face-to-face with a flock of about 500 sheep. A half dozen swarthy Humla men at the back were urging their charges across the precipice. There was no room to pass. Rather than retreat backwards to a wider place we stopped and made ourselves small by hugging the cliff face. When the lead animals saw us, however, they stopped still in their tracks and eyed us warily, baaing loudly. Eventually, two packmen worked their way to the front to push and pull the animals past us. To high-pitched whistles and the shrill calls of "*Drii! Drii!... Ah-ah-ah,*" and grunting pulls and tugs to keep them moving, the Humla packmen coaxed the animals on. Hesitantly at first and then in a rush the flock pressed by us in a cacophony of bleating and baaing, neck bells jangling, hooves clacking on stone, kicking rocks off the trail down the cliff in their wake. Dust and that peculiar musky odor of excited sheep and goats filled the air.

Eight..., ten..., fifteen minutes passed and still they came, a steady stream of sheep and goats laden down with bags of rice, emerging in a cloud of dust out of a scrub oak forest and onto the cliff track about 150 meters ahead of us, then passing us warily to continue on up the track towards Chainpur. It was a perilous but comical scene—the cliff dropping abruptly away at our feet and rising straight up over our heads, with packmen and their charges crowding by at our side. To top it off, while we clung tenaciously to the cliff wall, a band of mischievous rhesus monkeys began throwing sticks, stones and clods of dirt down on our heads from above! We were entangled in a wild, spirited and noisy traffic jam—sheep, goats, dogs, the packers, our porters, several foresters and this anthropologist, and a band of spiteful monkeys overhead.

As soon as the last animal was past, a second flock appeared coming up through the forest on the far side of the cliff, moving rapidly in our direction. I shouted to my companions that we would certainly be cut off if we did not move ahead quickly. I took the lead and ran, alternately ducking my head, steadying my balance against the cliff wall at my side, and bracing against the uneven track underfoot. Our goal was a wide spot in the trail ahead of us, where we could stand aside more comfortably while several hundred more sheep and goats passed us by. The track was no more than a foot or two wide in some places, treacherously narrow. All the while, the monkeys chastised us with their frenetic chatter, showering us with more dirt and twigs. We crossed quickly and without mishap, then stood bunched together where the trail widened to let the next Humla-bound pack train pass.

Then it happened! I had just turned into the forest to begin a gradual up-and-down passage through the scrub amidst the second flock. I was watching where to put my feet on a slight incline when I heard a deep WOOF!—and glanced up directly into the bearded muzzle of a magnificent dog. I, the dog and the flock stopped in our tracks. Here he was, the 'Humli dog' of my dreams, a KyiApso of outstanding characteristics, a highlander like none other—well-bred, well built, certainly to be respected but not feared. It was, perhaps, an archetype of the breed, an exquisite dog. Thankfully, he showed no aggression, only curiosity.

He was about two years old, cream-colored, with long hair badly matted, and a fully bearded muzzle. I fumbled in my pack for the camera. A moment later, the packman arrived after working his way up through the flock to see what had stopped forward progress. I persuaded him to hold the dog for photographs, and plied him with questions in Nepali:

"*Yo kukur—kaha paieko, daju?*," I asked. "The dog, where did you obtain it brother?"
"*Bhot baata, sah'b*" he answered. "From Tibet, sir (*sahib*)."
"How did you get him?," I continued in Nepali.
"While trading over the northern border above Humla," he replied.
"How much did you pay?"
"*Dherai mahango, sah'b*," he said. "Very costly, sir. Over one thousand rupees" (about $50 at the time, but a large sum even to a trader).
"Will you sell him?"
"*Nahi!*" was the firm reply. He was totally unwilling to part with his cherished guard dog.

This whole scene took place in a matter of one or two minutes, then it was over. The bunched-up animals were moving on, and after a few quick photographs of dog and master we stood aside to let them pass. Only after the flock and the packmen had gone did I realize that I hadn't asked for the dog's name, nor what the traders called this unique type.

Plate 5.5. *Bearded Tibetan mastiff (KyiApso) from Humla, far western Nepal Himalayas.*

We never saw another like it on that trip, and within a week we were back in Kathmandu, far away from the wild trails of Bajhang and Humla-Jumla. Short of mounting a special expedition back to the mountains of northwest Nepal along the border, or into western Tibet just a few miles north, to find and retrieve some of these rare and shaggy big dogs for breeding, I knew I might never see another like it again.[12]

Nine years later, however, a small expedition led by Dan Taylor did just that—set off towards Mt. Kailash in western Tibet, directly north of Humla, in search of KyiApsos to add to the small breeding stock already in the USA. The KyiApso fancier Diana Quinn was on that trip and later published the story.

## The Search for the Tibetan KyiApso*

In 1994, Daniel Taylor, founder of the Tibetan KyiApso Club of America, set out on a pilgrimage of sorts across Tibet to find new KyiApso breeding stock. He led a party of seven, including his ten-year-old daughter, an adventurous friend, and Diana Quinn. Three Tibetan drivers went with them, in two sturdy vehicles loaded with tents, sleeping bags, food and an extra supply of petrol for a long overland journey hundreds of miles west from Lhasa.

It took a week to cross Tibet, sometimes on roads that were little more than two ruts "across the dusty plains…, surrounded by distant mountains, under an intense sun and bright blue sky," writes Quinn. They imagined that they were "on another planet, or reliving Earth's first few million years," albeit by modern conveyance. "Huge, lumbering yak dotted the hillsides" and they passed "groups of tents belonging to nomads, pitched in their summer locations. Flocks of goats and sheep stayed close to the tents, and wild-looking dogs kept guard, barking when we got too close." It was several days before they encountered the first of the illusive shaggy dogs.

Kailash is one of holiest places on the planet, the center of the world, an *axis mundi* connecting heaven and earth. Mount Kailash is sacred to Hindus as the abode of Lord Shiva and his consort, to Tibetan Buddhists as the abode of the deity Demchog (Chakrasamvara) and his consort, to the followers of the ancient pre-Buddhist religion of Bön, and to followers of the Jain religion. Together, South Asian Hindus and South and Central Asian Buddhists comprise close to one-third of the world's population. Each year thousands of them come as pilgrims to Kailash, and to nearby Lake Manasarovar, from all over Asia. Their goal is to bathe in the sacred lake and circumambulate the mountain's 32-mile (52km) track. Buddhists and Hindus do their pilgrimage clockwise around the mountain, while the followers of the Bön and Jain religions do it counter-clockwise.

The highest point on this ritual *kora* is the pass called Drolma La, at nearly 18,500ft (5639m). While a few hardy souls trek each year around the mountain in one long day, the walk usually takes two or three, and longer for those pious Tibetans who prostrate themselves the entire way around. Tibetans believe that if one completes the circuit 108 times (a sacred number), he or she will achieve *Nirvana*, Paradise, and the supreme happiness that Buddhists believe comes when all passion, hatred and delusion cease and the soul is released from the otherwise endless cycle of purification through reincarnation.

---

*Dan Taylor's 1994 expedition to western Tibet in search of KyiApsos was written about by Diana Quinn and published later that year as 'The Search for the Tibetan KyiApso'. I've adapted much of their story here.

Taylor's unique pilgrim party felt that they, too, would achieve supreme happiness if only they could find some good KyiApso pups to take back to America. While Mount Kailash has always figured prominently in Tibet's sacred geography, for some Tibetan dog fanciers it also has this other 'sacred' significance.

After several days on the road, and many stops to ask about dogs, they met an old woman with "white hair in two neat braids, and wearing a huge belt made of seashells," who told them that she knew of a pup in a nomad's camp a few miles away. "We drove there at once," Quinn relates, "rolling up to the tent in a cloud of dust." Then, "A little girl in a red sweater ran out to greet us. Her parents followed, both wearing long braids, and smiling broadly. Yes, they had a beautiful male KyiApso puppy that they had bought only about a month earlier from a pilgrim passing by. About two dozen goats watched as we negotiated the price... 92 yuan, the equivalent of nine American dollars. Nine dollars is a great deal of money for a nomad family, and it made our new friends very happy."

"The little girl gave the puppy a hug, and the father petted him before handing him over. Dogs [here] are treated better than anywhere else in Asia. In Tibet, it is said that dogs are the souls of monks who have unsuccessfully transcended to the next level of being, and so are treated with respect... "

Taylor's expedition members named the pup Kang Rinpoche, another name for Mount Kailash. A few days later, at a tiny village called Hor, they found two more pups, both female and both severely malnourished. They paid twelve dollars for the pair and named them Mindu and Tashi. On the return trip, they drove out to Kathmandu holding the tiny pups on their laps and intermittently feeding them yak meat and *tsampa* (roasted barley flour) mixed with salt-butter tea. "And the puppies ate!" Quinn writes that "Mindu's nose was perpetually covered with the white barley flour. Nomads subsist on the sticky paste, and since it's all they have, they feed it to their dogs, along with the few odd scraps from the occasionally slaughtered goat, yak or sheep. Tibetan dogs have to scavenge whatever else they eat... "

A week later they were back in the United States with their newly acquired pups.

## Breeding KyiApsos

Among Tibetan nomads, a dog's *function* is its most important characteristic. But in the West, where the vast majority of dogs are bred for companionship or trained and groomed for the show ring, their *looks* are uppermost in the owner's mind. To Westerners, the KyiApsos' looks ("shaggy") and temperament ("laid back and mellow," as Dan Taylor describes it) supersede function. For Tibetans dogs such as these, which have 'come West' as imports, Western standards of physical characteristics and temperament are what guide some breeders and buyers.

In 1977, Daniel Taylor and Tibetologist Mel Goldstein imported their first pair of KyiApsos from the Himalayas. They called the male Thumdru ('Little Bear') and the female Singdru ('Little Lion'). According to Ann Rohrer and Cathy Flamholtz, Goldstein "managed to locate a bitch in Timi [sic: Limi], along the northwestern Nepal-Tibet border. Tibetans in the area told Dr Goldstein that Singdru, as he named the dog, was an excellent example of an Apso Do-kyi. They told him this was a rare and valued dog..."[13]

Both Thumdru and Singdru were physically large and very strong. When full grown, Thumdru weighed 110 pounds (50 kg) and stood 27 inches (68 cm) at the shoulders. Taylor worked with Thumdru under the tutelage of one of America's foremost dog trainers. The trainer, he says, was amazed at Thumdru's reflexes—"faster than a Doberman or a German Shepherd." Dan also noted that "Thumdru would start to react under his shaggy coat, and the bulk would camouflage the preliminary muscle action—he was quick and *smooth*. In races with other dogs, he outran most sporting breeds."

In the official Tibetan KyiApso Club standards, emphasis is placed on the following key features:

*Head*: large, including crown, eyes, and muzzle, and unusual facial hair, or 'beard.'

*Bite, muscular neck, body and curled tail*: all distinctively impressive.

*Forequarters*: "broad and powerful."

*Hindquarters*: built for "quick bursts of speed."

*Coat*: "weather resistant double," 3 to 6 inches long at maturity, profuse on its distinctive face and muzzle.

*Color*: black-and-tan, black-and-gold, black-and-silver, or all black.

*Gait*: "a trot with a marked bounce" exhibiting "exceptional speed and agility."

*Size*: the "most variable" characteristic, with males no larger than 28 inches (71 cm) nor heavier than 100 lbs (45 kg).

Some may think that the KyiApso looks a lot like a standard Tibetan mastiff, with exception of its shaggy muzzle and longer hair overall (and sometimes its smaller size). By comparison, however, breeders point out that KyiApso 'bone' is not as hefty as a Tibetan mastiff's. In a blind test, however, I suspect that it would take a good judge or an experienced breeder with educated hands to tell the difference. Most KyiApsos appear lighter and more athletic than the Tibetan mastiff, and lack the huge, sagging lips or dewlaps, facial wrinkles, or a lot of 'haw' (the fleshy red

*Plates 5.6 and 5.7. Thumdru, KyiApso import from Tibet.*

rim around the eyes) that characterize some of the extraordinarily large 'modern' Tibetan mastiffs.[14]

KyiApso breeders also note that their dogs are much more diverse in appearance than Western-bred Tibetan mastiffs (though on this point, some Tibetan mastiff owners would disagree, lamenting at times about the overabundance of diversity in the breed). KyiApso size tends to vary, from 60 to 100 pounds (27-45 kg), with heights of from 24 to 28 inches (60-71 cm). Judy Steffel, a Tibetan KyiApso breeder, notes however that "nobody's ever put any of them under a wicket to my knowledge. When pairs of KyiApsos—one heavy/powerful, the other lighter/faster—work together, they complement each others' capabilities. Their head shape, ear set, ear size and shape, tail set, amount of curl in tail, hair length, hair texture are all variable even amongst full siblings. There are some genetic differences in the coat colors, too, and some change with maturity. On bearded pups born black-and-tan, the tan often turns gold or cream colored, and the coat of a solid black pup will usually become sprinkled with lots of white or silver hairs, sometimes to the extent that the whole adult dog looks silver."[15]

The Humli KyiApso that I chanced upon in the mountains of western Nepal was the color of rich cream ("light gold" in breed standard jargon and in genetic terms), and Suptu, in England, was partly cream-colored but also had grey-black and brown (tan) markings. Singdru was black with reddish-tan markings. Minhsingh, Dan Taylor's KyiApso bitch, was bright gold. Thumdru, whose outer coat was about six inches long, was black with white flecking. The typical Kinnauri dogs of the north Indian Himalayas are similarly colored. The shaggy dogs bred by the Raja of Mustang are predominantly reddish in color. Apo, one of Judy Steffel's dogs, is black-and-tan with a lot of silvering, while her only genetically black dogs, Lishi and Badger, have real white on them.

Plate 5.8. Apo, a KyiApso.

The fine inner coat, a common feature among KyiApsos and Tibetan mastiffs, alike, is thick, soft and only one inch long. This silky cashmere-like fleece is sometimes called *pashmina* by Himalayan people, terms usually reserved for only the very finest quality spinning wool (for shawls). The outer hair on KyiApsos, however, is longer than on Tibetan mastiffs. Both types shed their under coat annually each spring and may also lose large amounts of guard hair. This under coat has a distinct adaptive advantage or the dogs, as protection against the biting cold and wind of winter. Their thick coats, and even the mats sometimes formed in their coats, also protect them from the teeth of other canids.[16]

Like some other early, more 'primitive' canine varieties, neither the KyiApso nor the Tibetan mastiff have detectable body odor, and the females of both types have only one estrus cycle annually, usually in late fall or early winter, like their wolf progenitors. The adaptive advantage of this timing assures that the pups are whelped in late winter, when the herds and nomads are relatively immobile, while staying in the climatically moderate and relatively safe environment of the lower valleys. Only after they are weaned and are physically active and strong do these dogs begin to encounter the severe stresses of life on the annual transhumance migration up to the highest mountain pastures.

## "The Most 'Laid Back' Creature You Can Imagine"

While their looks are of most interest in the West, how this landrace breed behaves instinctively 'on the job' is more important to the highland nomads and shepherds (though a few Westerners may argue that it is equally important for them, too). Though some KyiApso owners might have worded it a little differently, the dogs' traditional function and temperament are noted in the official 1995 standard of the Tibetan KyiApso Club, as follows:

> KyiApsos were traditionally used as guardians of homes, monasteries, and most importantly, flocks. They performed their duties by barking and threatening, not fighting. Hence, essential to the breed is their deep-throated, sonorous bark and double coat with long guard hairs which make these otherwise not large dogs seem formidable.
>
> Because KyiApsos tend to be assertive in their territory, an animal should not be approached suddenly by either human or canine strangers without the consent of the handler...
>
> The KyiApso is an energetic dog who enjoys running and activity. It has exceptional agility and fast start capability... Because of its energy, the dog may become destructive if confined alone for extended periods. In its 'off-duty' mode, the KyiApso is generally very laid back and mellow...

What most distinguishes KyiApsos from all other big Tibetan dogs is the length of the coat hair, especially that around the face and muzzle. "Hair completely covered Thumdru's face in the fashion of a Tibetan Terrier," Dan Taylor writes. The dog Suptu, imported from the Himalayas to Southampton, England, by Major Dan James also had a unique coat. "When he moulted," Major James has written to me, "his coat would peel off in enormous pieces almost like a lamb's fleece. We were told that this always occurred in the spring at the same time as the yaks moulted [in Tibet] and in a similar fashion. Underneath was a complete coat of new, soft, curly hair. The hair is used for spinning."[17]

In addition to the shaggy coat and bearded muzzle, Dan Taylor places great emphasis on the dogs' temperament. They display a very friendly, playful and largely peaceful approach to life, he says, contrasted with a protective, sometimes aggressive trait when aroused. Perhaps the two are complementary temperaments within a complex behavioral set that has allowed these dogs to adapt so successfully to the harsh and unpredictable conditions of life in the trans-Himalayan reaches of western Tibet. (Friendly, playful, peaceful behavior, of course, is reserved for pack, flock, family relations, and requires less energy than overt aggression.) A related trait, identical to that of Tibetan mastiffs, is their strong attachment to the place they consider 'home,' wherever that might be for the moment, and their absolute need to be outside the abode in order to guard it, especially at night. As one observer of KyiApso behavior remarked to me, the latter is an absolute need that can be modified by the owner. It helps if any inside dogs know that there is 'somebody' (another dog) outside on guard.

Judy Steffel, who has raised KyiApsos in America since 1994, describes the combination of their behavioral traits as follows. Her reference to smooth-faced KyiApsos (which she calls "smoothies") is to the dogs in a litter that carry the recessive genes for smooth faces and look much like standard Tibetan mastiffs when they mature. Most of what she describes fits within the standard range Tibetan mastiff behavior.[18]

"TK [Tibetan KyiApso] personalities are diverse," she writes, "just like individual people; but they all seem to have some strong similarities. They are fiercely protective, but discerning in their 'use of force.' They have 'boundaries'; i.e., they do not give 'unconditional love' and they don't accord automatic respect to just anybody with two legs. They will defer to a leader whom they have learned to respect, but there are some things they simply will not tolerate, intimidation being a big one!"

"In general," says Steffel, "the bearded dogs seem to be a little less 'serious' than the smooth-faced dogs. They seem to approach the world with a bit more whimsy and humor. They can turn deadly serious in a heartbeat if the situation warrants it, but they 'don't sweat the small stuff' and don't seem to be quite as distrustful as the 'smoothies.' I'm basing this observation on what I've read about other people's TMs [Tibetan mastiffs] and what I've observed in my own two smooth-faced girls [born in a litter of TKs]. Badger and Violet are *always* the last to stop barking after the UPS truck leaves and they are the last to approach a 'stranger' and allow petting. (It can take multiple visits for them to allow it!)"

The owners of Suptu, Singdru and Thumdru have each described their dogs as generally relaxed, affectionate and peaceful, but not without occasional surprises. "Laid back" and "mellow" are also terms used to describe them, based mostly on experience with Taylor's dog Thumdru. (But Thumdru also had a vicious side to his character, as Jennifer Ide notes below.)

Similarly, Major James writes from England that Suptu "used to love to have a game, always the same one. He would run around and around then go for one's legs, and with his tremendous power it was very difficult to keep standing." Suptu showed affection "by rubbing his head against us or lying close to our legs. He had a sense of humor and seemed to laugh with his eyes, which were most expressive. His manner was one of quiet dignity and pride..."[19]

After obtaining Suptu for a few rupees from a Tibetan trader in the mountains of northeastern Nepal, Major James learned firsthand what this sort of behavior meant. At first, while still in Nepal, he tried to keep the dog indoors at night. On the very first night he locked the dog in the main room of his small village hut. Almost immediately, he writes, Suptu "started whining and scratching at the door, so I brought him into my bedroom. But he continued his whining and when I eventually fell asleep he broke the door down and stayed outside. I realized later that the main role of these dogs is to guard and never stay in. From then on he stayed outside, where he fiercely threatened any person within sight—there is no doubting the serious damage he would have done if he were loose. He took to me and my wife and to guarding us at any cost."

Over time, Suptu changed his habits and started coming inside. After being imported to England he expanded his territory from the James' house to the front garden, and eventually to the whole street.

Taylor describes Thumdru as steadfastly guarding his wife whenever Dan was away from home for any length of time. Strangers were simply not allowed in or near the house, if Thumdru had his way. When left alone at home, the dog became highly nervous, anxious to break loose. "When we returned home once later than we had intended, we found Thumdru had methodically tried to gnaw his way out through a window frame," Taylor remembers. Standard Tibetan mastiffs have also been known to try to break out under similar circumstances—often quite successfully, and destructively, to the chagrin of their owners.[20]

The most sought-after characteristic of KyiApsos in their natural setting is their outstandingly protective behavior. This is also the key function of the standard Tibetan mastiff. Both the bearded and standard Tibetan mastiffs are prized guardians—the more aggressive in the face of danger, the better, according to local Tibetan/Himalayan people. In the yak and sheep pastures they are needed to warn the herdsmen of danger and to physically repel predators from camp (or, if necessary, to attack and kill predators such as leopards). Likewise, traders and pack animal handlers rely on them for protective purposes along the wild and dangerous mountain tracks and in their camps.

The strong protective instinct of the KyiApso is a point all contemporary owners whom I have queried remark upon. When approached on territory the dogs feel is theirs to guard, their normally quiet temperament rapidly changes and they become highly excited. Major James has noted that although Suptu was outwardly friendly towards people much of the time, he and his wife were never able to relax when Suptu was in a defensive mood.

Thumdru, about whom Dan Taylor has written the most, was affectionate towards most people, and could remember people he had not seen for even a year or more. Nonetheless, with strangers he could create serious problems. "I really think he might have killed to defend his master's home," Taylor says. By contrast, accounts of such intense protective behavior are few among Tibetan mastiff owners in North America where many of their dogs appear to have been bred to be calm but alert most of the time (though some are deliberately conditioned to retain an overtly aggressive guard dog behavior).

KyiApso owners in America and England have wondered if this dislike of strangers is idiosyncratic, or based on environmental conditioning. It is unclear whether it is in the nature of the breed or harks back, perhaps, to some early bad experience in Tibet. Jennifer Ide once cautioned that "It would be a mistake to treat Thumdru's defensive aggression lightly, both because it was so marked, and because it took an unusual pattern in his life. For the first year-and-a-half that we had him, he showed *none* of this behavior at all. Until roughly age two, Thumdru was the most 'laid back' creature you can imagine. While his light, athletic prowess was certainly displayed in romping, around the house he padded slowly and sedately about, modestly offering his head to a visitor for a pat. You couldn't ask for a better large dog house pet."

When Thumdru was about two years old, "he was attacked by a Bulldog, and fought aggressively back," Jennifer says. "From that time onward, he became increasingly 'touchy' around strangers. Over a period of several months he evolved a classic territorial behavior pattern. He remembered his friends for long periods, and once accepted by him, a person could be pretty relaxed about being accepted in the future. But Thumdru was breathtakingly vicious towards strangers, most particularly men, or anyone appearing suddenly or under unusual circumstances. The mildness of his early life and the coincidence of territorial behavior with maturity never suggested idiosyncrasy to us. We loved that dog and breed, but we would not want to play down the possibility of potent, inherited territoriality."

By comparison, the young KyiApso that I encountered on the remote trail in western Nepal displayed no outward signs of aggression nor dislike of me as a stranger suddenly encountered. If anything, he is best described as remarkably subdued, even shy, though I would not want to surprise him watching the herd in the dark of night in camp, when his guarding behavior would undoubtedly be more focused and intense.

Perhaps, as in most dogs, aggressive behavior comes from early socialization and conditioning. I hasten to add, however, that there is some evidence that a dog's 'mean streak' might be heritable. It is quite true that even the most gentle-natured dog may turn aggressive if restrained on a short chain for long periods (as they often are in the Himalayas and Tibet) or if mistreated in other ways, as in villages where rocks are thrown at them and they are sometimes beaten with sticks. I am quite certain, however, that mistreatment was *not* a factor in the dogs belonging to the Taylors, Major James or Judy Steffel. In the case of almost all dogs imported from Tibet, however, their treatment prior to being acquired by their Western owners is unknown.

KyiApso owners also comment on their dogs' seasonal restlessness. Dan associates the timing of their restiveness with the traditional yak migrations in Tibet, with which both KyiApsos and Tibetan mastiffs have long been associated. Similar seasonal patterns of transhumance are observed throughout the Himalayas of north India, Nepal and Bhutan. Taylor writes that his KyiApsos "became *very* restless in spring and fall. They would do everything they could to roam," he says, "disappearing for up to a week at a time and covering enormous distances by the phone calls that would come in from folks who saw the collar."

Whether the dogs' wanderings were truly associated with the traditional movement of Tibetan/Himalayan yak and sheep herds is an open question. There is disagreement among KyiApso owners on this aspect of their behavior. Jennifer Ide once wrote that while she clearly remembers Thumdru's restlessness, which "seemed more sporadic than seasonal. He would indeed roam, apparently non-aggressively, and would find his way home. I always felt it was stretching the point a bit, however, to fit Thumdru's behavior into a strictly seasonal pattern. It is certainly true that he did not always feel this urge. In other words, he was not just a roving dog."

Dan has wondered: "Is it the nomadic heritage? I can only guess so."

Thumdru and Singdru were born in the region of Mt. Kailash and Lake Manasarovar. The KyiApsos that I have encountered in Mustang and Humla-Jumla in northwestern Nepal, and those from Kinnaur in north India, may also have had Tibetan ancestry. And those I have seen on the way between Nepal and Lhasa undoubtedly had purely local Tibetan origins.

## Is the KyiApso a Separate Breed?

According to Diana Quinn and Dan Taylor, the Tibetan KyiApso Club was set up to support what they called an "exciting breed." The club leadership wanted to ensure that a responsible breeding program was established, in order to preserve the unique traditional features of the KyiApso as it developed in its native Tibet. The underlying assumption among the founders of the KyiApso Club is that this is a distinct breed of Tibetan dog. There is a debate, however, about whether it really is a distinctly "separate breed" from the Tibetan mastiffs, or merely a "variety" or "type," or the result of some other breed "intermixing" or "interference." Over the few years of its existence, the debate generated heated discussion and became controversial enough to be part of the reason for the demise of the Tibetan KyiApso Club in America in 1999. Lack of enough dogs to avoid inbreeding also contributed to it. Within a few years, despite acquiring a number of dogs from Tibet (Kang, Tashi and Mindu, along with Dzong, Singdru and Bhoté), relatively few KyiApsos are still in the club's (now outdated) registry, and perhaps only thirty or thirty-two KyiApsos are now found in the USA. To my knowledge, they are no longer being actively bred anywhere as a unique breed in the West.[21]

Those who fancy KyiApsos to be a separate breed, different from the Tibetan mastiff, do so on the basis of their purported origins and specific characteristics of temperament and physical build. They are keen on seeing KyiApsos gain more international recognition as a distinct breed. Nonetheless, some skepticism remains in America and abroad.

Many Tibetan dog fanciers consider the KyiApso a unique type of Tibetan mastiff, as one variety that "shares similar characteristics and forms a smaller division of a larger set," or breed. In support of this idea, they point to the fact that in many, if not all, litters of KyiApsos born in America, two varieties tend to occur: those with the characteristically long hair, and those with a short or smooth coat. The main difference is in facial hair. Many Tibetan mastiffs have body coats as long as those of a KyiApso, and KyiApsos are sometimes found to have relatively smooth body coats. It's the hair on the muzzle that distinctly sets the two types apart. Most KyiApso breeders in America routinely neutered the smoothies, to avoid perpetuating that particular (unwanted) trait, though it is unlikely to have any effect genetically on the production of the smooth-faced variety.[22]

There is some speculation that KyiApsos are a variation of the Tibetan mastiff created by crossing a Tibetan mastiff with a Tibetan terrier, which is smaller but also has a shaggy face and coat. Another idea is that the KyiApso may be the result of a Tibetan mastiff breeding with some other kind, such as one of the shaggier hunting dogs of Tibet. These are possibilities, and in my experience a Tibetan mastiff + Tibetan terrier cross is likely in some instances. It will take thorough genetic studies to make a firm determination, though I suspect that the results will be inconclusive or, at best, may indicate only that one or more strains or breeds of dog have been involved.

The leading expert on the Tibetan dogs in Nepal, Jay N. Singh, urges caution in designating the KyiApso as a new and distinct breed. "In all probability," he has said, the KyiApso "is not a pure breed and has no special function." He goes on to say that "I have seen quite a few of these dogs and they possess no common features to distinguish them as of the same breed which we find in the Tibetan mastiffs. There is no evidence of their existing as a distinct breed in Tibet for any length of time. We do find sporadic mention of long-haired dogs of Tibet, but it does not necessarily mean the existence of Apso Mastiffs. Relatively, the Tibetan mastiffs could be called long-haired in contrast to the English Mastiff. I saw one recent import from Tibet which has much longer hair than our dogs, but it is definitely not shaggy and an Apso Mastiff."[23]

One of North America's few dedicated KyiApso breeders has a similar opinion. Judy Steffel does not rule out classifying the 'original' KyiApso (whatever that might be) as offspring of a Tibetan mastiff crossed with some other Tibetan dog. "I don't think this is a separate breed," she writes, "but a bearded variety of the large, intelligent, independent, protective, hairy, deep-voiced, curly-tailed canine guardians that the Tibetans have been breeding for many, many generations—at least hundreds if not thousands of years."[24]

There is more to the story. Steffel takes credit (or blame) for coining the two terms that are associated with breeding KyiApsos in the West: "shaggy" for a pup born true to the bearded KyiApso morphology (i.e., its physical form and structure, including its coat), and "smoothy" for a pup from KyiApso breeding that looks, instead, more like a more common smooth-faced Tibetan mastiff. "Actually," she says, "I use 'shaggy' to mean a bearded one, whatever its size or structure." All shaggy-faced offspring were considered by KyiApso Club members as potential "show quality," while the smoothies born to registered KyiApsos were considered strictly as "pet quality" and not for show.

In considering the distinctiveness of the KyiApso, Steffel goes on: "It's a separate breed only in the Western sense" (as created under a specified breed standard in the West, that is, though not from the point of view of the Tibetans). Steffel's opinions are not popular among some KyiApso breeders and owners. As already noted, Daniel Taylor has said definitively that "The Tibetan KyiApso and the Tibetan mastiff are separate breeds."[25]

In support of her view that the KyiApso is a variant type of Tibetan mastiff, Steffel argues: "The TKC Breeding Guidelines called for the intense inbreeding (of only the shaggy-faced dogs—*all* smoothies were to be spayed/neutered) for at least three generations. We were told that the 'KyiApso'... was 'an ancient breed,' originating at the foot of Mt. Kailash in western Tibet—and that the dogs being used to establish the breed in the West will need to be genetically 'purified' to reform random genes from other Asiatic breeds. That was euphemistic for 'breeding out the genes that produce smooth-faced dogs'... Not an easy task at all since the gene (set) for the shaggy face appears to be dominant. Much easier to breed out the beards, as witness the fact that no TM breeding has ever produced a bearded pup..."[26]

As she began reading, breeding and learning about livestock guardian dogs, especially those of Tibet, and also about genetics, several things became apparent to her. The KyiApso, she concludes, is a dog that has been previously described as a "bearded Tibetan mastiff" (in the 1988 article I wrote) and that they are found all across Tibet, in adjacent northern Nepal, and perhaps elsewhere. (She also notes, as I have, that the Tibetan mastiff is not, strictly speaking, a true mastiff.) Steffel is concerned that founding a 'breed' on six dogs in the United States, some of which were siblings or half-siblings, was "doomed to failure." Everything she had read, by several authors, told her that the Tibetans do not delineate 'breeds' based on appearance (over function) as the American Kennel Club and Western dog breeders do. Her understanding of the Tibetan nomads, and other people who created working dog breeds, is that "they breed for what works." This is not to say that 'looks' (appearance) are not important, only that function is *more* important.

"Function," she says, "was a primary consideration because it had to be. The dogs were an integral part of the survival strategy of people living in an incredibly harsh, unpredictable ecosystem. A dog that couldn't or wouldn't work for its living didn't live long. Appearance of the dog was important primarily as a facet of performance: guardian dogs needed to be large and strong; needed a thick, weather-resistant coat; needed a deep bark that carried for miles; needed pendant ears and a curled tail so that they wouldn't be mistaken for wolves and shot at. Anything else on the outside of the dog was of secondary importance at most."

"Black dogs might be preferable because they were 'invisible' at night," she writes. "Dogs with 'spirit eyes' [tan spot over each eye on a black-and-tan dog] might be preferable because they could see the evil spirits their owners couldn't—or maybe just because their facial expressions were easier to read. There may have been individual or regional preferences regarding coat color and eye color—and shagginess of face—but a good *do-kyi* was a good *do-kyi* no matter what its color, whether it was bearded or not. (And some *do-kyi* were large and powerful while others were smaller, lighter, and faster—and that's the way the nomads liked it since different facets of the dogs' jobs required different physical abilities.)"

Considering comments she has heard about the apparent lack of any standard look in old pictures of big shaggy Tibetan dogs, Steffel says that "there is great variety in type amongst 'KyiApsos.' That is because there is great variety in type amongst the nomads' dogs in Tibet. *Intentionally.* Smaller, lighter, faster dogs can more effectively pursue predators while larger, heavier, more powerful dogs can more effectively hold and overpower predators once they catch up with the action. And the larger, heavier dogs would be more effective at terrifying human intruders and taking them down if they're not smart enough to go away. But note that the 'KyiApsos' imported... [as foundation stock] have all produced offspring larger than themselves, almost certainly due to the availability of greater and more reliable nutrition."

Judy Steffel's viewpoints shocked some members of the Tibetan KyiApso Club. After the subject came up at the club's final annual meeting in 1999, she witnessed the sort of intense politics that often contaminate the dog breeding fancy. The politicization of dog associations is common enough that several dog breeders I know have quit dog clubs in disgust and have left the 'looks only' show-oriented dog fancy altogether.

Steffel came to the 1999 KyiApso Club meeting "with reams of information in hand" (as she puts it) and outlined the case against the KyiApso as a separate and distinct breed. The question under discussion was: "*Should we continue with the intense inbreeding mandated by the Breeding Guidelines or should we start breeding the way the Tibetans always have?*" When it was decided to put it to a vote of the full membership of the club, there was a fiery reaction. As a result, the annual meeting was subsequently "voided," and club activities were discontinued. Since 1999, breeding KyiApsos has almost totally ceased in North America, "mostly for lack of new breeding stock due to nasty club politics."

Now, one might ask what difference it makes whether the Tibetan KyiApso is a separate *breed*, or a unique *type* of Tibetan mastiff that has been created according to a specific club standard in the West, out of breed stock that originated in one or another Tibetan locale, and possibly of one or another genetic combination of landrace breeds? Many great breeds or types of dogs have started out with a relatively uncertain or mixed provenance. That does not make them 'flawed' nor any less loveable as companions or show dogs.

In the end, the difference is that if the Tibetan KyiApso remains separate from the Tibetan mastiff, it will die out as a type. Without tapping into the much larger Tibetan mastiff gene pool, the Tibetan KyiApso cannot survive.

# 6

# Across Tibet

Are any of the big dog breeds and types that I have written about so far endangered, or in peril of extinction? According to some who are concerned about them the answer is yes. They point to the massive kill-off of dogs reported from Lhasa and other cities and towns of Tibet during the Chinese Cultural Revolution, to the interference of lowland dogs with the best of the mountain dogs of the Himalayas, and to deliberate cross-breeding. After I had examined everything I could read about Tibetan dogs and had trekked through the Himalayas of Nepal, Bhutan and India to find them, and after I had raised several, and trained, showed and bred my prize Champion Saipal Baron of Emodus, I felt strongly that to settle the question of their imminent extinction, I must look more deeply at the Tibetan scene to see for myself. To that end, I set out in April 2007 with one companion, a guide/translator and a driver, to cross Tibet in search of dogs. From that experience, I can confidently say that there is an intense interest in Tibetan dogs these days among upwardly mobile and wealthy urban Chinese and that neither the Tibetan mastiff, the shaggy KyiApso, the Himalayan mountain dog, nor even what some of my colleagues and I euphemistically called the *sha-kyi* ('hunting dog'), appear to be disappearing.

The Tibet trip was planned primarily to discover the current situation regarding big dogs, especially Tibetan mastiffs and KyiApsos. Our party was small, consisting of Judy Steffel (a noted American breeder of big Tibetan dogs), the author, and two Tibetan companions—our guide/translator Dorjé, and our road-wise and loquacious (entirely in Tibetan) driver, Lodenla. It was a team effort once the guide and driver understood our main purpose was to see dogs—and we saw many, several hundred in fact, over several weeks travel.[1] Dogs were commonplace along our route, at Nyalam, a busy transport town north of the Nepal border, then in Serlung, a village in Shishapangma National Park on the southern route to Mount Kailash. After Serlung we drove the Friendship Highway east, visiting Tingri (Dingri), Lhatse, Shigatse and Lhasa. The last part of the trip took us south into Lhoka Prefecture, then west through Gyantse to Shigatse and back the way we had come past Nyalam to the Nepal border.

We selected these locales, all of which are within the old administrative unit of Ü-Tsang, because of their significance in Tibetan dog lore. We were assured by the Tibetans that there are also numerous big dogs in all the places we did not visit. Each region and locale boasts one or more varieties of big dog, from Chomolungma (Mt. Everest) Base Camp and Rongphu

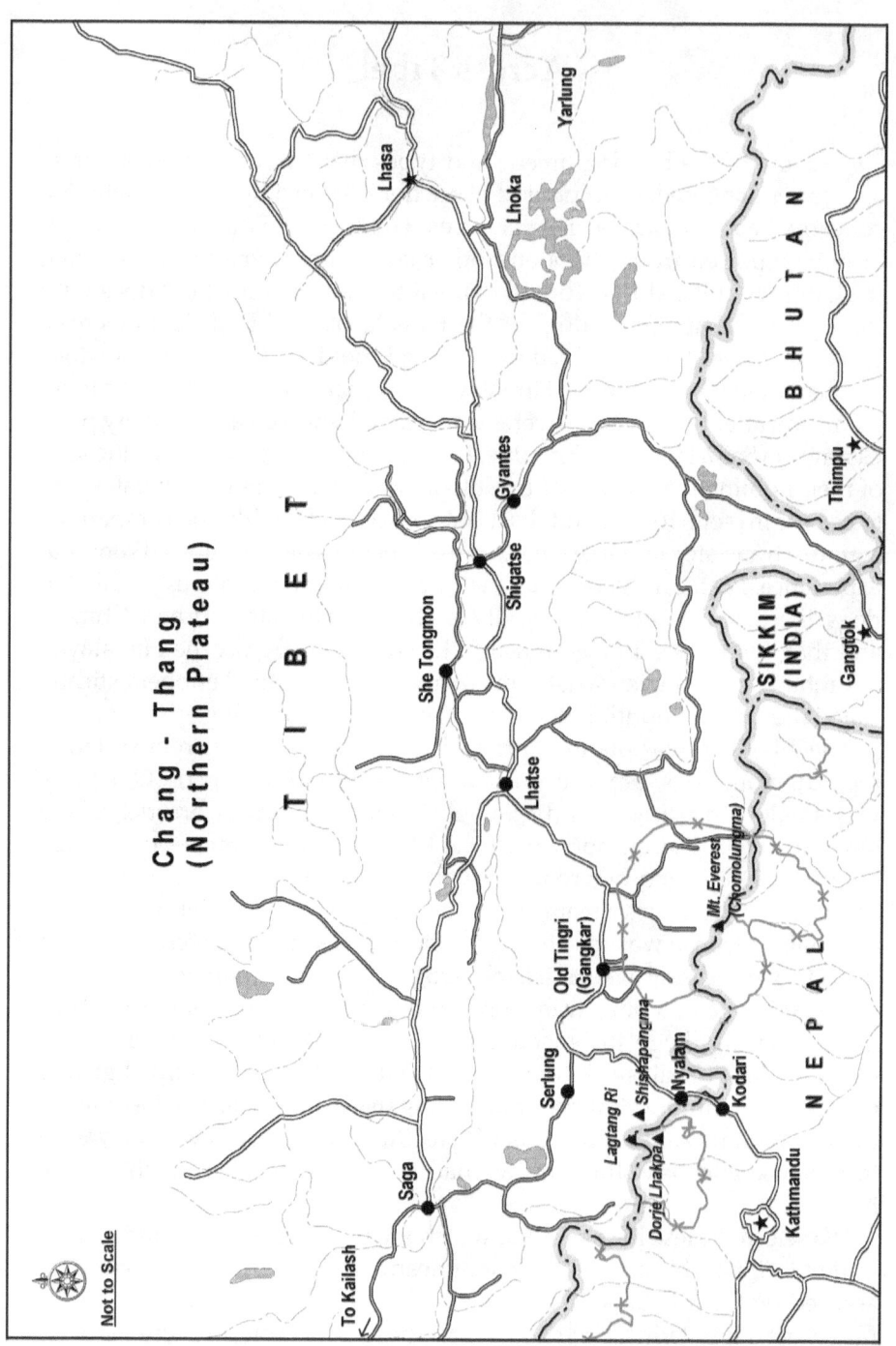

Map 5. Northern Nepal and Southern Tibet.

Monastery in the south, across the high dry expanse of countryside to Mt. Kailash and Lake Mansarovar in the far west, the Changtang at the north, and in old Amdo and Kham at the east. All dogs in Tibet are known as *kyi* (*phö-ky*, male dog), a generic category that includes the landrace *do-kyi*, *tsang-kyi* and *kyi-apso* (livestock guarian dogs), and *sha-kyi* (hunting dogs), as well as the smaller dogs that we in the West call the Tibetan spaniel, Tibetan terrier and Lhasa Apso.

On our travels, we saw big dogs accompanying nomads and their yaks, guarding monastery compounds, hotels and guesthouses, homes and factories, defending their turf on town and village streets and, often, just lying about (proper daytime behavior; their 'come alive' time is at night). We also saw numerous dogs in a commercial kennel on the outskirts of Lhasa and at Lhasa's weekend 'dog market' (for pets and guardians, not for eating). Black-and-tan and black dogs predominated, but there were many red and light golden dogs (sometimes characterized as 'cream' or 'white'), and a few bluish-grey in color. The most red dogs we saw in one pack was at the Pelkhor Monastery and Kumbum Stupa in Gyantse; but they were also abundant at the Lhasa street market. While smooth-faced Tibetan mastiffs were the most common, we were pleased to see over a dozen shaggy, or bearded, KyiApsos, several of very fine quality.

We saw so many dogs, in fact, that I can only conclude that they are ubiquitous in Tibet.[2] Depending on what one is looking for in a Tibetan mastiff—be it conformity to a Western breed standard for show, or conformity to traditional breeding for not-so-showy guardianship—there are many in Tibet, in both categories, some of which are quite noble looking. We encountered dogs that were loud, but few were as "ferocious" as they are often described. Most could be cautiously approached if we moved slowly and spoke softly. When they wagged their tails, they let us pet them. Most of the aggressive ones were tied well out of our reach inside locked, gated compounds, being guard dogs. It's what they do best. One blue-eyed critter, chained outside a restaurant at the Lhasa dog market, guarding a bicycle, acted like he would kill me if I got any closer, and the only other one that gave us both an anxious moment was a black-and-tan do-kyi at the entrance of a police compound on a back street in the roadside town of Lhatse. As we walked past, he 'rushed' us, barking and snarling, but to our amazement he stopped in his tracks at the command of the officer stationed at the open gate. We were impressed, both at the dog's behavior (appropriate) and at the officer who controlled him (skilled).

## Nyalam's Little 'San-San'

We travelled during April, a good month to look for dogs in Tibet. There are many pups to be seen then, after the winter whelping season, including whole litters for sale at the Lhasa market. The first that we encountered, however, were in the town of Nyalam on the second morning of our trip.

'Nyalam' in Tibetan is translated as 'gateway to hell,' reminding travelers going out of Tibet that the way south drops rapidly to lower elevations and into the much warmer climate of Nepal. In earlier times, Nyalam was called Kuti (or Kutti), and was well known as a trans-Himalayan trade entrepôt.[3] The elevation of Nyalam is 12,303ft (3759m).

For modern travelers going north into Tibet from Nepal, Nyalam is the first overnight stop, about thirty kilometers beyond the border town of Zhangmu (also called Khasa). Zhangmu (at 7,546ft, or 2300m) is a busy place precariously perched on a steep hillside above the 'Friendship Bridge' linking Tibet with the Nepalese town of Kodari (6,145ft, or 1873m). North of Zhangmu the mountain road parallels the Bhoté Kosi River, a whitewater torrent (popular with rafters and bungee jumpers on the Nepal side of the border) that cuts its way torturously through an awesome gorge emboldened with high cliffs and dramatic waterfalls. For centuries, all trade and travel between Kathmandu and Lhasa was on foot via this difficult track. In the early 1720s, when the Jesuit Ippolito Desideri (whose description of the big dogs we read in Chapter 2) descended by this track from Kuti (Nyalam) down into Nepal, he wrote that "...the road [meaning the foot track] skirted frightful precipices, and we climbed mountains by holes just large enough to put one's toe into, cut out of the rock like a staircase. At one place a chasm was crossed by a long plank only the width of a man's foot, while the wooden bridges over large rivers flowing in the deep valleys swayed and oscillated most alarmingly."[4] Today, the motor road is cut across those same cliffs high above the turbulent river, so that while the passage is much safer, it is no less spectacular.

We found Nyalam unattractive, cold and uncomfortable. Because its narrow main street is also the highway, it was crowded with heavily loaded freight trucks going to and from Nepal, most of them driven by men of Uighur ethnicity.[5] When we were there, the hotels were full up with international mountaineering expedition members acclimatizing before setting off to scale Chomolungma, Cho Oyu, Shishapangma or one of the other high peaks along the Tibet-Nepal border. We, like they, also stayed over in Nyalam while adjusting to the high altitude.

The morning after our arrival, after being rudely awakened too early by truck and bus drivers revving engines and honking horns at an ungodly hour, we took a walk through the town. There were many big dogs about, all of obvious Tibetan mastiff heritage. One big black-and-tan was especially approachable, and we took a number of photographs. In front of a fresh produce shop we were pleased to see a plump little black-and-tan pup tied up next to a bin full of cabbages. The Chinese storekeeper spoke no English so Dorjé translated. "The pup's name is 'San-San'," the shopkeeper told us, "a true *tsang-kyi*." He had paid 1,300 Yuan for it (approximately US$170), he said, and was willing to sell it for whatever profit he could make.

We explained that we did not want to buy one, but were interested in photographing dogs and hearing what people said about dog origins and associated cultural traditions and beliefs. I asked where he had gotten San-San. "From a village in Lhoka," the shopkeeper said. Then, with conviction, he repeated it to impress us: "He's a true tsang-kyi, from Lhoka." It was a refrain we would hear often during our trip; whenever we admired a nice looking dog we were inevitably told "It's a tsang-kyi, from Lhoka"!—no matter where it was actually from.

*Do-kyi vs. Tsang-kyi*
On our second day in Nyalam, over cups of green tea in a restaurant, my companion Judy Steffel and I 'talked dogs.' Throughout the trip we held many discussions like this over tea or Lhasa beer, yak stew or *thukpa* (noodle soup) or *momo*s (meat-filled dumplings), mulling over and jotting down what we were learning about the big dogs, and framing new questions to ask. Anthropologists call these sorts of informal discussions in the field "debriefings." To us they were simply "dog talk."

Reading back through my journal, one of the first notes reminds me that it was in Nyalam that Dorjé first cautioned us *not* to believe all who claim their dogs are "tsang-kyi from Lhoka." "They think that's what foreigners want to hear," he said. Apparently besides Lhoka Prefecture being the cultural birthplace of the Tibetan people it is also considered to be the historic provenance of the Tibetan mastiff. On that basis, some people have promoted the belief that the tsang-kyi is somehow superior to all other varieties of Tibetan mastiff.[6]

The question of purported tsang-kyi superiority among big Tibetan dogs, specifically the tsang-kyi *vs.* do-kyi debate that rages in the modern Tibetan dog fancy, was something we wanted to examine during our travels. But, it was Dorjé, a Tibetan, who brought it up first.

We told Dorjé it was not only some Tibetans who believe the tsang-kyi to be a distinct and superior type, but that some Chinese, European, American, Australian, and other breeders also believe it. Not only do they say that the tsang-kyi is the "original" Tibetan mastiff (the progenitor), but that, historically (some claim) it was once a consistently larger dog than it now appears; indeed, the largest of all varieties of the breed. This conviction is apparently based on descriptions in old travelers' accounts of large fearsome dogs, beginning with Marco Polo's dogs "the size of donkeys" that we met in Chapter 1. The notion of the tsang-kyi's superior looks and conformity, as a super-dog from the past, is now widespread in the Tibetan big dog fancy. It is a notion that seems even to have reached the shop keeper in Nyalam and breeders and owners far removed from Tibet and the Himalayas.

Our inquiries and observations over the next few weeks, and our wider appreciation for cultural traditions and customs (both Tibetan and Western) concerning these dogs, did little to confirm the alleged 'superiority' of the Lhoka tsang-kyi. We voiced a mutual opinion several times, and Dorjé agreed, that it is not so much the *looks* of these dogs, but their *performance*—their *function*—that should be the standard by which they are judged. As we continued to discuss the merits of various dogs that we saw, we agreed that the operative question should be: "Are they doing their job as guardians?" This is what Tibetan nomads look for in them and has been a focus of my research over the years. Unfortunately, by taking the landrace Tibetan mastiff out of its original natural and cultural habitat, and by turning it into a show dog, pet or status symbol, *function* has been eclipsed by *form* (looks), and as the definition has changed so has the dog.

I am not saying that Tibetans do not concern themselves with what their dogs looks like. Indeed, they know full well what they should *look like* and what to *look for* in a dog—just ask them! They *know*, for they've been breeding them successfully for centuries. Among the nomads, however, function traditionally takes precedence over looks when selecting a working dog pup. I must qualify this slightly, however, by noting that 'looks' and 'function' are sometimes intertwined in Tibetan belief. For example, black-and-tan mastiffs with tan spots above the eyes are said to have "four eyes," a trait some believe gives them superior night vision and certain supernatural capabilities, all helpful in fulfilling their guardian functions. Someone told us, for example, that the best of the four-eyed dogs could sense trouble approaching from a distance of 300 kilometers!

Tibet's big landrace dogs are bred and raised locally to perform the important economic functions of protecting both fixed and moveable assets, land as well as livestock (anthropologist Robert Ekvall calls nomad livestock "fields on the hoof"). Among the nomad folk, having a big dog that does its traditional job remains a necessary condition for cultural and economic survival.[7]

There are two ways to select dogs: either as pairs for breeding, or as pups regardless of breeding. It is mostly the latter that guides Tibetan nomads in their choice of dog; that is, their main interest is determining which pups in the litter will do the best job for them. Any others in the litter, regardless if they are bigger or more handsome, are better traded or sold off.

As a long time owner/breeder of big Tibetan dogs, Judy Steffel has given considerable thought to the matter. Reflecting on what she has read (and what I interpreted to her on the trip) about Tibetan history and Himalayan area culture, as well as what she has learned about other landrace livestock guardian breeds elsewhere, "The landrace do-kyis," she says, "were developed by nomads over many centuries, generally not by the selection of breeding pairs but by 'postzygotic selection'; that is, the culling of pups and dogs that didn't or couldn't do the work required of them in a subsistence culture in a very harsh climate. Selection for ability to function kept the dogs 'correctly' structured and healthy. The nomads *certainly* weren't breeding to conform to a written (or even verbal) 'standard'! They supported the dogs that worked, culled or sold the ones who didn't/couldn't."[8]

"There was plenty of outcrossing," she points out, given the peripatetic nature of life in Tibet, "because the nomads and their dogs were nomadic. The dogs moved with the flocks and herds from one seasonal pasture to another and from the pastures to the villages on trading expeditions, etc." Then, expanding on the logic of her argument, she says: "Bitches that were fertile in the late fall were exposed to numerous dogs en route to and in the villages, and made their own mate selections. (Multi-sire litters may have been the rule instead of the exception, depending upon a bitch's opportunities and her 'tastes.') The pups were born in winter, when the nomads were stationary, allowing the bitch to den up and raise the pups—for a couple of months at least—in one place, not while trying to follow the herds or flocks."

Steffel goes on to say that "It has been noted in the literature on working dogs that most subsistence dog 'breeders' do not select for either consistently large or consistently smaller dogs, but for some of each—because they have uses for both. Lighter, leaner, faster dogs have the speed and agility to put the fear of dog into predators. The heavier, more powerful, but not-quite-so-fast dogs can take on larger predators (humans?) and provide effective backup for the speedsters."

"Any *really* large pups/dogs in a litter, however, wouldn't be of much use to a nomad directly, for a variety of reasons," she concludes. "They are too big to keep fed. In hot weather they too easily succumb to heat stroke. Their greater size and weight reduce their speed and agility to the point that they aren't much good against four-legged predators. But they are impressive; so they are valued by people who *can* afford to feed them and who have no need of quick, agile dogs: People who live in palaces and monasteries. People who can afford to shell out real money (or many blessings) for a dog they want."

Think about it: this implies (and we are convinced) that big Tibetan dogs have long been selected by the nomads for their essential working potential, over and above their looks. In their popular book, *Dogs: A New Understanding of Canine Origin, Behavior, and Evolution,* Raymond and Lorna Coppinger note the same thing (well outside of traditional Tibet)—that in today's dog fancy many working dog breeds are "no longer... chosen for the way they behave, but for the way they look..."; that is, for coat color, size and (in some breeds) ear carriage. The Coppingers call these "superficial traits" that are only cosmetically related to their previous functions and survival.[9]

"It is ironic," the Coppingers say, "that the village dog, well suited to survive, is rejected as 'just a mutt' for those same traits"—that is, he's suddenly "inferior." Among some Tibetan dog fanciers who raise them for show, the corollary to the Coppingers' village dog-as-mutt is the do-kyi, which they consider to be distinctly smaller, hence inferior, to the allegedly superior looking tsang-kyi. The interesting thing is, of course, that both are Tibetan mastiffs and, from my experience, there is very little to tell them apart.

The Coppingers go further to raise an even more serious point about what happens when the selection of breeding pairs in the modern world for non-working traits becomes uppermost, and working dogs become the object of breeding for show and pets, or as status symbols. The result, they say, is ultimately "bad genetics" that "dooms any breed that gets caught in that physically isolating trap..." By closing the stud book to the traditional nomad dog population (such as the do-kyi from western Tibet and the Himalayas) in order to promote specific non-working dog traits, inbreeding inevitably occurs. Inbreeding, the Coppingers and others remind us, decreases genetic diversity and vitality. This, in turn, leads to the rise and perpetuation of debilitating genetic diseases.

Let's look again at the attitude of some Tibetans we encountered in our travels, specifically those who made a big thing out of the relative superiority/inferiority of particular dogs. Their comments to us were usually based not on the dogs' looks or functions, so much as those traits in combination with *place* of origin (e.g., dogs of Tsang *vs.* Amdo or Kham, or dogs of central or eastern Tibet *vs.* those from western Tibet, or the northern plateau, or the Himalayas, and so forth). Depending on what region or locale one's dogs are from in this simplistic dichotomy, all others are considered inferior.

Why so? We anthropologists tend to explain such behavior as part of the common human cultural trait called 'ethnocentrism.'

Ethnocentrism is defined as an ingrained tendency among humans to judge other people and cultures, or specific cultural characteristic of others, against the practices and standards of one's own ethnic group, social set or culture—that is: them and theirs *vs.* us and ours. Usually, our own is considered superior, while 'others' are viewed with disfavor or contempt. For example, Tibetans often point out to foreigners the obvious differences between Lhasans and people from Amdo or Kham, or the difference between sophisticated city folk and their country cousins. Their observations are usually genial, with a focus on uncontentious styles of dress, accent, cuisine or physical size. We were told that Khampas, for example, tower over men of other regions; that Amdo is noted for some of Tibet's best cooking; and that some rural people talk with a "funny accent." When it comes to comparing dogs from one region or another it takes little imagination to translate regional rivalries into emotional 'brag dog' behavior. It makes no difference if the braggart's dog is from Amdo, Kham or Lhoka, the Changtang or Mount Kailash area, Bhutan or

Mustang, or from a commercial kennel. Such rivalries are commonplace in human society. The worldwide dog fancy is rife with them, within and between breed clubs and owners. (It has been said that 'brag dog' behavior is especially intense and sometimes downright 'dirty' within the Tibetan mastiff fancy in the West).

In Tibet, where traditional guard dog functions are essentially the same everywhere, what's left then but to compare and judge their physical features (size, shape, color), temperament (relative ferocity or gentleness) and other characteristics (even their bark), one against the other?

Some breeders outside of Tibet, for example, may say that the "inferior" do-kyi is more prevalent on the western plateau and in the adjacent Himalayas, and that it typically has a smaller head and smaller body than the tsang-kyi. We had trouble confirming this, however, as many of the dogs identified to us in Tibet as do-kyi were no smaller than some of their tsang-kyi cousins. Many claims that dogs from one region or another are "superior looking" were visibly contradictory. I can only conclude that much of the regional rivalry over dogs that we encountered reflects more of an ethnocentric braggadocio than anything of great substance.

During our travels, I kept a list of the major distinctions described for each type of Tibetan mastiff. Based on my list, and copious notes accumulated during years of research, I have prepared the accompanying table for comparison (Table 3, p.142). The list reflects regional bragging, and from my perspective not all of it is true. Some of the terms used are mine, some are my Tibetan colleagues', and some are Tibetan mastiff breed standard nomenclature. (See also my 'Adaptive Traits of Tibetan mastiffs' in Appendix 3, pp.218-20.)

Physical features not listed on the table are more or less consistent with Tibetan mastiff breed standards from Europe and America, with some variation. For example, not all of the dogs we saw in Tibet were (to quote the FCI breed standard) "powerful, heavy, well built, with good bone," although those in factory compounds and at the Lhasa dog market (well fed and cared for) were exemplary in these regards. Note also about color that reds and golds (sometimes called cream) are closely related genetic variations, and that many Tibetans like a black dog to have a white spot or 'blaze' on the chest.[10]

The so-called "true Tibetan mastiff" on the table needs an explanation. It is what those who promote the tsang-kyi super-dog call their newly redesigned breed. This variety has highly exaggerated features, to the point of gigantism. It is entirely kennel-bred, and though genetically based on the Tibetan *phö-kyi* (which includes all landrace Tibetan mastiffs), no natural landrace parallels of the same elaborated size are known, either now or in the past, other than in some exaggerated travelers' accounts (and in novels). It is also alleged (and is sometimes quite obvious) that

| Big Tibetan dogs (generic *phö-kyi*): | *Do-kyi* (western Tibet and Himalayas) | *Tsang-kyi* (*do-kyi* from Tsang) | *Tsang-kyi* as "the true Tibetan mastiff" |
|---|---|---|---|
| Origins | Tibet's western plateau (old Ü-Tsang); Bhutan, Nepal, India | all of Tsang, including Lhoka Prefecture south of Lhasa | kennel-bred in America, Europe and Asia |
| Local economy | yak/sheep/goat herding | farming, sheep/goat herding | kennel sales; expensive exports |
| Function | livestock protection, sentinel | livestock protection, sentinel | pet sales, show dogs; ostentation |
| Temperament | assertive | assertive | *less* assertive? |
| Body | solid, bulky | slightly larger, bulkier | much bulkier, oversize, gigantism |
| Mane & hair | thick | thicker | much thicker and longer |
| Head | blocky | blockier | very blocky |
| Muzzle | moderately blocky, short | slightly longer, wider | equally long, heavy, blocky |
| Jowls | tight to slightly pendulous | pendulous | very pendulous, deep wrinkles; a major feature (hound-like) |
| Haw | none, or minimal | slight (in some dogs) | obvious, deep, intense; a major feature |
| Colors | black-and-tan, black, red; occasional 'blue' (blue-grey), and variations | black-and-tan, black, many red; occasional 'blue', and variations | black-and-tan, black, red/gold/cream, 'blue'-&-tan, chocolate-&-tan, other dilute colors and variations |

Table 3. *The major distinctions described for each type of Tibetan mastiff.*

crossbreeding has been done in some Chinese kennels in order to achieve the exaggerated traits.

There has been considerable research on the factors that give some dogs their large size. Some early explorers in Tibet described extremely large dogs, and it is now thought that perhaps what they described can be attributed to the Tibetans having castrated some dogs. It is well known among veterinarians and other experts that sterilization of any male animal by gonadectomy before puberty results in increased long bone growth *after* puberty. While sterilization is probably not in the best interest of guardian dogs among the nomads, it is known to have happened. Tibetologist Robert Ekvall has noted, for example, that some of the nomad's watchdogs "are so fierce that some of them are castrated to heighten their vigilance but also to lessen their ferocity and make them less likely to charge into a spear point or wildly swinging sword, and get hurt."[11] Watchdogs at monasteries may also have been sterilized to tame them down a bit and keep them from roaming, as well as to make them larger and more intimidating.

According to recent studies, neutering a young dog can, indeed, affect long bone growth and result in a detectable increase in the dog's height. "Testosterone plays an important role in closing the growth plates, so if testosterone is missing (the testes are a major source of testosterone in males) the bones will grow longer." The younger the dog when neutered, we are told, the more dramatic the bone growth effect, though—and here's an important point—"it should not be envisioned as a 'King Kong' type of effect."[12]

Plate 6.1. Example of gigantism in a Tibetan mastiff.

I said above that there are no natural Tibetan landrace canines with truly gigantic features. But wait! There is a 1905 observation that sounds something *like* that, especially those 'hound-like' jowls under "*Tsang-kyi* as 'the true Tibetan mastiff'" (in the right-hand column on table). Perceval Landon, a reporter for *The Times* (London), accompanied Colonel Younghusband's military incursion into Tibet in 1903-04. In his 1905 book, *The Opening of Tibet*, Landon describes four varieties of Tibetan dog, of which "the finest," he says, is the "so-called" Tibetan mastiff. "This," he writes, "is a great shaggy creature, with a very massive head. It is usually black-and-tan in color, and has a very thick, rough coat. Its eyes show some 'haw' *like a bloodhound,* and it has *the pendulous lips of that breed"* (my emphasis). Landon concludes that "No monastery of any pretensions in Southern Tibet is without at least a pair of these fierce dogs chained up on either side of the entrance."[13]

That last observation about monastery dogs is important. But first, Landon is *not* equating the *overall* looks of these dogs to bloodhounds. He says only that they have "*some* haw *like* a bloodhound" (my emphasis again). Photographs of Tibetan mastiffs brought back to England after the Younghusband mission do nothing to enlarge the image: a little haw and some pendulosity of jowls, yes; but no shaggy long-haired bloodhounds. By comparison, however, some dogs on today's market (at the Lhasa dog market and in promotional materials from dog clubs and kennels in China and elsewhere) look a bit *too much* like furry hounds... (suspiciously so, methinks).[14]

Landon's remark about dogs at monasteries reveals an even more important point. He is talking about big dogs used for one specific purpose—chained at the gates of Buddhist monasteries to impress and intimidate visitors, and let loose at night to keep out thieves (for there is a lot of gold on the temple altars). Judy Steffel observes the following about Landon's description of those early twentieth-century dogs, and about what was probably going on then, and now. First of all, remember that the landrace Tibetan mastiff was primarily a livestock guardian dog, born in the camps and villages of livestock herders, and was used only secondarily as sentinels in homes and monasteries. With that in mind, she writes, "Landon was talking about 'monastery' dogs. I believe that there might, indeed, always have been a few tsang-kyi in the nomads' litters, and that the nomads learned early on how to recognize which pups would be good with livestock, which would be good as village guardians—and which would be good for nothing other than 'looking' impressive; i.e., temple/monastery/estate guardians. The nomads weren't 'just' herdsmen; they were tradesmen," she points out. "They bred more livestock than they needed for subsistence, harvested more salt than they could use, etc. They needed things to trade for what they couldn't produce themselves. They needed livestock/camp guardians; other people needed village dogs; and others needed (and might pay handsomely for) big hairy beasts to scare (and impress) the pants off visitors to the monastery or palace."[15]

Steffel then makes an interesting analogy. "The big, extra-hairy, heavily jowled and hawed, often poorly-structured tsang-kyi," she writes (if there ever was, historically, a dog of such extremes), "is a palace guardian, a 'Beefeater'—fancy uniform, stands around a lot, may roar on occasion, but probably couldn't actually 'do' much if the need arose." The functions of the Beefeater, or Yoeman of the Guard, who stands as a sentinel in front of the Tower of London, like that of a big dog stationed in front of a Tibetan monastery, are essentially the same: guard, impress, titillate one's curiosity, stand still for photographs, and (in the dog's case, for sure) instill fear and second thoughts among persons of ill-intent. One difference is that, unlike the dog, the Beefeater is not chained to his post (though he is effectively 'leashed' by obedience to strict rules).[16]

In a comparative vein, Steffel goes on to equate the nomad's do-kyi to "a good beat cop." (She knows what she's talking about here. She's been a cop, she's been to Tibet, and she owns fifteen big Tibetan dogs that guard a pasture full of goats.) "And, like a beat cop," she says, "he or she does the job mainly just by being there and being seen and heard (and smelled). No reason to chase or bite or beat up or shoot at somebody if one's mere presence will do the job." On the guard dog's beat, amidst the nomad's livestock, "the 'uniform' isn't fancy; its purpose is to allow quick identification by friend and foe, not to make anybody gasp in admiration!" she says. "These guys know their 'beats'—what's normal, what's out of place, what can be ignored, and what needs a second look. They are slow to anger, patient with kids, suspicious of strangers, not eager to expend a lot of energy, but perfectly willing and 'able' to catch and kill if that is what's necessary."

There is another set of concerns about the super-dogs, one that has been the subject of discussion on the Internet. If the slobbery jowls, deep haw (open 'red eye'), and massive bulk of the super-dog were standard characteristics of the original, historic tsang-kyi, how well would the dogs have fared in the high, arid, dusty, windy and bitterly cold environment of Tibet? Fluid loss from drooling, inevitable eye infections from irritation of the haw, overall heaviness and their often poor structure would, each, have posed a serious challenge to dog survival and, together, would have been be deadly.

So, what are these new super tsang-kyis all about? First of all, they are bred for show and might even make wonderful pets (for the right kinds of owners, bearing in mind that these independent-minded dogs are not lap dogs, nor are they submissive like labs and retrievers, for example). They are also used as 'heavies' (guards) around the estates of their wealthy owners. (To my knowledge, however, none of the new breed has been sent to nomad camps to guard yaks.) I believe the super-dog has been developed ostensibly to conform to overstated travelers' accounts going as far back as Marco Polo in the thirteenth century and in some of the exaggerated nineteenth- and early twentieth-century writings that we encountered in Chapter 2, as well as to highlight conformity with ancient Molosser breeds to which some contemporary breeders claim Tibetan mastiffs are related. They are also being bred to make money. Some have recently sold for outrageous sums (to honor Marco Polo as *Il Millione*,' perhaps?). The result, the so-called "true Tibetan mastiff," has become widely known as an ostentatious acquisition among relatively rich Tibetan mastiff aficionados worldwide, in China, Taiwan, Europe and North America.

While avidly sought after by some (the rich), the super-dog is also the focus of controversy. Some breeders favor, promote and staunchly defend its development. Others are strongly opposed, saying (as I do) that there never was such a breed in the past. Dogs the world over are the creation of human culture, and through both formal and informal breeding great liberties have been taken to modify gene lines to develop new types, usually without much fuss. (The Leonberger and the Irish Wolfhound come to mind. The Leonberger was 'invented,' genetically speaking, barely a century ago, and the Irish Wolfhound was essentially 'reinvented' at about the same time by using outcrosses to similar breeds, including a Tibetan KyiApso, in an attempt to resuscitate a breed fading into extinction.) In fact, some supporters of the Tibetan super-dog openly speak now of two separate breeds. No problem; but doesn't that diminish the available genetic diversity of both? And, wouldn't much of the opposition and controversy go away if the super-dog was given a different breed name? (though, of course, that would reduce both gene pools). There already *is* a well-recognized Tibetan mastiff breed which includes both the do-kyi and the tsang-kyi as localized types.[17]

The breeding and sale of these modern super-dogs is big business, especially in China. The China News Service reports that the collective value of a few Tibetan mastiffs at a recent dog show was over Y10 million, equivalent to US$1.3 million, and elsewhere one reportedly sold for Y600,000 (almost US$74,000) and another, as recently as September 2009, for four million yuan (over $585,000)! That's no exaggeration. In Beijing alone, there are said to be over a hundred Tibetan mastiff breeding centers and more than a million pet dogs, among which the Tibetan mastiff is considered one of the most highly prized, and pricey. "Not surprisingly then," a reporter recently wrote in Beijing's *China Daily*, "most Tibetan mastiff breeders are celebrities or wealthy personalities." It's been reported that in one recent two months' period a well-known Chinese breeder earned close to a quarter million US dollars (over two million Yuan) from his dogs. They have become such popular status symbols that it has been said of China's new urban elite that "They want Hummers; they want Tibetan mastiffs," and their dogs have become so valuable that they are now the target of thieves.[18]

### The Dogs of Lhasa

A few months before we went to Tibet, I met some Lhasa Tibetans in a Kathmandu restaurant. We discussed dogs over beer, and I jotted down these cryptic notes:

> "Tsang-kyi is the offspring of a cross between a do-kyi (dog) and a tiger."
> "The dog has red eyes (bloodshot). Very big. Very expensive."
> "The ground bone (after a dog dies) is used in Tibetan medicines."

"Many big do-kyi in Tibet (esp. Lhasa) are not true Tibetan mastiffs (i.e., mixed genes)."

The closer we got to Lhasa on our cross-country trip the more excited we became. Along the way we saw many dogs, all doing their jobs. For example, some were seen protecting yak herds grazing the hills (where there are tents and yaks there are dogs). Some were seen accompanying heavily loaded yaks and their masters transporting goods and supplies cross country (the old way). Some were guarding monasteries and homesteads in towns and villages along our route. And some were seen guarding factories in the cities. Most of them were black-and-tan or all black, and some were red, both tsang-kyi and do-kyi (we were told, though it was difficult to tell them apart). Everywhere we went there were dogs. Some were loose, but many were tied, or ensconced safely behind a gate, or uncomfortably confined to a small cage with iron bars, securely shut in until nightfall when they are let loose to roam the premises. The caged dogs made the most menacing noises at our approach (or were they just being very vocal about being locked up?).

On our second afternoon in Lhasa I visited a factory where colorful dyes are made for use in the Tibetan carpet industry. A friend took me there, saying that the factory owner had two beautiful big Tibetan mastiffs. Inside the compound I was assailed by loud, deep enraged barking from two cages near the door. Despite the ruckus, both dogs, black-and-tans, wagged their tails vigorously at my approach. I took that to mean that they were friendly and could be approached, with caution. In minutes they were licking my hand as I petted them through the bars of the cage. I returned at dusk with Judy and Dorjé to see the dogs again. With the dog handler present they were let loose and were soon romping about playfully, happy to be out of confinement. They recognized me from earlier and after proper introductions to my friends they were soon licking our hands and nuzzling us, begging for the small snacks that the handler had given us to use as doggy treats.

Those two dogs conformed in all respects with European and American Tibetan mastiff physical standards. Each stood over 66 centimeters (26 inches) high at the withers. Each had a heavily plumed tail curled up over the back. Their heads were heavy and broad, with deep, square muzzles. Their limbs were muscular, with deep chests, thick manes and hair, and they had those imposing deep, loud barks. They did not, however, have excessive haw or extremely pendulous hound-like jowls. We also saw dogs of almost identical physical and behavioral characteristics at a factory in Shigatse. Some observers may have said that they, too, were "tsang-kyi from Lhoka," but in these guard dogs we knew we were seeing some of the best of the traditional Tibetan do-kyi (which includes tsang-kyi) at work.

That evening at the Lhasa dye factory we also saw a small, blue-grey KyiApso pup. He was a marvelous dog with an attractive fuzzy bearded muzzle. He was one of about twenty KyiApsos that we saw during our trip along the Friendship Highway and in Lhasa. That so many were seen in southern and eastern Tibet surely dispels the belief among some KyiApso fanciers that these special dogs are found only around Mt. Kailash.

Lhasa, of course, has more to see than big dogs. We photographed the Potala, which dominates the main modern business street of the city, and we visited the sacred Jokhang temple, the very 'heart' of traditional Tibetan religion and culture. As part of that experience, we circumambulated the Jokhang via the Barkhor, along with hundreds of pious Tibetan pilgrims thumbing their rosaries and mumbling their prayers: "*Om mane padme hum...,*" "Ode to the jewel in the lotus" (Lord Buddha). We were moved by the religious fervor of the devout of all ages who come daily from every corner of Tibet to worship a panoply of revered deities and saints. Entering the Jokhang temple was like stepping out of China back in time into the 'real' Tibet.

Our time in Lhasa was short, and we declined touring the Potala and the nearby monasteries of Drepung, Sera and others, partially in the knowledge that these famous sites remain today only as shells of a part of Tibetan culture that has been seriously wounded. We were disheartened to see how the face of this fabled and ancient city and surrounding countryside has changed under modern circumstances. Besides the fact that the Tibetans do not govern themselves now, as they did for centuries before the 1950s, one of the worst offenses is an attempt by the authorities to force nomads to settle in houses and graze their yaks on private land. In several places along the Kathmandu to Lhasa highway we saw rows of newly constructed houses for the nomads—identical looking, one after another, like 'company towns,' empty. Village life in a mud brick house, with a mongrel dog and a few yaks in a fenced paddock are anathema to the nomads, given their age-old transhumant lifestyle in tents, with their livestock guardian dogs, tending vast herds of livestock on the wide open plateau. If the government succeeds in forcing such a radical lifestyle change, what will become of the nomads and their famous dogs?[19]

Much of what remains of old Tibet's physical culture (monasteries, shrines, temples and old fortresses) has been turned into tourist attractions, including those that have had to be rebuilt or replaced after the destructive excesses of the Cultural Revolution. Since we did not consider ourselves to be typical tourists, we avoided many sites hyped by expensive tour agencies and their guides. Our main interest was dogs, livestock, nomads and villagers in the hinterland, and the dogs in the Lhasa dog market.

## The Lhasa dog market

The major highlight for us in Lhasa was our visit to the weekend 'dog market.' There, on a back street near the Kyichu river, Tibetan mastiffs are displayed, traded, bought and sold with great enthusiasm every Saturday and Sunday. We saw a few of what the locals called do-kyi, and dozens of others they identified as tsang-kyi, and a half dozen or so KyiApsos. (Judy labeled some of them—'Blue Lady,' 'Gold Puppy,' 'Red Scruffy Boy,' and 'Black-and-Tan Scruffy Boy.')

That there were so few of what some observers consider to be the small do-kyi was to be expected, for two reasons. Their traditional homeland is many days' drive in the west of old Ü-Tsang, while many of the reputedly larger tsang-kyi come from Lhoka, only a few hours' drive south of Lhasa. And, it is now the *big* dogs that sell best to city dwellers, a reflection of the insidious influence of the West. Ultimately *all* are do-kyi, of course.

There were also several examples of the new, hound-like "true Tibetan mastiff" (the *super* tsang-kyi). Their presence and the considerable hype about them in the market were quite obviously influenced by some of the big Beijing kennels and by some apparent (to us) cross-breeding to achieve the large hound-like looks.

I was surprised to see so many red dogs (actually more of a deep orange). Red is apparently a popular color, especially among the super-dog breeders. The majority, however, were black-and-tan. Because it was April, there were many adorable pups to be seen. (Both Judy and I briefly considered smuggling one or two back to Nepal with us.) The largest dogs were tethered, and the pups were in open boxes. Some of the dogs wore bright red yak-hair ruffs, which made them look very imposing.

Plate 6.2. *KyiApso in the Lhasa dog market.*

The shaggy KyiApsos that we saw included lithe black-and-tans, with the typical bearded muzzle. The one attracting the most attention and highest price was a magnificent blue-grey one-year-old, with hair all over her muzzle and covering her eyes (like a Tibetan terrier, but much larger). The price was 30,000 Yuan (nearly US$4,000) even after Judy and I light-heartedly haggled it down (with Dorjé's help) from over twice that amount. It was tempting...

As the only Western foreigners at the market, we attracted attention. We took many photos and asked many questions, and were shown many more dogs in small kennels down a side alley. Posted high over the marketplace were colorful signboards advertising some of

*Plate 6.3. Tibetan mastiff with wool ruff, in the Lhasa dog market.*

Lhasa's Tibetan mastiff kennels, with pictures of their big best dogs.

We were told, of course, that many of the big dogs in the market were "tsang-kyi from Lhoka," the reputed heartland of Tibetan mastiffs. If some western Tibetan nomads had been present they would undoubtedly have disputed that claim. There is a distinct Lhasa-centric or southeastern Tibet bias in this business, one that has spread worldwide.

In conversation with some dog sellers and bystanders we learned that the search for puppies in Lhoka had become a problem. Because so many of the "best" pups in Lhoka are "sold to Lhasa" each year, they told us, it is feared that the number of really good dogs in the hinterland is going down. We were also told that the residents of Lhoka are under considerable pressure from Chinese breeders to supply them with good pups. Consequently, some of the locals have become resentful and hide the litters, or at least the best of the pups, from the buyers and middlemen who come looking for them.

Everywhere we went, both in Lhasa and on the road, we interviewed big dog owners and interested bystanders. We carried a photo album of pictures depicting a range of dogs by size and color, both so-called do-kyi and tsang-kyi, along with some shaggy KyiApsos, and several snapshots of non-Tibetan dogs for comparison. A crowd gathered whenever and wherever we opened the album. Dogs are a subject of great interest and debate in Tibet. Inevitably, after our new-found friends had leafed through all the photos and discussed some of them with obvious enthusiasm, we asked them to tell us which dogs, and especially which colors, they preferred. Invariably, they turned back to pictures of black-and-tan Tibetan mastiffs, declaring them "best." It was not a scientific sample, and we were not always sure what "best" meant, but black-and-tan was the consistent first choice. They also commented on the red- and cream-colored dogs and were impressed by the KyiApso photos; but after a few minutes, most people turned back to the black-and-tans.

## *The dogs of Gangaling Kennel*

A short drive outside of Lhasa we visited the Gangaling Tibetan Mastiff Kennel where we saw dozens of dogs, some quite large. Most were identified to us as tsang-kyi, and some as do-kyi. The kennel owner made a special point of showing us his best bitches and their pups, especially those he said were descended from breed lines previously raised at the Dalai Lamas' Summer Palace on the outskirts of Lhasa and by the Panchen Lama at Shigatse.

Plate 6.4. *Tibetan mastiff (golden/cream color) from the Mt Kailash area of Western Tibet, at Gangaling Kennel, Lhasa.*

In 1946 Charles Bell wrote about the importance of big dogs to the 13th Dalai Lama whom he visited in the Summer Palace, the Norbulingka, near Lhasa. The dogs he saw there are probably an earlier generation of those shown to us in the Gangaling kennel. Bell wrote: "Large Tibetan dogs—used for herding yaks and for guarding houses and tents—with their long black hair, and collars of thick hanging wool, dyed a dull red, and chained up here and there. Incessantly they tug at their chains, as they try to spring on you, and bark with the deep, low note which Tibetans tell you should be like the sound of a well-made copper gong."

*Plate 6.5. Tibetan mastiff at Gangaling Kennel, Lhasa.*

"The Dalai Lama is fond of this breed. To be able to give such a dog to him people will pay a great sum of money, the earnings of several months. One of these dogs in the enclosure is a particularly fine specimen; it comes, as the Dalai informs me, from his own district, Takpo. He is proud of that dog, but indeed he is fond of all animals..."[20] Somewhere in the old literature is a picture of one of the 13th Dalai Lama's dogs—a bearded or shaggy KyiApso.

One dog we were shown at Gangaling (one not from the Dalai Lama's breed line) was extraordinary. He was a long-coated dog from somewhere near Mt. Kailash whose color appeared pure white (more correctly, 'light gold' genetically), like a Great Pyrenees. It wasn't a bearded KyiApso as one might expect from that far west, but a variation of the regional do-kyi, and the only one of its kind in the kennel.

In discussion with the Gangaling kennel owner and his staff, several distinct kinds of dog handlers were identified in Tibet. The list is not much different from one which could be drawn up in America or Europe. There are those whose legitimate aim is to preserve and save the breed from corruption or extinction; those who run 'puppy mills' strictly for profit (Tibetans call it "dirty money"); middlemen and profiteers who buy up dogs from anywhere to sell at the Lhasa dog market; and super-dog promoters allied with big Chinese kennels who are in it strictly for profit and for pride (ego). The Gangaling folks assured us that they take special care to avoid inbreeding. They asked to be remembered as legitimate breeders concerned with protecting and promoting the breed.

## To Lhoka, and Westward

From Lhasa we drove south to Lhoka where we spent several days and, very briefly, played tourist by visiting several historic and sacred sites in the ancient Tibetan heartland. We divided our overnight stays between a big Chinese hotel in Tsedang, and a typical Tibetan farmers' home in a rural village (by far the most interesting). We saw many dogs, but were not greatly impressed by them. None fit the image of the noble tsang-kyi that we expected after seeing such fine dogs in Lhasa on the weekend. Perhaps there really is no significant difference between the various types of Tibetan mastiffs, after all. Or, maybe the locals are, indeed, adept at hiding their best dogs from outsiders. Or, perhaps the best dogs from Lhoka had already been sold away to breeders and middlemen. There was scant evidence to the contrary.

After Lhoka we turned west from Tsedang 140 miles to Shigatse via Gyantse, then the remaining 300 miles back to Nepal. We crossed the Khampa La, one of many high passes on our route, then drove along the large, high, sacred Yamdrog Tso ('Scorpion Lake'). The mountain scenery was awesome, one panorama after another of unsurpassed geological entertainment. Gyantse was also impressive, with its great mountaintop fortress, the Gyantse Dzong, and the city's Pelkhor Monastery and Kumbum Stupa. It was there that we saw the red monastery dogs described earlier.

Gyantse is famous (*infamous*, that is) in Tibetan and British history as the site of the 1904 assault on the Dzong (the walled fortress that dominates the city skyline), led by Colonel Francis Edward Younghusband and his British Army of mostly Indian Sikh and Nepalese Gurkha soldiers. During the British invasion of Tibet that year, hundreds of ill-equipped Tibetan foot soldiers were massacred south of Gyantse and at the Gyantse Dzong, in what is now considered to have been a most unseemly military action by the British Raj.[21]

Three days later we passed through Nyalam on our way down to Zhangmu and Kodari, where we crossed back into Nepal. On the street in front of the Nyalam fresh produce shop we saw little San-San again, the pup we had been assured was "a true tsang-kyi, from Lhoka." Since then, however, somebody has surely bought him, and he may well be guarding a compound in Kathmandu, the nearest large city and a great market for big dogs from Tibet...

## Hunting Dogs

No story of big dogs of Tibet and the Himalayas is complete without mention of *sha-kyi*, the hunting dog. The sha-kyi is sometimes loosely compared with the more well known and popular sighthounds (like Saluki), but it is not recognized in the West as an established breed. Little is written about this type except some snippets in travelers' accounts, brief observations by researchers, and a few photographs.

There is a popular legend in Tibet and northern Nepal about a hunter and his dog. It relates the famous confrontation between the great eleventh-century sage Jetsun Milarepa and the hunter Gonpo Dorjé. Milarepa is renowned in Tibet as the first ordinary person to attain Enlightenment in one lifetime. There are countless stories about him and his religious teachings. The one about the hunter, his dog and a stag who came upon Milarepa meditating in a cave is so popular and so widespread that, in fact, *many* caves have been identified all across the Himalayas as the very *one*. There are also several versions of the story, each of which come to the same conclusion: the renunciation of killing.

Here's one version from Nyeshang (Manang District) in northern Nepal, bordering Tibet, where the story of Milarepa's cave is called Khyira Gonpo Dorjé (literally, 'The dog and Gonpo Dorjé'). Located high in the mountains about a half day's walk from the village of Braga, it is a popular pilgrimage destination for Tibetan villagers each summer. When I visited the cave in 2005, the resident monk told me that he was there to keep up the site including a small temple for worship. He had recently overseen the dedication of a large image of Milarepa, which attracts many pilgrims. There is even an exaggerated likeness of the hunter's bow wedged into the cliff above the cave where, it is said, the hunter tossed it.

The story goes like this, quoting an anthropologist's account:[22] "A hunter in Nyeshang with his dog was trying to hit a deer with his arrows, but failed. He chased the deer up a steep hillside and arrived at a hermit's cave. The deer entered first and saw Milarepa, meditating. Milarepa sang to the deer, inviting it to seek refuge with him. The deer was calmed and sat beside him. Then the hunter's dog entered the cave, and it too was calmed and sat on the other side of the saint."

"The hunter charged into the cave, became enraged by what he saw and shot an arrow at Milarepa. It could not touch him. Then Milarepa sang to the hunter, asking him to kill the passions within rather than kill sentient beings without. He invited the hunter to remain and practice the dharma."

"The hunter threw his bow and repented of the sin of killing sentient beings. The deer and the dog were sent to a Buddha field, and the hunter himself stayed with Milarepa until he had obtained liberation."

Travelers' accounts of Tibetan hunting dogs are less illustrious and imaginative. The best we can do with them is to quote a few. For example, Gabriel Bonvalot wrote in *Across Thibet* (1891) that "Now and again we meet with hunters carrying matchlocks, forks, and lances, with powerful dogs in leash, long-haired, like our shepherds' dogs, and with broad heads shaped like that of a bear. Many of these dogs are black, with reddish-brown spots, this latter being generally the colour of their chests and paws..." These sound suspiciously like do-kyi, but if was used exclusively for hunting, which the author implies, they would have been called a sha-kyi (following the Tibetan custom of 'naming' the dog according to its function, not its appearance).

In 1910 John Hedley noted hunting dogs while traveling in Mongolia. One evening a hunting party arrived on horseback, accompanied by a pack of dogs: "Eight or ten dogs were with the party," he writes, "and while the village dogs snarled and snapped at their heels, these hunting dogs behaved with absolute indifference, sticking close to the heels of the horses as they were led about to cool off. These dogs were big and bony, and save that they had long hair, somewhat resembled our grey hounds. They were an entirely different breed to the average house-dog, evidently well trained, and kept for sport only."

Hedley goes on to describe the style of Mongolian rabbit hunting with dogs: "The Mongols are accustomed to assembling three days each month for hunting, and at this meet there are said to be fully a thousand men present. Their methods are primitive and crude, and consist of chasing the frightened hare at a gallop, and flinging, when within reach, a curved club about 2 feet long, heavily weighted with lead at the curved end, and thrown with such unerring accuracy that they rarely fail to bring down their game. Their dogs retrieve the quarry, and the trophies are hung... to the crupper of the wooden saddles, remind one of the scalps secured and carried in pride by the Red Indian of former days."[23]

Douglas Carruthers also noted hunting dogs in Mongolia and wrote in 1914 that "The dogs we saw... were of a peculiar breed, and were said to be remarkable for their hunting capacities. They were lean, undersized animals of the lurcher type, with prick-ears and very pointed noses..."[24]

When some officers of the Raj, and diplomats and their wives, returned to Great Britain with exotic Asian dogs, a keen interest in them was raised among dog-loving British elites. Writing in 1932, Will Hally, the popular author of a column on dogs in England's *Our Dogs* magazine, penned the following armchair observation: "The Northern Tibetan Hunting Dog is a breed that is completely new to me, a photograph of which has been very kindly sent to me by the Hon. Mrs. McLaren Morrison..." It had "a magnificent body, beautiful legs and great bone, and with the most intelligent expression. The head contains a hint of the Labrador, also of the Golden Retriever and also of the Mastiff, but at the same time it is quite distinctive, with nothing of the cross breed or mongrel about it. Indeed, the whole animal looks like the veritable picture of pure breeding." Hally goes on to note that 'Shakya,' the dog's name (Sha-kyi?), was purchased near Gyantse and that its coat was "fluffy from muzzle to tail, and the tail is also profusely coated, but nowhere is it what we could call a long or even a feathered coat. The colour is sandy, but gets darker in the summer

Plate 6.6. *Tibetan hunting dog, early 20th century*

months." (It always amazes me how from one specimen a 'dog expert' such as Will Hally can find so much to say, though his comments here are entirely from the perspective of the show ring, with nothing about the dog's function.)

After living in Lhasa early in the twentieth century, Irma (Mrs. Eric) Bailey described the hunting dog as "about the size of an Airedale. In color it is a creamy grey with a thick coat. The tail may be carried curled over the back, but also some times down. The head is long and is a smoky black shading into the creamy grey of the body. The ears hang forward."[25]

She goes on to say that "The dog is used for killing game. He is taken on a leash to within sight of the game—Bharal (wild sheep) musk deer, serow, etc.—and slipped."

"When the quarry is pursued, it adopts its natural defense against a wolf by getting into a cliff, where it turns to bay and attempts to butt the dog over the precipice."

"This is where the quarry is wrong, for the dog does not go in and attempt to kill as a wolf would presumably do, but keeps barking in complete safety and distracts the attention of the quarry while the hunter comes up and shoots the animal at close quarters with his primitive matchlock." Mrs. Bailey concludes that "These dogs are very keen sighted..." Indeed, they are sometimes described as 'sighthounds.'"[26]

There is another description, of Khampa dogs, that sounds remarkably like hunting dogs. It was written by Heinrich Harrer in *Seven Years in Tibet* (1953), where he describes crossing the northern Tibetan plateau to Lhasa during the winter of 1944-45. Harrer and his companion, Peter Aufschnaiter, were occasionally confronted by dangerous brigands, mostly Khampa ('men of Kham,' from eastern Tibet). Some of them had big dogs. Harrer writes: "We often saw, happily in the far distance, men on horseback, whom we knew to be Khampas from the unusual type of dogs that accompanied them. These creatures are less hairy than ordinary Tibetan dogs, lean, swift as the wind, and indescribably ugly. We thanked God we had no occasion to meet them and their masters at close quarters."[27]

There are, unfortunately, no in-depth studies of these special, rare and probably endangered canines in the literature. There are, however, a few short accounts by scholars on Tibet. One of them, for example, is the Tibetologist Robert Ekvall who describes them briefly in his classic 1963 article on 'Role of the dog in Tibetan nomadic society' (see Appendix 2, pp.208-17). Professors Melvyn Goldstein and Cynthia Beall also write about them with a few photos in their 1990 book, *Nomads of Western Tibet*. The geographer Alton Byers mentions hunting dogs in his essay (in Chapter 3) on the Tibetan mastiffs of Mustang and Dolpa on the Nepal-Tibet border. Byers' account is especially interesting for the discussion about conflict between the nomads' need to hunt and the Buddhist proscription against killing.[28]

Ekvall wrote that "In some areas, a few rangy prick-eared dogs—mostly fawn or brindle—are kept for tracking stag or musk deer. They are called SHa KHyi (meat dog(s)) which is a somewhat general term for hunting dogs, and sometimes more specifically SHwa KHyi (stag dog(s)). They vary greatly in pelage, conformation and size for the breed is not at all pure. In a Golok tribe which I visited... I saw quite a number of slender greyhound type dogs which had certain resemblances to both borzois

*Plate 6.7. Tibetan nomads with hunting dogs.*

and salukis... They were considered useless as watchdogs but, as their name Wa KHyi (fox dog) indicates, were kept for hunting foxes."[29]

M. Hermanns, in *Die Nomaden von Tibet*, describes a hunting dog from Koko Nor. It was a long-bodied, wiry-haired canine, he says, and was so valuable to its owners that it fetched a price equivalent to a good riding horse.[30]

Sylva Simsova refers to a Tibetan hunting dog based on an article she found entitled 'Kattuk the Tibetan Hunting Dog,' by A.J. Sellar of the UK. The dog was picked up by Sellar while traveling near Gyantse, in southern Tibet, and was eventually taken to England. The description of this dog is sketchy, other than that the author insists it should not be confused with the larger, more solidly built Tibetan mastiff.

On a German website labeled 'Tibetan Hunting Dog,' we get this interesting description of the Kunlun hound: "Found in remote rural regions of Tibet, the Kunlun Hound has remained virtually unchanged throughout the centuries. Closely related, yet very different than the Kunlun Mountain Dog, this rustic breed is oftentimes seen as a variant of the Tibetan Mastiff. Unlike other dogs of the region, the Tibetan Hunting Dog is neither a livestock herder nor protector. Its main role has always been and still is that of a hunter, although it can make a capable watchdog as well."[31]

A rigorous and rather cruel training regimen is then described (on the same website) for Tibetan hunting dog puppies (a description that appears to be borrowed directly from Irma Bailey's 1937 article). Young dogs are tied to their mother's collars and forced to learn from her by being dragged alongside during a hunt, it says, then concludes that "Only the strongest and smartest puppies are able to run and keep up with their mother without getting choked or trampled underfoot. This ensures that

*Plate 6.8. Tibetan hunting dog.*

*Plate 6.9. 'Shikari' Himalayan mountain dog.*

only the best working dogs get bred and pass on their excellent working genes to the next generation."

The Tibetan hunting dog/Kunlun hound is portrayed as having a "lean, muscular and long-legged body... covered with a dense, harsh medium-length coat, coming in solid shades of fawn, brown and grey," with an average height of 23 inches. The author concludes that they are extremely rare and often misidentified as Tibetan mastiffs. While it is an interesting description, it is undocumented.

Goldstein and Beall's brief remarks about hunting in western Tibet describe how dogs are used by the nomads to track blue sheep, wild yak, gazelle and antelope. As brief as they are, their observations as anthropologists provide us with perhaps the best description of these dogs and how they function. Some nomads still maintain the hunting tradition, the authors say, although hunting and hunting dogs were far more important in the past than now. In their book, Goldstein and Beall quote a hunter describing how he uses his dogs to hunt *na*, blue sheep: "I always travel with my hunting dogs and rifle," he begins, "so that if I spot some *na* I can go after them. But you've seen my matchlock rifle... With just our Tibetan rifle we are no match for the *na*. That is why we always hunt with our hunting dogs. They tilt the odds in our favor. As you will see, as soon as I spot blue sheep on a mountain slope I turn loose my dogs. Their job is to corner one of the *na* among the crags, and bark loudly to lead me to the spot. The best dogs will even try to run back a ways to make it easier for me to find them, all the time, of course, keeping their prey at bay. Once I get there, I have plenty of time to set up my rifle and shoot..."[32]

Several times during our travels across Tibet in 2007, Judy Steffel and I saw what looked like hunting dogs. We called them, euphemistically, "sha-kyi" in preference to other terms that came to mind. (Dorjé, for example, called them "scrawny mutts" and "mongrels.") They were thin and rangy, some dun-colored and others of the same hues one sees in the more common do-kyi and tsang-kyi. Some were undoubtedly the offspring of Tibetan mastiffs that had been genetically 'interfered with' by the lowland dogs that have made their way in recent decades to rural Tibet along with Chinese migrants. We were not impressed. None had the noble look of a mastiff nor the sleek racy form of a Saluki, and none paid any attention to us beyond a glance or, at best, a high-pitched yip.

In 1925, William Montgomery McGovern described the same sort of scavenger dogs as "all hopelessly hungry-looking animals, usually of a light-brown color. They are left free to prowl about at will but seem to have been divested of every form of moral or physical courage and with furtive eye and drooping tail slink around the family courtyard. They are really despicable curs. The only thing one can find to say in their favor is that as scavengers they are really effective, as there is nothing, no matter how filthy, which they refuse to eat."[33]

Compared to the scavengers, McGovern found the watch-dogs to be "of an entirely different build, larger and stronger, with a much longer coat, generally black. They are always chained up in front of the great gateway, which leads into the courtyard, and thus lead a life of perpetual captivity. They bark vigorously at the approach of any stranger and generally make a blood-curdling attempt to bite as well. These dogs are purposely underfed in order to keep them in a savage temper."

A recent article in the *Wall Street Journal* (of all places!) describes the nondescript smaller big dogs of Tibet "mixing with mutts and other breeds like German shepherds."[34]

The only hunting behavior exhibited by the ones we saw was scavenging in the streets, sometimes alongside free-roaming cattle and *dzo* (yak/cow crossbreeds), at village dumps, and even in dumpsters, a behavior to which dogs naturally revert when no longer cared for by humans in houses, tents, monasteries or nomad camps.

*Plate 6.10. Tibetans with yaks and Tibetan mastiff guardian dog.*

# 7

# Summing Up

It's time to tie up some loose ends, now that we have discovered the big dogs of Tibet and the Himalayas with Marco Polo, in the journals of various eighteenth to twentieth century explorers, through my own research, while raising Kalu, and from travels across Tibet. For example, what do we really mean by 'dog' and 'breed,' and what is a 'purebred' Tibetan mastiff?

We must also address the persistent myths and malarkey that have grown up around the Tibetan mastiff, the best known and most popular of big dogs from the Asian 'Roof of the World.'

## What is 'Dog'?

'Dog,' according to John Ayto in his *Dictionary of Word Origins*, "is one of the celebrated mystery words of English etymology." In Old English dog was *dogca,* meaning a powerful breed of canine and the source of our modern term. But *dog,* "has no known relatives of equal antiquity in other European languages," says Ayto. We know that the term 'dog' proliferated in the thirteenth century, but not until the sixteenth century did it more fully replace English *hound,* a term that originally applied to all dog breeds. The Indo-European root of 'hound' is *kuntos,* which becomes *hund* in German, Danish and Swedish. 'Dog' was eventually borrowed from English into other European languages as German *dogge,* French *dogue,* Spanish *dogo* and Swedish *dogg.*

Then, while researching other English roots, I found an intriguing connection between dogs and canary birds.

### *The canary and the cynic*

There are many derivatives in English and other European languages from the Latin and Greek terms for 'dog.' Out of Vulgar Latin *canile,* for example, we get French *chenille* and English *canine, kennel* and—oddly—*canary.* As the story goes, canary birds (yellow or green finches) were introduced to England in the sixteenth century as caged pets from the Canary Islands. The Canaries are a group of Spanish islands in the Atlantic Ocean so named because in Roman times one of them was known for a large breed of dog. To the Romans, *Insula Canaria* meant 'Island of Dogs.' In time, the local finches became popularly known as "birds of the dog islands," or "canaries."[1]

The Latin term for dog is *canis* and the Greek is *kuon* (also spelled *cuon* or *cyn*), from which the English terms *canine* and *cynology,* and *canary* and *cynic* are derived. There are also probable links between Greek *kuon/kyn* and 'dog' in several Asian languages. Eugene Cotter observes, for example, that the Indo-European term for dog "was probably *kuon or *kwon, with a shorter form *kun, the root being perhaps *ku." It is not much of a stretch to consider *kutta* in Hindi, *kukur* in Nepali, *nakyu* (or *nagi*) in northern Nepal's Gurung language, Tibetan *kyi*, and Chinese *kou* (or *gou*) to all be related to the *ku* <*kun* <*kuon* root.[2]

Etymologically speaking, when tracing word origins, *kuon/cuon* transmutes to *kyn/cyn*, the root of *cynology*: the 'study of dogs.' Fair enough, but also consider it more surprisingly as the root to *cynic*. A 'cynic' is "one who believes that people are motivated purely by self-interest." Hmm. If you thought the origin of canary bird was interesting, the root meaning of cynic is all the more inspiring! According to linguist Victor Mair: "A *cynic* may be pardoned for thinking that this is a dog's life. The Greek word *kunikos*, from which cynic comes, was originally an adjective meaning 'doglike,' from *kuōn* (dog). The word was most likely applied to the Cynic philosophers because of the nickname *kuōn* given to Diogenes of Sinope, the prototypical Cynic. He is said to have performed such actions as barking in public, urinating on the leg of a table, and masturbating on the street."[3]

Moving on (with appropriate cynicism) we note with a chuckle that Ambrose Bierce in *The Devil's Dictionary* (1911) defines dog satirically as "A kind of additional or subsidiary Deity designed to catch the overflow and surplus of the world's worship. This Divine Being in some of his smaller and silkier incarnations takes, in the affection of Woman, the place to which there is no human male aspirant. The Dog is a survival—an anachronism. He toils not, neither does he spin, yet Solomon in all his glory never lay upon a door-mat all day long, sun-soaked and fly-fed and fat, while his master worked for the means wherewith to purchase the idle wag of the Solomonic tail, seasoned with a look of tolerant recognition."

## What Do We Mean by 'Breed'?[4]

The word "breed" also appears often on the pages of this book. Ah, yes—that mischievous term *breed*! The concept of breed has been bandied around a lot in dog talk. Raymond and Lorna Coppinger come close to defining it in a way that makes good sense in our context. "There are hundreds of breeds of dogs," they say. "Each breed has a definable conformation, a uniformity of size, shape, and behavior—a breed standard, if you will..."

Within those hundreds of breeds, they go on, "breed differences suggest a distinctive genetic program for each. Yet unbelievably, there is no appreciable genetic difference between them. One would think that there was some sequence of alleles that would act as a genetic marker that would identify a breed. Yet to date none has been found... I doubt there ever will be a genetic test of breed..."[5]

There are even more ways of defining the term. One meaning of 'breed,' for example, is *a group of organisms having common ancestors and certain distinguishable characteristics; a special kind of domesticated animal within a species*. Under this loose definition we can speak of *landrace* dogs, such as the Tibetan mastiff as it has existed in Tibet and the Himalayas for a very long time prior to or aside from the relatively recent establishment of breed standards outside of Tibet.

Another meaning of 'breed' is *a stock of animals or plants within a species having a distinctive appearance and typically having been developed by deliberate selection*. This definition better fits our image of *purebred* dogs. The "deliberate selection" part is what leads to breed standardization in the Western sense. In fact, however, landrace Tibetan mastiffs in Tibet and the Himalayas were also subject to "deliberate selection," though not to the extreme nor with the rigor required of Western breed standardization.

Some people talk about the Tibetan mastiff breed in the generic sense, as a type of big dog from high Asia. By this they mean the *landrace breed*, regionally developed and locally adapted. We often use the term breed, however, in its more restrictive (more modern) sense as a group or type within a species that has been *carefully and purposefully developed* under a high degree of human control and selection. This does *not* mean that nomads and other Tibetans are not careful or purposeful in their management of Tibetan mastiff breeding. It only means that compared with Tibet, Western standardization requires much more attention to type and conformity, and especially to looks (often with far less attention to function and health). In the West, this typically results in an artificially standardized breed or what we call a "pure breed," one that conforms to a standard set by one or another kennel club or breed association. Selective breeding for a standard type usually focuses on *show feature conformity* largely ignoring the traditional functions of the landrace forebears. Tibetan mastiff breed standards have been written by and for contemporary commercial breeders with little concern for the dog's life and purpose in Tibet.[6]

## Landrace dogs

'Landrace' is *a local or regional type that has sprung up and become relatively uniform by virtue of local selection for a specific purpose or function adapted to its original natural and cultural environment*. Thus, a landrace is a type that fits the 'land'; i.e., the geographic area and both the physical and cultural environment of its origin. The Tibetan mastiff and other big Tibetan/Himalayan dog types began as landrace dogs that were selected naturally in that region of the world with minimal assistance or guidance from humans, following traditional rather than modern breeding methods; that is, without interference or artificial selection for standardized appearance as in 'purebred' breeding outside of Tibet. In their original form, Tibetan mastiffs and other associated types were adapted to guard livestock as well as other assets like monasteries, houses and factory compounds and to withstand the harsh climate and persistent lack of resources. Tibetans call them *do-kyi* ('tied dog') or *yun-kyi* ('loose/untied dog'), or *tsang-kyi* (a do-kyi from Tsang), or by the generic term *phö-kyi* (a male dog, but loosely any dog of Tibet).[7]

### Purebred dogs

'Purebred' dogs are *types of dogs that are purposely created through focused, human-guided processes usually well away from their landrace origins*. The cultural predilections that guide standardized breed development are often far different from those of the derivative culture. This is particularly true of Western standardized Tibetan mastiffs vis-à-vis landrace Tibetan do-kyi. (The adaptive traits of Tibetan mastiffs, natural and cultural, landrace and purebred, are listed in the Appendix 4, pp.221-226.)

Breed standardization is a form of 'gentrification,' meaning a changed or improved (it's relative) condition designed to conform to Western or Westernized middle and upper class tastes.[8] Gentrification occurs when a breed is removed from its place of origin or its indigenous niche; from the land and the livestock of Tibet, for example. Under Western-style controlled breeding, an animal is created that may or may not reflect its original functions or even its original appearance. Many modern or standardized versions of traditional or landrace breeds are recognized in the registries of established breed associations. When recognized as such they are considered to be purebred dogs. In the modern context, only those dogs whose parentage consists of other registered purebred examples of the breed (to a specified number of generations) are regarded as belonging to that breed.

The concept of a 'pure breed,' however, is controversial, for three reasons. The first is *the difficulty of regulating breeding* with complete accuracy and honesty. The second is *the potential for unwanted genetic consequences* from inbreeding with a limited population of dogs. The third and most distressing is *the loss of original landrace cultural characteristics* of function and purpose. In the case of standardized Tibetan mastiffs, we find that all three of these problems arise to one degree or another.

### Registration

The American Kennel Club (AKC) demands a three-generation pedigree in writing from a recognized registry. Other clubs have similar requirements. It is not difficult, however, to fake a pedigree and registry. Some foreign-bred dogs exist where there is no formal registry. One of the problems with dogs originating in China, including Tibet for example, has been the lack of a recognized registry, though the Chinese are now working to perfect one. In 2007, while visiting a Tibetan mastiff kennel near Lhasa, my colleagues and I were met with a look of incomprehension when we asked if individual dogs were registered with a national or provincial kennel club. In fact, however, the kennel owner assured us that he took great care to maintain a pedigree of sorts, as a breed line. He just didn't know quite how to formally standardize or officially register it.

Furthermore, while in many countries breeding is ostensibly regulated by breed clubs and associations, other deliberate (and sometimes questionable) breeding regimens also may occur. Some modern breeders, for example, are striving to produce a larger, heftier, more 'jowly,' wrinkle-faced dog that is *thought* to *look* more like the *mythical* large dog of Tibet—i.e., Marco Polo's dogs "as big as donkeys." Some want to call it *tsang-kyi*, but that term is both questionable and contentious.

On our trip across Tibet, my companions and I saw dogs for sale in urban Lhasa that evinced *highly* exaggerated and suspiciously hound-like morphology, with great sagging, slobbery jowls and considerably more haw than is usual or expected in Tibetan mastiffs. These *non*-traditional characteristics reflect deliberate breeding towards a form of gigantism uncommon in the rural landrace breed. Both characteristics would appear to be highly detrimental to the dogs' survival under the typical climatic conditions of the high, cold, dry, dusty, windy Tibetan plateau.[9] (See Chapter 6.)

*Inbreeding*
Sometimes inbreeding causes further deleterious complications. A prime example is the purebred Tibetan KyiApso, which suffers (outside of Tibet) from a very small population and limited gene pool (discussed in Chapter 5).

*Loss of landrace characteristics*
In virtually all instances of creating a modern breed (like a Tibetan mastiff) for show or companionship (*sans* livestock, *sans* harsh environment), the original functions of landrace dogs as they adapted physiologically and culturally over many centuries to guardianship and protection functions may be diminished, lost or radically changed. Thus, the attention of Western breeders and buyers is focused on phaneroptical traits (looks), morphology (form) and phenotype (observable characteristics resulting from interaction of genotype with environment), whereas pastoral nomads were (are) most interested in the roles and functions of their working dogs. In short, as the modern standard Tibetan mastiff has been bred to be "massive," "noble," "pretty" or "as big as donkeys," its original protective nature, its physical stamina and durable structure and its guardian dog behavior are played down (though some of the original temperament may remain). The change to standardized breeding began in the 1800s when European aristocrats began selecting dogs for a very specific purpose; that is, for *looks* that would win at dog shows. That's when the dog's original function took a back seat.

## "Myth and Malarkey"[10]

When we examine what has been said about Tibetan mastiff origins it becomes clear that some members of the popular Tibetan dog fancy have bred not only canines, but also canards ('canard' means 'duck' in French), often with wistful, troublesome and unsubstantiated breed histories.

In addressing the fictions and fancies that have been perpetuated, we must ask ourselves the following questions:

Is the so-called 'Tibetan mastiff' a true mastiff?

Are we really expected to believe that the dog depicted in the bas-reliefs at Assyrian Ninevah (*c.* 640 BCE) is a Tibetan mastiff, or that Alexander the Great collected Tibetan mastiffs on his sojourn in the heat across Southwest Asia in the fourth century BCE?

Should we continue to call them Tibetan mastiffs, or some other more reasonable appellation like Himalayan mountain dogs? Or, should we split the breed and use both terms, considering the latter to be a variant of the former?

Have the big dogs of Tibet truly disappeared at the hands of the Chinese?

Was there once a "gigantic" Tibetan mastiff that is now lost to the dog world?

Was the so-called tsang-kyi a huge but long gone landrace breed or merely a localized name for one type of the traditional Tibetan guardian dog called do-kyi (i.e., a do-kyi from Tsang)?

Is the long-haired, bearded or shaggy kyi-apso (the Tibetan KyiApso) a breed apart?

By raising such questions do I appear to be a little cynical? Yes, of course—but healthily so, in the true sense of the word 'cynic,' as a skeptical or 'dogged' inquirer. As Raymond Triquet has said (he's the French cynologist whom some call the father of the modern Dogue de Bordeaux): "...because Doubt is the essence of science and research, we need first to stop affirming without giving evidence."[11]

## Mastiff / No Mastiff

The definition and derivation of terms can reveal a great deal about things, especially things as subject to conjecture and controversy as the origins, history, definitions and nature of dog breeds. First, the Tibetan mastiff is not a true mastiff. Second, many of them should probably be called Himalayan mountain dogs, a notion that has been suggested by such respected Tibetan mastiff fanciers as the American Kristina Sherling and the Scandinavian Kåre Konradsen.

So, where does the term 'mastiff' come from? When was it first applied to large Tibetan dogs?

For sure, Marco Polo did not call them Tibetan mastiffs, though after reading many of the books and Internet sites that discuss the history of this dog it seems that a lot of people believe he did. The Internet holds a treasure trove of stories about Tibetan dogs. I recently Googled 'Tibetan mastiff' and got nearly a half million hits. When I narrowed the search to 'Tibetan mastiff' + 'Marco Polo' I still got over 700 hits.

Marco Polo, who lived in thirteenth century Italy, did not call them mastiffs because that term was not invented until the nineteenth century, in England. More commonly and for centuries, the big dog of Tibet and the Himalayas was known to English readers and writers as the "Thibet [or Thibetan] dog" and sometimes, early on, as the "Nepal [or Nepaul] dog" or "Bhotean" dog.

Here is some of the early terminology *before* the term mastiff came into popular usage for these dogs (I've highlighted the key words in bold):[12]

1121BCE "Cynologists ransacking the ages for evidence concerning the early breeds, have discovered... ancient testimony to the antiquity of **the dog of Thibet**, contained in Chinese writing in a record of the year 1121 B.C., in which it is stated that the people of Liu, a country situated west of China, sent to the Emperor Wou-wang, **a great dog of the Thibetan kind**. The fact is also recorded in the Chou King (Chapter Liu Ngao), in which the animal is referred to as being four feet high, and trained to attack men of a strange race" (Robert Leighton, 1916).

1298CE "They have **dogs of the size of asses**..." (Marco Polo, in his chapter on Tibet, Thomas Wright translation, 1880), or "**dogs as big as donkeys**" (Yule-Cordier translation, 1903).

1712 "Many of the **Thibetan dogs** are uncommon and extraordinary..." (Ippolito Desideri, written in the 1700s, published in 1939).

1800 "...**huge dogs... natives of Tibet**..." Samuel Turner (1800).

1847 "...**large dog from Tibet**..." In 1847 Lord Hardinge, Viceroy of India, sent this large dog named Bhout, from Tibet, as a gift to Queen Victoria (A. Croxton Smith, 1931).

1852 "...**Thibet dog**... large dogs, fierce, strong, and noisy... impetuously furious" (William Youatt, 1852).

The anonymous author of one Internet site maintains that "...the breed name 'Tibetan Mastiff' was assigned by cynologists of the 1800s to a number of closely related breed strains from the Himalayan Mountains. No distinction was made according to size or function."[13] The nineteenth-century British naturalists Brian Houghton Hodgson, William Charles Linnaeus Martin and Joseph Dalton Hooker were among the first to *call* the Tibetan dog a "mastiff", but they were not the first to *compare* the big Tibetan/Himalayan dogs with mastiffs. The distinction is important.

*Earliest use of the term "Tibetan mastiff"*
After an exhaustive search I have determined that the earliest use of the term "mastiff" in association with (but not yet *for*) these dogs was published in 1820 by James Baillie Fraser whose *Journal of a Tour through Part of the Snowy Range of the Himālā Mountains* is one of the earliest sources on the Himalayas written by a European. Fraser toured the mountain districts of the India colony for two months in 1815, before the monsoon. His travels took him across what are now the north Indian states of Himachal Pradesh and Uttar Anchal. He accompanied his brother, William Fraser, a political agent for the British army, along with an escort. The Frasers were the first Europeans to make such a trip. During the trek, they reached the sacred site of Gangotri on the upper Ganges River, and thinking they were at the source of the river went no further. Fraser's journal is a superbly written, detailed primer on the geography, military outposts, religious sites, wildlife, people, local economies, domestic animals including dogs, and a great deal more. Fraser's descriptions have great historic value, and because his book was the first authoritative account published about the region it was read with great interest by other explorers, naturalists and diplomats of the day.[14]

Fraser referred to the Himalayas as "Himālā" (an early spelling) and to Tibet as "Bhootan" (not to be confused with modern Bhutan). In those days, Tibet was known as Bhot, and he called the Tibetans whom he encountered near the border "Bhoteas" (a common though slightly derogatory term still heard today, though nowadays it is usually spelled 'Bhotia'). On his travels he encountered a "large and fierce species" of dog that, with no other handy reference, he proceeded to *compare* with the European mastiff. Here's what he wrote: "...we saw a dog of a large and fierce species, highly valued through these countries... They tell wonderful tales of their strength and activity; and it is commonly said that, of the fiercest sorts, two will kill a tiger. As no true tigers are founded in the hills [sic], it is not probable that the trial has been made..." He assumed there were no tigers in the region; but there were. He thought that the locals meant leopards, which were also there in relatively large numbers.[15]

Fraser remarked that the finest of the dogs he saw were noted for their size and hardihood and that those "that came under our observation *bore a considerable resemblance to a mastiff*, but retained a good deal of the cur" (emphasis added). "Their color in general," he went on, "was black and white, with a little red occasionally; their hair is long and thick, and the tail long and bushy, curling up behind..." By "a little red" he undoubtedly means the typical reddish-tan of a black-and-tan dog. He concludes that they are "often very fierce, and sometimes attain a considerable size, but are seldom so large as a full sized mastiff." By "full sized mastiff" he undoubtedly means the *English* mastiff; but, more importantly, note that nowhere in his account does he *directly* call the dogs he saw *mastiffs*. He only *compares them with* what one may assume was *the English mastiff*.

Fraser described one feature of these dogs that anyone who has owned a Tibetan mastiff knows full well: "These animals are furnished with a down under their long shaggy hair, which is as fine and soft as shawl wool; this comes off easily in warm weather, and is regularly shed with the hair. Every animal is similarly furnished in this cold country." He saw them at the beginning of the hot season when they were blowing their undercoat.

Finally, he remarks, "We found that the natives used these dogs as sheep-dogs, in the same way as those of other countries, and also for hunting all sorts of game, even birds which they tire out in flying; and some were valued at a very high price."

It was Brian Hodgson of the British Residency in Kathmandu (and a gifted naturalist), who *first* called the big dogs of the Himalayas and Tibet by the term "mastiff." In an article published in 1832 'On the Mammalian of Nepal,' Hodgson describes the big dogs of the Himalayas as a "noble beast usually denominated the Nepal dog," In a section of the article subtitled '*Genus Canis*' he writes that the Nepal dog "is found only in the Kachár, where alone in Nepal he can live. It was introduced into the Kachár from Tibet, in which region it is indigenous, and in various parts of which there are several varieties. That of Lassa [Lhasa] is the finest, and is almost always black, with tan legs, and a false or 5th digit before and behind."

Kachár means 'foot of a hill' or 'hillside' in Nepali, though Hodgson uses it to mean the high hills and mountains adjacent to or on the edge of (at the foot of) Tibet. He also calls that high mountain region the "juxta-Himalayan." He writes that "For half the year the summits of these mountains are buried under snow; and, *near* to the Æmadus, their sides and basal intervals also..." (his emphasis). (Æmadus, Emodus, and sometimes Imaus, are interchangeable ancient Greek geographers' terms for the "Snowy Range"; i.e., the Himalayas.)

Hodgson describes three varieties of Tibetan mastiff: the so-called "Nepal dog" of the Kachár highlands, the red dog of Mustang, and the black Lassa or Lhassan type (his spellings). Referring to a noted British painter of his day, Edwin Landseer, Hodgson writes that "Landseer has excellently figured a male and female of this dog, which were taken from the residency [in Kathmandu] and presented to the king of England. The *mustang* variety", he goes on, "is rather smaller, of a bright red colour, with wall eyes; and he wants the 5th digit behind..."

It is here that Hodgson classifies the big "Nepal dog" as a mastiff for the first time: "This dog is justly placed in Curvier's 3rd section of the

*Plate 7.1. Tibetan mastiff by the English painter Edwin Landseer (1802–1873).*

*caninæ;*" he writes, "but he ought surely to be classed under the variety mastiff, not bull dog. His superior size, moderately truncated muzzle, long fur, sunken eye, perfectly pendant ears, and 5th claw on the hind foot (in the Lhassan animal at least) seem decisive of this point. The chief character of the skull consists in the great development of the longitudinal and transverse cristæ."[16]

Some of the earliest uses of the terms "mastiff" and "Tibetan mastiff," including Fraser's and Hodgson's, are listed below (with key words highlighted by me):

- 1820 "...the breed of Bischur is noted for its size and hardihood. The finest that came under our observation **bore a considerable resemblance to a mastiff**, but retained a good deal of the cur... They are often very fierce, and sometimes attain a considerable size, but are seldom so large as a full sized **mastiff**" (Fraser, 1820, pp.354-5).

- 1832 The big dog of Tibet and Nepal "**ought surely to be classed under the variety mastiff, not bull dog**... His superior size, moderately truncated muzzle, long fur, sunken eye, perfectly pendant ears, and 5th claw on the hind foot... seem decisive of this point. The chief character of the skull consists in the great development of the longitudinal and transverse cristæ" (Hodgson, 1832, p.342).

- 1843 "**The Mastiff of Tibet**... [is] larger than the English, with a thick head, elevated occiput, very pendulous lips, the skin from the eyebrows forming a fold towards the outer edge of the eyes, and ending in the jowl; the ears are round and

# 7. Summing up

drooping; the neck remarkably full; the back slightly arched; the tail, turned over the back, is well fringed, and, together with the very rugged hair of the body, deep black, with the sides somewhat clouded; over the eyes, about the muzzle and the limbs, there is some tawny..." (William Jardine and Charles Hamilton Smith, 1843).

1845 "...dogs of gigantic stature, which were no doubt **Thibetan mastiffs**..." (W.C.L. Martin, 1845).

1854 "...beside these stalks the **huge, grave, bull-headed mastiff**..." (J.D. Hooker, 1854; Hooker's mentor in the Himalayas was Brian Hodgson).

1868 "...**Thibet mastiff**" (Charles Darwin, 1868, after corresponding with both Hodgson and Hooker).

1878 "...**Thibet dog**... **mastiff**" (John Henry Walsh, aka 'Stonehenge,' 1878).

1886 "...the **mastiff of Tibet and Bhután**" (William Wilson Hunter, 1886).

1886 "...the finest **Tibetan mastiff** which I saw in all the Himáliya. It was a sheep dog..." (Andrew Wilson, 1886).

1891 "...a magnificent **Tibetan mastiff**" (William Woodville Rockhill, 1891).

1891 "...excellent watch-dogs... One of them, a **mastiff**... keeping on the watch all night" (Gabriel Bonvalot, 1891).

1894 "...**mastiffs**...in Eastern Tibet and **pure mastiffs**...in Lhasa" (William Woodville Rockhill, 1894).

1894 "...a huge, savage **Tibetan mastiff**" (Isabella Bird, 1894).

1906 "...gigantic black **Lhasa dogs**... These **Tibetan Mastiffs** are unpleasantly savage" (Thomas Hungerford Holdich, 1906).

1908 "...**the larger dog of Tibet**, the so-called **mastiff**... The **Tibetan Mastiff**" (H.W. Bush, 1908)

1927 "...**Thibetan watchdogs**... **mastiffs**" (Alexandra David-Neel, 1927).

1937 "...Speaking of names, **mastiffs in Tibetan are called Do-Kyi**, which means "a dog you can tie up".... **Tibetan mastiffs** are usually black in color with tan points..." (Irma Bailey, 1937).

1968 "...**encampment dogs**... more or less of **mastiff** breed" (Robert Ekvall, 1968).

In a few instances in the literature on these dogs, the term "mastiff" and "Tibetan mastiff" are used indiscriminately as if they are identical. For example, in Ann Lindsay Wynyard's *Dogs of Tibet* (1982, Chapter 4), a book mostly about Tibetan spaniel, the author tends untidily and without rigorous documentation to confuse the history and spread of the Tibetan mastiff with the *true* mastiffs of Europe. "The dogs of Tibet traveled all over the world," she writes, "not only along the trade routes, but also as a result of war and invasion. When Julius Caesar landed in Britain he found Mastiffs here, brought, in all probability, by the Phoenicians on their trading expeditions..." Such an incredible leap of faith connecting the big Tibetan dogs with ancient Mediterranean and British mastiffs is so outlandish that it needs no further elaboration.

Finally, for a long time it was thought that first 'official' reference to the Tibetan mastiff was in the *Kennel Club Stud Book* (UK) of 1873. An exhaustive search, however, has recently revealed this to be wrong. While it appears later in Kennel Club documents, it is *not* found in the first *Stud Book*. In May 2006, Eric Holliday, a British Tibetan mastiff owner, asked the staff at the UK Kennel Club "to try and see if they had any references to a TM anywhere, but they have nothing until much later. I had always believed that in 1873 the breed was included because it was a relatively well known breed then but as the rules changed it was removed. This is not the case though, and the TM by that name or any other does not appear in that first stud book. In fact the TM will not be included in the Stud Book until we get Championship status and a final Standard," he writes.

## *More on the meaning of 'mastiff': Take your pick*

Some commentators now consider the term "mastiff" to be a misnomer for any of the big Tibetan/Himalayan dogs. They ask: Is the Tibetan mastiff a true mastiff or something else? If something else, is it a molosser as some believe? The mastiff/molosser question has been debated for a long time, and it goes on, as wishful thinking.

A mastiff is generally defined as an old breed of dog that probably originated in Asia (a huge place). The evidence points to countries in Asia Minor bordering the eastern Mediterranean. True mastiffs are described as smooth-coated dogs (in most accounts) and occasionally as rough-coated (in a few), but never as heavy-coated with an undercoat (as in the Tibetan mastiff). They are especially noted for the combination of large size, strength, courage, deep chest, short muzzle and loud, sonorous voice. Given their instinctive guarding behavior, they have traditionally been used as watchdogs for the protection of livestock against large predators and poachers, and of property against intruders and thieves. There are also stories of their being war dogs, particularly in Asia Minor and in Kublai Khan's China.

So, what *is* a mastiff? The origins and definitions of the term are highly variable and not entirely conclusive.

# 7. Summing up

Here is a representative selection of standard definitions from current dictionaries and encyclopedias: The *Encarta Encyclopedia*, for example, presents a rather sparse definition of mastiff as a "large dog with a short brownish coat." The *Oxford Dictionary* expands on that only slightly by adding some facial features: a "dog of a large, strong breed with drooping ears and pendulous lips." The *American Heritage Dictionary* adds probable origin—"an ancient breed of large strong dogs, probably originating in Asia and having a short, often fawn-colored coat." The *Merriam-Webster Dictionary* includes a hint of function, describing the mastiff as "a breed of large smooth-coated dogs used especially as guard dogs."

The *Hutchinson Encyclopedia* provides a slightly more expansive reference to the dog's size and function with a little history thrown in, as a "breed of powerful dog, usually fawn in colour, that was originally bred in Britain for hunting purposes. It has a large head, wide-set eyes, and broad muzzle. It can grow up to 90 cm / 36 in at the shoulder, and weigh 100 kg / 220 lb."

The *Columbia Encyclopedia* gives a long definition that includes additional historical background, a broader discussion of function, and more about the dog's appearance. The mastiff, it says, is "a breed of very large, powerful working dog developed in England more than 2,000 years ago. It stands from 27 to 33 in. (68.6–83.8 cm) high at the shoulder and weighs from 165 to 185 lb (74.9–83.9 kg). Its coarse, short, close-lying coat may be silver, fawn, apricot, or dark fawn brindle in color, with a black muzzle, nose, and ears and black around the eyes. The mastiff was first bred as a fighting dog and guardian... throughout the entire history of the breed in England its greatest popularity has derived from its widespread use as a guardian of home and family..." Then, this surprising afterthought appears, suggesting that the term 'mastiff' "is also applied to a general type of giant dog whose origin has been traced to Asia and of which the modern Tibetan mastiff, infrequently seen in the United States, is representative." This is the only definition I have found that specifically mentions the Tibetan mastiff as an example and the only one that refers to a "giant dog."

By delving further into the word's European origins we find a broad, speculative and confusing range of possibilities. Etymologists have established that the term 'mastiff' most likely entered the English language around the mid-fourteenth century, from one of several sources. It may be related to the word *massive*, from French *massif* (<Old French *massiz*, <Latin *massa*), meaning "heavy," "bulky," "huge," "monstrous" or "forming or consisting of a large mass," like a mountain. Or, it may have originated from the Old French *mastin* and ultimately from Vulgar Latin *canis mansuetinus* (meaning 'domesticated' (<*manus* 'hand' + *suēscere* 'accustom'); hence a dog "accustomed to the hand"—i.e., domesticated, tame, or gentle.[17] There is also speculation that the Old French *mestif*, for 'mongrel' in the sense of vulgar, may be involved. Note that vulgar implies "lacking cultivation or refinement, offensive to good taste or refined feelings"; i.e., just plain common. (I'm not sure that many mastiff dog fanciers will be keen on this source of origin, given that they are unlikely to consider their canine charges as 'vulgar' or 'common.')

In conclusion, these various definitions are contradictory enough, and their speculative historical origins cover such a wide range of possibilities, that almost anything goes. Take your pick. Confusion reigns.

*The molosser disconnect*

The molosser group, among which 'true mastiffs' are often classified, includes short-haired, fawn, beige or brindle colored, smooth-coated breeds such as Boxer, Bulldog (various), Bullmastiff, Dogue de Bordeaux, Great Dane and Neapolitan mastiff. The term is derived from the name of a northwestern Greek region called Molossia, noted for its big dogs. Does the longer-haired rough-coated Tibetan mastiff (black, black-and-tan, or other colors) fit this description or classification? It comes down to two issues: Is so-called Tibetan mastiff a *true* mastiff? Is it a molosser?

The Norwegian breeder Kåre Konradsen, on molosserworld.com, makes one of the strongest arguments *against* the molosser connection. He convincingly argues against the following two well-accepted and interconnected beliefs: (1) that "the Tibetan Mastiff is the first of all Molosser breeds" and (2) that it is also "the ancestor to all other breeds known today as Molossers... This you can read in all dog books," he says, "but think about it, which [proofs] have been presented to you that are placing giant dogs in Tibet before anywhere else?" There is no archaeological or paleontological evidence, he says, placing "the Molosser dog in Tibet before anywhere else. No pictorial evidence, no folklore evidence, nothing!" (Recent DNA evidence has the first proto-dog arising from East Asian wolves.)[18]

Konradsen cites several other European dog experts who also question the Tibetan mastiff/Molosser connection. He then concludes that there are simply no scientific facts to support the thesis that Saint Bernards and other Swiss cattle dogs, for example, are derived from Tibetan mastiffs via the molossians—"No written documents, no pictures and no osteological proof...," he says.[19]

## Sheepdog, Bhutia, Bhotia, or Mountain Dog

Besides Tibetan mastiffs, some of the dogs along the southern fringes of Tibet have been called by other names, such as "Himalayan sheepdogs," "Himalayan mountain dogs," "Tibetan mountain dogs", "Bhutias" and "Bhotias."[20] (Sometimes the Mongolian mastiff/sheepdog is included on this list.) These types of names are sometimes used for dogs in the high mountains of India, Nepal and Bhutan that do not measure up to the 'ideal' as it is defined by contemporary Western show dog standards. H.D. Brooke noted early on, in *The New Book of the Dog* (by Leighton, 1907) that "On the borders and outskirts of Thibet, the size and type of the dog deteriorates; the marked properties disappear, and an ordinary looking animal of sheep-dog type is reached. But the true type is unmistakably Mastiff." Thus, some speak of these "lesser" or "smaller" dogs in terms of one or another *type* (not *breed*), different in various ways (most notably in size) from the "true type" Tibetan mastiff.

The challenge comes when we attempt to nail down the distinctions between the various types more clearly. Do we really "know one when we see one," as we sometimes say? Juliette Cunliffe has cautioned that "Although the dividing line between various breeds in the Himalayan regions can be thin, we must continually take care not to confuse other dogs with the Tibetan Mastiff."[21]

A century earlier, in *The Dog Owners' Annual 1901*, L. Jacob made the following cogent observation about our inability to clearly and faithfully distinguish among types of big Tibetan dogs: "A group of twenty Thibetan dogs might be taken at random, and if arranged in line, the most mastiff-like animal at one flank and the most sheepdog-like at the other, the rest being sorted in between by gradation, it would be *absolutely impossible to say where the one variety ended and the other began*" (emphasis added). Jacob's challenge is provocative.[22]

The following is a partial but representative list of references to the Tibetan mastiff as "sheep dog," "Bhutia," "Bhotia." "Himalayan" or "mountain dog," from the nineteenth century to the present:

1820     "We found that **the natives used these dogs as sheep-dogs**, in the same way as those of other countries..." (James Baillie Fraser, 1820).

1886     "...the finest Tibetan mastiff which I saw in all the Himáliya. It was a **sheep-dog**" (Andrew Wilson, 1886).

1891     "The **sheep dogs** kept by the Himálayan shepherds are warmly spoken of by their owners, who say that when the mountain paths are hidden in mist, they are infallible guides..." (John Lockwood Kipling, 1891).

1908     "...**Bhutia Sheepdog**..." (H.W. Bush, 1908)

1931     "It may be that **the word 'mastiff' is a misnomer**, and that **'sheepdog'** would be better" (A. Croxton Smith, 1931).

1976     "**The (Tibetan) Himalayan Mastiff**... are scattered throughout the Himalayan hills and are not and *were* not particular only to Tibet" (Amar Rana, 1976; original emphasis.)

1981     "One very big black and tan **Bhotia dog** I bought in November, 1919... Today it would be called Tibetan-Mastiff" (Mukundi Lal, 1981)

1983     "To me the **so-called Tibetan Mastiff** is a mountain dog… the **Tibetan Mountain Dog… The title 'mastiff' is a misnomer**" (David Hancock, 1983).

In 2002, Kristina Sherling summed it all up: "There is some controversy about what kind of dog the Tibetan Mastiff really is/was, Some say it is not a Mastiff breed, and should be called Himalayan Mountain Dog instead of its current name. Others says that the history clearly shows that this is of real Mastiff type. Other claim that the original Mastiff of Tibet is extinct, and that the current version of the breed is a cross of several types of dogs coming from Tibet." Her conclusion: "It is a difficult issue to decide…"[23] Yes, indeed it is.

## "When all is said and done…"

Regardless of the controversy over what it is and what it isn't, the big dogs of Tibet and the Himalayas are all fine dogs to know, to raise, to enjoy and, tapping into their natural instincts, to put to work as property or livestock guardians. It is fitting at the end of this long journey of discovery to quote what I think is one of the best celebrations of these big dogs in print, from well over a century ago:

> "Perhaps the noblest specimen of all dogs in Asia is the Thibet mastiff. He inhabits the Himalaya Mountains, as that other noble dog, the St Bernard, inhabits the Alps. But he is a fierce, savage animal, and is used for the purpose of repelling strangers, instead of rescuing them from the snow, as does our good friend the St Bernard. The Thibet dog is larger and stronger than an English mastiff, and with a heavy black coat. He has a peculiar overhanging upper lip, and a general looseness of skin about the face that imparts to him a strange, forbidding expression. His very look implies a terrible threat, and seems to bid the approaching stranger, 'Beware!'…"

> "This dog has been called the 'Guardian of the Himalayas,' by travelers who have seen him standing guard on some rocky eminence, warning the stranger in deep, hoarse tones, on peril of his life to come no farther. At such times the imaginative traveler has likened these dogs to black canine sentinels stationed there to guard the rugged Himalayan passes from the advance of civilization."[24]

# 7. Summing up

## *The last word*

*Plate 7.2. HAGAR by Chris Browne.
(Hagar © King Features Syndicate, with permission.)*

# Endnotes

## Preface

1. Jigme Wangchuk Taring (1908-91) was a Sikkimese prince and aide to the 14th Dalai Lama, former director of the Council for Tibetan Education (1968-74) and member of the Tibetan *Kashag* (Cabinet) in exile. Taring was well acquainted with the big dogs in Tibet.
2. Eugenia Shanklin, 'Sustenance and symbol: Anthropological studies of domesticated animals' (1985).
3. Michel Peissel, *The Last Barbarians* (1997, p. 84).
4. The choice of 'American Tibetan Mastiff Association' for the club name was mine, in part for its acronym ATMA which means 'heart,' 'mind' or 'spirit soul' in Sanskrit (sometimes also spelled *atman*). Ann Rohrer, ATMA's first president, agreed that it was a fitting name for the club. After awhile, however, I let my involvement in ATMA and other dog organizations lapse, for a variety of reasons. Though I oversaw the breeding and sale of several litters of Emodus Tibetan mastiffs from Kathmandu, I did not become a well-known breeder because my transient lifestyle did not allow it. Frequently moving from one country to another on consultancies puts a damper on such ambitions. I also lacked the financial resources necessary to establish a commercial kennel. And, quite frankly, I developed a distaste for the egotistical infighting and political machinations for which the Tibetan mastiff fancy has a reputation, especially among competitive breeders and clubs. I have always been interested in dogs more from an ethnographic point of view and for companionship than as show dogs or objects of profit.

## Chapter 1

1. E. Kostova, *The Historian* (2005, p.26).
2. R. Latham, *The Travels of Marco Polo* (1958, p.1).
3. Paul Smethurst, a Marco Polo scholar, reminds us Polo wrote in the 13th century to a rational and scientific world that had maps and compasses and taxonomies, and while his *Travels* strikes a chord of authenticity and reality in the modern geographical imagination, it is easy to ignore his fabrications and exaggerations based on travelers' tales and hearsay (Smethurst, 'Travel writing—Marco Polo and Sir John Mandeville,' 2001).
4. The *Z Codex* appeared nearly a century after Marco Polo. It's full title is the *Zelada Latin Text*. It was discovered in 1932, lying unnoticed in the library of Cardinal Francisco Xavier de Zelada at Rome, based on an original in the Cathedral Library of Toledo (Spain). It was brought to public attention in 1935 by A.C. Moule and Paul Pelliot as volume II of their edition of Marco Polo's *Description of the World*. To big dog aficionados the famous and misunderstood section appears in the description of "Thebet": "*Habent insuper magnos canos molosos qui sunt magnitudinis asinorum optimi ad feras*

*quaslibet capiendum & abiles sunt ad capiendos boues siluestres. Boues sunt maximi & feroces & multi sunt ibi... Prouincia thebet est subdita dominio magni can*" (p.xxix). In English: "They have many large molosos dogs which are as large as asses. They are excellent at capturing game and are able to catch wild cattle... The province of Thebet is under the dominion of the Great Can [Khan]."

Note that the Latin *molosso* refers to the region of Molossia in northwestern Greece, known for its large dogs, and that *asinorum* is derived from Latin *asinus*, the small local ass or burro. For more on the importance of the Z Codex to Polo studies, see Dana B. Durand (1939).

Note that since the 14th century, the Greek *molloso* dog has taken on far greater meaning among dog fanciers than it had then. The comparative size issue is discussed later in this book.

5. *Marco Polo*, Komroff translation (1926, pp.130-131).
6. The distinction here between physical and ethnographic Tibet is an important one, first raised by Sir Charles Bell in his authoritative books published in the 1920s and '30s. Tibetan culture, language lifestyles, and dogs, for example, have spread far beyond its physical or political boundaries to include adjacent high altitude parts of Western China at the east and the Himalayas at the south and southwest.
7. In her book *Tibetan Mastiff*, Juliette Cunliffe (with S. Elworthy) makes the same mistake as Rohrer and Flamholtz by stating that Polo "first encountered such dogs in China's Szechuan province..." (2007, p.10).
8. Rohrer and Flamholtz make a telling observation (1989, p.28) that "Marco Polo's stories of the Tibetan dog, first set in print in the 1300's, are remarkably similar to accounts penned in this century." And why not? Much that has been penned about them in this century repeats the most popular mis-translations of Marco Polo's account.
9. It was common during the Han Dynasty to place ceramic 'spirit goods' in tombs, in this case a glazed clay sculpture of a powerful but stout looking dog, to stand guard in the afterlife. Some may think the tomb dog *looks like* a Tibetan Mastiff, but *is it?* We can only speculate. There is no proof. But, look—the ears are wrong!
10. On the gift of the dog Ngao ("of Tibetan breeding") to the Chinese Emperor in 1121 BCE (the first year of the Zhou Dynasty) see Robert Leighton, *The New Book of the Dog* (1907), and Paul Strang and James Giffen, *The Complete Great Pyrenees* (1981). The Zhou originated among the nomadic Xirong and Rongdi peoples of western China, bordering Gansu Province adjacent to Tibet. Perhaps the earliest account of this gift dog in Western writings is in M. Pauthier, *Livre de Marco Polo (The Life of Marco Polo)* (1805). Citing Pauthier, Max Siber took up the story in *Der Tibethund* (1897), which comes down to us in *The Venerable Tibetan Mastiff*, Cathy J. Flamholtz's 1995 translation of Siber's book. The relevant passage reads: "In the year 1121 before the beginning of our chronology, a people named Liu living in the west of China sent a dog of Tibetan breeding called the 'Ngao' to Wou-Wang, the Emperor of China. According to Chinese historians of Chou-King, this dog was four feet tall and trained to track down human beings, as was often the case

at that time in western India" (Flamholtz translation, 1987, pp.52-53).

Adolf Krassnig, in an article entitled 'Do Khyi—mythos und wirklichkeit—killing legends!' (2006) throws cold water on the story when he writes: "The beauty of Chinese sources is that there is hardly anybody who can scrutinize them." He then asks, rhetorically, "why these 'exotic' sources are so willingly cited by people who cannot distinguish Chinese characters from crow feet!"

11. Many China travelers before the Polos are described by Hugh Murray in the beginning of his translation of *The Travels of Marco Polo* (1852). Some scholars question *if Polo really visited China*. See Frances Wood, *Did Marco Polo Go to China?* (1995); also Adolf Krassnig (2006, quoting Dietmar Henze at http://de.wikipedia.org/wiki/Marco_Polo, in German), and the brief discussion in Frances Wood, *The Silk Road: Two Thousand Years in the Heart of Asia* (2002). Skeptics aside, Stephen G. Haw has written the definitive narrative in favor of Polo reaching China in *Marco Polo in China*. Haw points out, however, that the Tibetan country that Polo claims to have penetrated to a distance of five days' walk was not Tibet at all, but parts of western China far to the east of the high plateau and at a considerably lower elevation (Haw 2006:99).

For further study of early European travelers to the Far East, see Henry Yule, *Cathay and the Way Thither* (1866); Sir Charles Beazley *Dawn of Modern Geography* (1897); Willem van Ruysbroek, *The Journey of William of Rubruck...* (1900 [13th century]); Manuel Komroff, *Contemporaries of Marco Polo* (1928); Christopher Dawson, *The Mongol Mission* (1955); Nigel Cameron, *Barbarians and Mandarins* (1989); John Larner, *Marco Polo and the Discovery of the World* (1999); and Laurence Bergreen, *Marco Polo: From Venice to Xanadu* (2007). The detailed 'Exploration in the Medieval period' (with extended bibliography) is also useful, at www.win.tue.nl/~engels/discovery/medieval.html.

About Tibetan dogs on the Silk Road, there is vague reference to a Tibetan mastiff on a mural by the renowned Tang Dynasty painter Yan Liben (*ca*. 600–673 CE) in *The Silk Road* by Frances Wood (p.77; after Edward G. Schafer, *The Golden Peaches of Samarkand: A Study of T'ang Exotics*, 1963). See also Tonia Eckfeld, *Imperial Tombs in Tang China, 618–907: The Politics of Paradise* (2005). The Yan Liben mural is entitled 'Emperor Taizong in a Sedan Chair Greeting Three Envoys from Tibet,' a copy of which is in the Palace Museum, Beijing.

12. The description of Battuta as "one of the greatest travelers" is by Jennifer Speake in her *Literature of Travel and Exploration* (2003, p.577). For mention of the dogs that the Chinese ate, see H.A.R. Gibb, *Travels* (of Ibn Battuta) (1994).

13 Four centuries after Polo the Jesuit Father Ippolito Desideri visited Tibet (1712–27) and described the big dogs he met. Compared with Polo's writings, Desideri's description is clearly based on firsthand experience (in *An Account of Tibet*, 1932, p.126; and below, in Chapter 2). The Greek historian Megasthenes (350–290 BCE) also wrote (vaguely) about what sounds like Tibet, though it is unlikely that he actually went there.

14. A number of sources were used to inform the discussion, including Thomas Wright, *The Travels of Marco Polo, a Venetian* (1880); Paul Smethurst,

'Travel writing: Writing the East' (2001); Mary Campbell, *The Witness and the Other World* (1988); Stephen Greenblatt, *Marvelous Possessions* (1991); and Italo Calvino, *Invisible Cities* (1974). Other sources are noted in the text.

15. From Agatha Christie's *The Body in the Library* (1970).

16. Doug Brown speaks of refraction "through the prism of translation" in 'The Bible delusion,' reviewing Bart Ehrman's *Misquoting Jesus* (2005).

17. Note that the term 'donkey' did not enter the English language until the late 18th century. To Marco Polo it was an ass (Italian *asino*, Latin *asinus*), referring to a rather small burro. The relatively recently invented term 'donkey' appears only in translations, many centuries after Polo. Etymologists tell us that 'donkey' is probably derived from the dull brown color *dun* + the diminutive *–ki*; literally, a small dun-colored critter.

18. Rohrer and Flamholtz (1989, p.10) also consider "mastiff" to be a misnomer, similar to the erroneous classification of the Tibetan terrier as a terrier and the Tibetan spaniel as a spaniel. "The Tibetan Mastiff has likewise suffered the same misclassification," they write. "By virtue of his size alone, the 'mastiff' label was hung on the large dog of Tibet. We must remember, however, that in earlier days, the word 'mastiff' was a much less specific appellation than it is today. Western explorers, catching their first glimpse of the powerful dogs of Tibet, doubtless, were anxious to explain the dogs in terms that their readers would understand. And so, they referred to the dogs as 'mastiffs,' knowing that the reader would immediately envision a large, powerful dog... The name stuck and is with us still."

19. See Wright *The Travels of Marco Polo, a Venetian* (1880, p.287, n.12; also pp.268-69); M.B. Wynn, *The History of the Mastiff* (1886, p.18); and http://nmbe0.unibe.ch/deutsch/531_5_1_4.html (by anonymous; emphasis added). In an interesting bit of terminological invention (now forgotten) Wynn (in Chapter 2, 'The Mastiff Type') classified the Thibet mastiff as "Mastivus Thibetanus."

20. One Internet writer, however, has turned this skepticism on its head by stating that it is absurd "to believe the quantum leap of assuming that Marco Polo just happened upon [in Tibet] the smallest of miniature Mediterranean donkeys" (kesangcamp.com). Of course, if he didn't go to Tibet then he didn't "happen upon" any kind of 'donkeys' (or asses), tame or wild, there.

My point is that Polo most likely chose a comparison that made sense *to his European readers*. Thus, he chose to equate the dog's size to that of a Mediterranean ass *because that's an animal with which his European readers were familiar*. Comparing them to a large wild Tibetan ass—an animal unknown to his readers—seems far more bizarre.

Polo (in translation) also described his "lions" to look like tigers. He wrote that "The Emperor... hath also several great Lions, bigger than those of Babylonia, beasts whose skins are coloured in the most beautiful way, *being striped all along the sides with black, red, and white*" (in the Yule-Cordier edition, 1903, v.2, p.397; emphasis added). In an interesting footnote earlier in the same edition (1903, p.93), Yule speculates that if Polo thought in the Persian language (and there is considerable evidence that he did), then the ambiguity inherent in the Persian term *sher* (meaning both 'lion' and 'tiger') may be the cause of the confusion.

21. Lions are not native to China, though their images, especially statuary, commonly appear all across China, traditionally placed as if to guard the entrances to homes, palaces, religious buildings and shrines (the famous 'temple lions'). Today they are even found at the entrance to banks and restaurants. There is a belief that they fend off evil and protect truth. (Now, *that's* appropriate!)

22. There are many other versions, editions and translations of Marco Polo's *Travels*, most relying extensively on previous translations (a type of translator inbreeding). They are listed in Smethurst's 'Introduction' (2005, pp. xxvii-xli) and Bergreen's 'Notes on Sources' (2007, p.367-81). See also Durand (1939). Many of them are listed in References at the end of this book.

23. The unconvincing assertion that there were 30,000 Tibetan mastiffs in Kublai Khan's army is by Li Qian in *China's Tibetan Mastiff* (2006, p.21); it appears again, online, in the anonymous article 'Tibetan Mastiff' at http://hi.tibetwindow.com. (Perhaps the same Li Qian is the anonymous author?)

24. Elsewhere in this book (Chapter 3) Alton C. Byers describes Tibetan mastiffs that help with hunting in the high Himalayan region of Dolpa. Their being "accomplices" in the hunt is from Ekvall (1963, p.166); a similar description appears in M.C. Goldstein and C. Beall, *Nomads of Western Tibet* (1990).

25. Yule (1903, p.126).

26. Juliette Cunliffe has perpetuated the myth. In her 2007 book *Tibetan Mastiff* (with Susan Elworthy), the following appears in a box entitled 'Believe It or Not' (we don't): "It is believed that Marco Polo was the first Westerner to own a Tibetan Mastiff. So impressed was he by the breed that he had one of his own for protection on his journeys. This dog accompanied him on his homeward-bound journey, though it is not thought likely that the dog could ever have reached Polo's eventual destination in Italy" (p.12).

27. The source of the article describing the Tibetan mastiff "as sturdy as black bear" is a news release from the China Tibet Information Center (CTIC) dated January 2005 (online at www.tibetinfor.com.cn/English/). Such exaggerations seem to appear often in Chinese sources on the big dogs. One states that "He [Marco Polo] says that their lips were raised upward and there were deep wrinkles around their eyes." But, did he say that? *Where?* The source of this spurious quote is Li Qian in *China's Tibetan Mastiff* (2006, p.24). Li Qian's book supports the current trend towards overstatement about Tibetan mastiffs in print and gigantism in their breeding.

    The account of killing "three wolves" and of seeing "a two-meter long tiger-like Tibetan mastiff" is from the unsigned article from China noted earlier, entitled 'Tibetan Mastiff,' June 14, 2007, online at http://hi.tibetwindow.com.

28. Henry Yule (1903, v.2, p.49n.).

29. Stephen Haw (2006:99). Laurence Bergreen, in his book *Marco Polo: From Venice to Xanadu* (2007) agrees that Polo traveled widely in and near China,

but that he did not reach Tibet. He notes, for example, how difficult it is to interpret Polo's descriptions in modern geographical terms and that, at best, Polo may have visited Yunnan in southern China, and perhaps even what we know today as Vietnam and Burma, but not Tibet (2007, p.173).

30. Gavin Menzies describes the systematic erasure of knowledge about the Outside by Chinese leaders in the early 15th century, in his popular book *1421: The Year China Discovered America* (2003).

31. That Marco Polo was known as *"Messer Milione"* for "talking big" is noted by Paul Smethurst in his Introduction to *The Travels of Marco Polo* (2005, pp. xxvii-xli). Apparently, Polo also endured the nickname "*Messer Mentitore*" ('Mr. Liar') by his contemporaries who did not believe his stories, nor that he had gone to China (Krassnig 2006).

32. Paul Smethurst (2001) considers Polo's writing far less exaggerated, for example, than that of Sir John Mandeville, his 14th century near contemporary. The description of how Polo entertained his readers is from Smethurst (2005).

33. The complete comment by Yule: "It is a great book of puzzles, whilst our confidence in the man's veracity is such that we feel certain every puzzle has a solution" (1903, v.1, p.1).

34. Paul Smethurst confirmed the lack of originals to me in a personal communication (2006). He wrote that "you will not find anything in the manuscripts that has not found its way into the modern editions, which are usually a sum of existing manuscripts." In his 'Introduction' to the 2005 Barnes & Noble Library of Essential Reading Series edition, Smethurst writes: "The original manuscript of *The Travels of Marco Polo* gave rise to numerous copies, translations and editions. There are about eighty-five surviving manuscripts in different languages preserved in museums and libraries around the world. Most modern editions are based on either the medieval French edition (the Paris manuscript or F text) written in the early 14th century, or the much fuller Italian version by Ramusio (1559), which is generally regarded as the first printed edition. The F text is usually considered to be closest to the original, and it is generally accepted that it is written in the same language as the original, i.e. medieval French. Ramusio claims that his edition is based on a Latin text of great antiquity (it is certainly partly based on the Latin translation made in Polo's lifetime by Francesco Pipino, a Dominican of Bologna, who claims to have worked from an Italian, not French, manuscript). This may contain original material omitted from F, or possibly contains additions made by Polo himself in later life. The origin of Ramusio's 'additions' is a matter of conjecture, but it is conceivable that some of these came from a fuller translation of the original manuscript..."
See also the so-called *Z Codex*, note 4, above.

35. Italo Calvino's post-modernist book, in which he invents the conversation with Marco Polo, is entitled *Invisible Cities* (1974).

36. In *The Life of Samuel Johnson, LL.D.* Boswell wrote (based on a conversation with Samuel Johnson) that: "It is more from carelessness about truth than from intentional lying, that there is so much falsehood in the world" (1900, p.256).

37. The Canine Information Library and Dogbreeds.bulldoginformation.com (2003–2009) states that "No other group of dog breeds seems as difficult to define as the Molosser dogs" and that their true origin is not known. "They are one of the few breeds that appear in reality as well as in legends, adding to the difficulty of separating truth from fantasy."
38. M.B. Wynn, *The History of the Mastiff* (1886, Chapter II: The Mastiff Type).
39. Kåre Konradsen, 'Molosserworld's origin & history of the Molossers' (2002).
40. Yes, even the Bible is not spared the effects of translator error and misinterpretation. There are interesting parallels between versions of the Bible and Polo's *Travels*, including lost originals and errors in translation—so much so that we cannot be sure of 'the gospel truth' in either case. Regarding the Bible, Doug Brown writes that "So many hands have altered and edited the now lost originals... [that] we will never know for sure what [they] said." There are many versions and they have been so "accidentally and intentionally altered over the centuries by other men with agendas of their own" that some of the original intent may be there, but how sure can we be. He concludes that "In many respects, the Bible was the world's first Wikipedia article." Marco Polo's *Travels* has suffered the same fate; i.e., it's the Marco Polo Effect all over again.
41. It may well be that my own (few) comments about Marco Polo in earlier publications have also been misinterpreted by others. In my article entitled 'The Tibetan mastiff: Canine sentinels of the range' I wrote that "Marco Polo thought they were as large as donkeys" (*Rangelands*, 1983, p.173). One American breeder has informed me that the note on his website stating that Marco Polo described the Tibetan dog as being "tall as a donkey with a voice like a lion" came from my 1987 article in *Dog World*, 'The mysterious mastiffs of Tibet: Finding the dogs of Kesang Camp' (Ron Bombliss, personal communication, 2006; his kennel's name, 'Kesang Camp,' comes from the same article). On checking, however, the "voice like a lion" quotation does *not* appear in my article, and neither of us has been able to find the original source.

    Similarly, captions to photographs of big Himalayan dogs and written descriptions of them attributed to me in earlier publications have also been misinterpreted. One correspondent has recently pointed out, in relation to the ongoing debate in Tibetan mastiff circles about appropriate and acceptable coat colors, that I identified a cream/light gold colored dog in one of my photographs from northern Nepal as a Tibetan mastiff. Yes I did, but I have also pointed out that (a) white, cream or light gold colored dogs are *not* common and that (b) *that* particular fine-looking dog should probably be called a 'Himalayan mountain dog.' I clearly imply this where I refer to such dogs as "'near'-TMs" and as "Himalayan shepherds" in a note 'On varieties of TM dogs' published in the *ATMA Gazette* in 1987.

    For further discussion of the genetics of coat color, see Charles W. Radcliffe and Matthew J. Taylor, 'Coat color genetics in Tibetan Mastiffs,' reprinted here in Appendix 4, pp.221-6, from the *TMCA Sentinel* (n.4, 2001; revised 2008). Radcliffe and Taylor point out that the color "cream" is not recognized in all international Tibetan mastiff breed standards and, in their revision they favor the term "light gold."

42. A contemporary example (from the Internet) of blatantly uncritical repetition appears in a written reference to the recent reclassification of dog from *Canis familiarus* to *Canis lupus familiarus*. In a Google search I found 16 exact copies of this garbled statement: "The designation was presented by a group of scientists in the article 'The naming of Wild Animals and their Domestic Deriatives' [sic] in the Journal of Archeological Science # 31 in 2004." That misspelling of Derivatives as "Deriatives" is repeated over and again on Internet dog sites. It is amazing that so many folks neglected to question and correct the mistake. The original article—with Derivatives spelled correctly—is by Anthea Gentry, *et al*, on pages 645 to 651 of the journal. As Adolf Krassnig (2006) has put it: "everyone copies from the other one!"

43. The description of Kirkpatrick's writing as "antiquated and cumbrous in form" is from William Blackwood and Sons (1852, p.88). All quotations about Orfeur Cavenagh and Thomas Smith abusing Kirkpatrick's writings are from the same source (pp.86-89).

44. Aguirre's observations are quoted in 'The truth about dogs' by Stephen Budiansky (1999).

45. The quotation was attributed to Mark Twain by Rudyard Kipling in 1890 (see Ralph Keyes, *The Quote Verifier*, 2006, p.232); emphasis added.

## Chapter 2

1. L. Austine Waddell, *Lhasa and Its Mysteries* (1906, p.89).

2. Ippolito Desideri, *An Account of Tibet* (1932, p.126, from his travels of 1712 to 1727).

3. George Bogle's account is in C.R. Markham's *Narratives of the Mission of George Bogle to Tibet...* (1999 [1876], p.68). Samuel Turner's encounters are described in *An Account of An Embassy to the Court of the Teshoo Lama in Tibet* (1800, Part I, p.155).

4. Captain William John Gill, *The River of the Golden Sand* (1880, pp.247-8).

5. All references to Bela Széchenyí are from a chapter on 'Die Tibetaner Dogge' by Richard Strebel, *Die Deutschen Hunde* (*The German Dog*) (1904), quoting Széchenyí's original writing in *Die Neue Deutsche Jagdzeitung* (*The New German Hunter's Magazine*) (1882, pp.183-204). My source is Christina McFarland's 1990 translation. (By "buffalo herd" the author undoubtedly means yaks.) Széchenyí's dogs are also described by Ludwig Beckmann in *Leipziger Illustrierte Zeitung* (*Leipzig Illustrated Newspaper*) (1880); and see Max Siber (1897/Cathy J. Flamholtz, *The Tibetan Dog*, 1995).

6. Gustav Kreitner, quoted in McFarland (1990, p.7). See also Kreitner, *Im Fernen Osten: Reisen des Grafen Bela Széchenyí* (*In the Far East: Travels with Count Bela Széchenyí*) (1881).

7. Traveling to and through Tibet today is relatively easy and safe. The modern traveler can ride on well-constructed highways (part of the time) from the Nepal border east to Lhasa, or fly by jet to Lhasa (then hire a car, with driver and guide), or ride the new Qinghai-Lhasa Railroad (highest in the world). The train trip from Beijing to Lhasa via Golmud, Qinghai Province, takes only forty-eight hours. Because of the altitudes involved,

the passenger carriages feature aircraft-like pressurization including oxygen masks.

Harrer and Aufschnaiter entered Tibet in 1944 after a daring escape from a British internment camp in India where, at the outbreak of World War II, they were incarcerated as Nazi sympathizers. They were in India on a German expedition to Nanga Parbat, for which Aufschnaiter was the leader. See Heinrich Harrer, *Seven Years in Tibet* (1953), and *Peter Aufschnaiter's Eight Years in Tibet* edited by Martin Brauen (2002).

8. According to the *Guinness Book of World Records*, the highest town in the world is (since 1955) Wenzhuan, China, at 16,730ft (5488m), on the Qinghai-Lhasa road. Lhasa is second. By comparison, La Paz, Bolivia, is only 11,913ft (3631m) above sea level. There are some villages on the Nepal/Tibet border as high as 14,000 ft.

9. António Andrade's account appears in L'Abbe Huc's *Christianity in China, Tartary and Thibet* (1897).

10. The visit to Lhasa in 1661 by D'Orville and Grueber is well documented, but they may not have been the first Europeans to enter the city. The Franciscan friar, Ordoric (Oderico) of Pordenone, is known to have visited Tibet over three centuries earlier, in 1327, but because he did not leave a written account of his journey the claim is disputed; see Bill Buxton, *Dramatis Personae* (2005).

11. The Joseph Dalton Hooker quotations are from *Himalayan Journals* (1854, vol.I, pp.189, 191, 193).

12. A.E. Brehm, *The Animals of the World* (1895, p.221).

13. Sarat Chandra Das, *A Journey to Lhasa and Central Tibet* (1902, p.131).

14. See Hamilton Bower's *Diary of a Journey Across Tibet* (1893); Henry Savage Landor's *In the Forbidden Land* (1898) and *An Explorer's Adventures in Tibet* (1910); and Gabriel Bonvalot's *Across Thibet* (1891).

15. The article by Robert Ekvall, on the 'Role of the dog in Tibetan nomadic society' (1963), is referred to often in this book and in writings on Tibetan mastiffs by many others. Because it is one of the best descriptions ever written and is so insightful about the place of dogs in Tibetan culture, it is reprinted here as Appendix 2, p.208. One of the few other studies of dogs in Tibetan culture is by Siegbert Hummel, 'Der Hund in der religiösen Vorstellungswelt des Tibeters' ('The dog in the Tibetan religious worldview,' 1958/1961).

16. See Giuseppe Tucci and E. Ghersi, *Secrets of Tibet* (1935) and Tucci's *Tibet: Land of Snows* (1967) and *Transhimalaya* (1973); Colonel Sydney Burrard, *Records of the Survey of India* (Part 8: *Exploration in Tibet and Neighbouring Nations 1879–1892*), and S. Burrard and Horace Hayden, *A Sketch of the Geography and Geology of the Himalaya Mountains and Tibet* (1930 [1908–09]). L. Austine Waddell wrote *Lhasa and Its Mysteries* (1906), *Among the Himalayas* (1906) and *The Buddhism of Tibet or Lamaism* (1895). Frank Kingdon-Ward wrote *On the Road to Tibet* (1910), *Land of the Blue Poppy* (1913), *Mystery Rivers of Tibet* (1930) and *A Plant Hunter in Tibet* (1934). (A good introduction to Kingdon-Ward's work is his *Himalayan Enchantment*, an anthology of his writings published in 1990.)

17. L. Austine Waddell, *The Buddhism of Tibet or Lamaism* (1895, p.406).
18. Alexandra David-Neel, *My Journey to Lhasa* (1927, p.188).
19. On women adventurers in the Himalaya and Tibet, see Luree Miller, *On Top of the World: Five Women Explorers in Tibet* (1976); along with discussions in William Carey's *Travel and Adventure in Tibet, Including the Diary of Miss Annie R. Taylor's Journey* (1983); and Peter Hopkirk's *Foreign Devils on the Silk Road* (1980) and *Trespassers on the Roof of the World* (1982). See also Isabella L. Bird, *Among the Tibetans* (2004); Alexandra David-Neel, *My Journey to Lhasa* (1927) and *Tibetan Journey* (1926); Nina Mazuchelli, *The Indian Alps and How We Crossed Them* (1876); Susie Carson Rijnhart, *With the Tibetans in Tent and Temple* (1904); and Fanny Bullock Workman, *Ice-Bound Heights of the Mustagh* (1908) and *Peaks and Glaciers of Nun Kun* (1909). Ms Workman also co-authored, with her husband William Hunter Workman, *The Ice World of Himalaya* (1900) and *Two Summers in the Ice-Wilds of Eastern Karakoram* (1917).
20. All quotations from Irma Bailey are from her article 'Dogs from the roof of the world' (1937; reprinted here in Appendix 1, p.203). Before Irma joined her diplomat husband F.M. (Eric) Bailey in Lhasa he had been a renowned and clever spy. At one point in his career, while playing the 'Great Game' in central Asia, the well-disguised Bailey, operating under a pseudonym, was hired by the Russians to track down an especially notorious British spy named Eric Bailey! See *Mission to Tashkent* (1946). See also Bailey's *China-Tibet-Assam: A Journey* (1945 [1911]) and *No Passport to Tibet* (1957). Eric Bailey is also credited with discovering the Himalayan blue poppy (*Meconopsis baileyi*); see Michael Mcrae's *The Siege of Shangri-La: The Quest for Tibet's Sacred Hidden Paradise* (2002).
21. On their purported immense size, see Li Qian, *China's Tibetan Mastiff* (2005, p.35) along with photographs of highly exaggerated facial features (p.25). The gigantism that some breeders seek these days is frequently seen in catalogs and brochures about contemporary Tibetan mastiffs in China.
22. See Thomas Hungerford Holdich, *Tibet, the Mysterious* (1904, p.3). The account of Frank Kingdon-Ward's encounter with the menacing dogs is from Tom Christopher's *In the Land of the Blue Poppies* (2002, p.198).
23. Giuseppe Tucci, *Tibet: Land of Snows* (1967, p.133).
24. J.B.L. Noel, *Through Tibet to Everest* (1927, p.52).
25. The description of the bark "most peculiar: not sharp and crisp" is from R.F. Johnston, *From Peking to Mandalay* (1908, p.162), and the allusion to a copper gong is from Charles Bell, *Portrait of the Dalai Lama* (1946) and repeated in Douglas Oliff's *The Ultimate Book of Mastiff Breeds*, (1999, pp.166 and 174).
26. Lawrence Swan, *Tales of the Himalaya* (2000, pp.99-100).
27. J.B.L. Noel, *Through Tibet to Everest* (1927, p.248-9).
28. William Moorcroft and George Trebeck, *Travels in the Himalayan Provinces* (1841, p.261). Amarnath cave is located in the northwestern Indian state of Jammu and Kashmir, at an elevation of 13,000 feet (the average elevation of Tibet, across the border a few miles north). Amarnath is famous for a

pillar of ice, a giant stalagmite, thought to resemble a phallus or lingam, the symbol of Lord Shiva. This (super)natural wonder annually draws thousands of Hindu pilgrims each summer, who come to worship it.

29. The observations by Andrew Wilson on these pages are from his book, *The Abode of Snow* (1886, pp.135, 172, 226-9, and 273n).

30. By *"mustang"* Wilson apparently meant the Himalayan kingdom of Mustang, a culturally Tibetan district in northern Nepal long renowned for its big red dogs.

31. All quotations of Hedin are from his books *Through Asia* (1899, pp.106 and 144) and *My Life as an Explorer* (1925, 496-525 *passim*).

32. Some historians date the Great Game as starting earlier and lasting longer. See the 'Great Game Timeline' online at www.oxuscom.com/greatgame.htm. The 'Great Game' has inspired many books on the subject, including Peter Hopkirk's *Setting the East Ablaze* (1984), *The Great Game* (1992) and *On Secret Service East of Constantinople* (1995); K.E. Meyer and S.B. Brysac's *Tournament of Shadows: The Great Game and Race for Empire in Central Asia* (1999); Tatiana Shaumian's *Tibet: The Great Game and Tsarist Russia* (2000); and Sir Martin Ewans' *The Great Game: Britain and Russia in Central Asia* (2003).

    For a highly readable biography of Colonel Francis Younghusband, see Patrick French's *Younghusband: The Last Great Imperial Adventurer* (1994). See also Charles Allen's *Duel in the Snows: The True Story of the Younghusband Mission to Lhasa* (2004).

    Looking back on the story of Kim and the Great Game, Peter Hopkirk has written a contemporary travelogue tracing Kim's adventures, *Quest for Kim* (2001). Even Sherlock Holmes got involved in novelizing the Great Game, in the company of at least one main character from Rudyard Kipling's *Kim* (1901), in Jamyang Norbu's prize-winning *The Mandala of Sherlock Holmes* (1999). But, alas, neither Tibetan mastiffs nor Holmes' (that is, Arthur Conan Doyle's) Baskerville hound are mentioned in Norbu's otherwise remarkable mystery novel.

33. Charles Allen, *Duel in the Snow* (2004, p.294), quoting Mark Synge (aka Powell Millington), *To Lhassa at Last* (1905, p.195). Place name spellings, like Lhasa (for 'Lhassa') and Tibet (for 'Thibet') were not settled until later in the century.

34. It has been reported that a second dog named 'Lhasa' was imported to England by someone from the Younghusband mission, but little is known of this particular dog.

35. Juliette Cunliffe, 'Early history of the Tibetan Mastiff' (2002, pp.31-32).

36. H.W. Bush, 'The dogs of Tibet,' *The Kennel Encyclopedia* (1908).

37. Colonel Sir George Everest (knighted in 1861) insisted that his name, and the mountain named after him, be pronounced '*Eve*-rest,' despite the more popular pronunciation of '*Ever*-est.' The naming of Mount Everest as Sagarmatha by the Nepalese is described in 'An autobiography of Sagarmatha's height' by Tirtha Bahadur Pradhanang (n.d.).

38. L. Austine Waddell, *Among the Himalayas* (1906, p.345). See also John Keay, *The Great Arc: The Dramatic Tale of How India was Mapped and Everest was Named* (2000).

The early ban on traveling in Tibet was routinely broken. The British colonial authorities in India were especially adept at infiltrating Tibet with secret emissaries, mostly notably by specially trained undercover Indian surveyor-spies whom they sent to map the country. Each of these men, with names like Kishen Singh, Nain Singh and Kintup, was known euphemistically as a *pundit*, a 'learned man' in Sanskrit. The pundits carried sextants (well hidden) with which they made clandestine solar and stellar observations. They measured distances by counting their paces. They hid their findings and smuggled them out to their controllers in India inside innocuous prayer wheels that every pilgrim carried. Their story is told by Peter Hopkirk's *Trespassers on the Roof of the World* (1982), Derek Waller's *Pundits: British Exploration of Tibet and Central Asia* (1990), and Jules Stewart's *Spying for the Raj* (2006).

39. On the recently recalculation of the height of Mount Everest see NASA, 'New Height for Mt. Everest: 29,035 ft' (1999). That 1999 mapping expedition is also described in *Bradford Washburn: An Extraordinary Life* by Washburn and Lew Freedman (2005). Recent GPS readings indicate that Everest grows taller by about 0.16 inches, or 4 mm, each year.

40. Sir Francis Younghusband, quoted in *Mount Everest: The Reconnaissance 1921*, by C.K. Howard-Bury (1922: Introduction).

41. One of the most comprehensive books on the history of Mount Everest conquests is by Walt Unsworth, *Everest* (2nd edition, 1989). See also Maurice Isseman and Stewart Weaver, *Fallen Giants: A History of Himalayan Mountaineering* (2008).

42. The story of Rawlings and other Alpine Club members planning for Mt. Everest is told by T.L. Blakeney in 'The First Steps Towards Mount Everest' (1971).

43. The various excuses for failure are noted by Jon E. Lewis in *The Mammoth Book of Eyewitness Everest* (2003, p.xviv).

44. On the mystery surrounding Mallory and Irvine's ill-fated attempt on the summit in 1924, see J. Hemmleb *et al*, *Ghosts of Everest* (1999); D. Breashears and A. Salkeld, *Last Climb* (1999); C. Anker and D. Roberts *The Lost Explorer* (2001); and Reinhold Messner, *The Second Death of George Mallory* (2002). Hillary's famous remark at the conclusion of his summit success is from 'Quotes from Everest' at www.mnteverest.net/quote.html.

45. The story of Tendrup on Mt. Everest is from Hugh Ruttledge, *Everest: The Unfinished Adventure* (1937, pp.62-63, 139-40 and 145). The "dump" is where the expedition equipment and food were kept and Tendrup stood guard.

46. The description of Police-ie as "one-eyed and suspicious" is by Hugh Ruttledge in *Everest 1933* (1934, p.74). The bulk of the expedition's dog story, however, is told by E.O. Shebbeare later in the same book (Part III: Transport, pp.273-4).

47. Ekai Kawaguchi's travels are described in Scott Berry's *A Stranger in Tibet: The Adventures of a Zen Monk* (1990, p.100), and in the monk's own *Three Years in Tibet* (1909, p.80 and 199).

48. Peter Matthiessen, *The Snow Leopard* (1978, p.50).
49. John Hedley, *Tramps in Dark Mongolia* (1910, p.245-6).
50. Snellgrove's travels with Pasang Sherpa and the pup Nying-kar are told in *Himalayan Pilgrimage* (1961, pp.xiii-iv, 2-3 and 225-74 *passim*) and in *Asian Commitment* (2000, p.156). (In the first of these, Snellgrove misspelled the district of Gorkha, as 'Gurkha,' the famous Nepalese soldiers.)

## Chapter 3

1. 'Landrace' is defined as a local or regional type that has sprung up and become relatively uniform by virtue of local selection for a specific purpose or function adapted to its natural and cultural environment of origin. (The concept is further discussed in Chapter 7.)
2. All quotations from the French expedition are from *Annapurna* by Maurice Herzog (1952).
3. The Swiss climb and the crash of the 'Yeti' are described in *The Ascent of Dhaulagiri* by Max Eiselin (1961) and at www.flymicro.com/everest (History>Emil Wick), www.porter-yeti.org, and www.pc-6.com/history/337.htm.
4. A few years after our reconnaissance several other Peace Corps Volunteers succeeded in climbing Tukché Peak by the north face, an accomplishment that was neither sanctioned nor registered, and that remains undocumented and unrecognized (on official rosters). The first 'official' ascent of Tukché Peak, also by the north face, was by a Swiss team (the Schweizerische Nepal-Expedition Thakkhola) in May 1969 (after the Peace Corps climb). The leader, Georges Hartmann, later described Tukché Peak as "one of the most beautiful peaks he had ever seen in all his experience climbing in the Andes, Hindu Kush and Alps" (personal communication from Elizabeth Hawley, January 2006; see also Hawley 2004).
5. The description of a dog's 'rush' behavior is from an article on 'Livestock protection dogs' by Andrews and Kidd (1982).
6. The Count's story is quoted here from Christina McFarland's 1990 translation of Richard Strebel's *Die Deutschen Hunde* (1904, pp.183-204).
7. For more on the history of Mustang as part of western Tibet see Jackson 1976, 1978, 1984, and Vinding 1988, 1998. The observation about "Moostang" as "a place in upper Tibet" is from Colonel William Kirkpatrick's *An Account of the Kingdom of Nepaul* (1811) based on his travels in Nepal in 1793.
8. Though many partisans were from districts other than Kham in eastern Tibet, all members of the secret resistance army were often referred to by outsiders simply as Khampas ('men of Kham'). Their organization was called the 'National Volunteer Defense Army,' so as to include all districts of Tibet; but it was more popularly known as *Chushi Gandrug* in Tibetan.
9. The story of the Tibetan resistance movement and CIA involvement, including a brief account of the life of the leader, Wangdu (sometimes spelled Wangdi), and his operations in Nepal's Mustang District up to his death in 1974, is told by Mikel Dunham in *Buddha's Warriors* (2004, Chapter 10: 'Last Stand'). See also 'Tibet's Cold War: The CIA and the Chushi

Gangdrug Resistance, 1956–1974,' by Carole McGranahan (2006).

10. The "last strongholds of Tibetan culture" is a quotation from D. Nelson, *et al*, in *The Major Ecological Land Units and Their Watershed Condition in Nepal* (1980).

11. The description of the fierce dogs that "will attack any stranger" is from Thubten Jigme Norbu and Colin Turnbull, *Tibet* (1968. p.39). Robert Ekvall, also describes them as "noted for ferocity" in *Fields on the Hoof* (1968, p.40).

12. The quotation from Ekvall is from 'Role of the dog in Tibetan nomadic society' (1963, p.170). (See Appendix 2, p.208 in this book.)

13. Inter-regional warfare in Mustang is further discussed in Alton Byers, 'Resource management in the arid Himalaya' (1985). Michel Peissel describes the training of dogs by the Khampas in *Mustang: The Forbidden Kingdom* (1968, p.103-4).

14. Robert Ekvall, 'Role of the dog in Tibetan nomadic society' (1963, pp.165-9), reprinted here as Appendix 2, pp.208-17.

15. George Schaller, a wildlife zoologist, describes the high mountain predators of this region in his book *Mountain Monarchs* (1977). For more about reducing depredation through the greater use of trained dogs see the discussion in Rodney Jackson, *Threatened Wildlife* (1990) and Jackson, *et al*, 'Reducing livestock depredation in the Nepalese Himalaya' (1994).

16. Nepali terms like *gothalo* (for 'herder') are commonly used by Tibetan and Tibeto-Burman speaking mountain people in this region.

17. On the Tibetan predilection for hunting, see Ekvall 1963, 1964, and 1968. In his 1964 study (p.75) he writes that "The doctrinal ban on killing has left its mark on many aspects of Tibetan life, although often it would seem to be more honored in the breach than in the observance. It is not, of course, completely effective, for the Tibetans must eat. A large segment of the Tibetan population is nomadic and engaged in animal husbandry.
Their subsistence depends on livestock raising. The local production of grain is small; there is a lack of vegetable fibers and textile production is limited; and the harsh climate dictates the use of sheepskins and furs for clothing. Of necessity, the Tibetan must kill to live, and against that necessity, explicitly or implicitly, his creed says he should not kill. The conflict and the compromises or subterfuges that result affect much of Tibetan life."

Note that while the founding of Buddhism in South Asia dates to the 6th century BCE, it was only during the 19th century CE it was introduced and formally adopted in Tibet.

For a description of hunting in the nomadic subsistence economy of western Tibet, see Melvyn C. Goldstein and Cynthia Beall, *Nomads of Western Tibet* (1990). On hunting by Mongolian sheepdogs (similar to Tibetan mastiffs), see G. Yavorskaya, 'The Mongolian sheep-dog—the most ancient breed' (n.d.).

18. On the Tibetan hunting dog as a type of sighthound, see Juliette Cunliffe, 'Special breeds from "The Roof of the World" (2005).

19. M.C. Goldstein and C. Beall, *Nomads of Western Tibet* (1990, p.125).

## Chapter 4

1. The discussion of language and geography is from S.G. Burrard and H.H. Hayden, *A Sketch of the Geography and Geology of the Himalaya Mountains and Tibet* (1933, p.7).

2. Luis Vaz de Camões, *Os Lusiadas* (1572, canto VII, verse 17), translated by Yule and Burnell, in *Hobson-Jobson* (1903, p.436).

3. Henry Yule and A.C. Burnell in *Hobson-Jobson* (1886 [1903], p.414, with original diacritics and italics). By the "roving Tartar" of the "Himālăya" the authors meant the Tibetan people. In the 13th century Tartary was the name that Europeans gave to the greater Mongol Empire; i.e., from central Asia east of the Caspian Sea and Ural Mountains including Turkic and Mongol peoples, all the way to what is now Mongolia, Manchuria, Siberia, Turkestan, and sometimes Tibet (as here).

4. Thalar was one of a group of 22 ancient principalities of far western Nepal. After the Gorkha Conquest of Nepal by Prithvinarayan Shah (unified Nepal's first king) in 1768, Thalar was incorporated into the larger kingdom. Descendants of the Rajas of Thalar and neighboring Bajhang, Salyan and other old principalities, are related to Nepal's elite Shah and Rana families. For more on this period see John Whelpton, *A History of Nepal* (2005).

5. Reminiscences of Jay N. Singh to the author, 2005–07. The Byansi shepherds and traders in the story are an ethnic group indigenous to the far northwest corner of Nepal bordering western Tibet.

6. From my short article 'On the "ideal" Tibetan mastiff,' in *ATMA Gazette*, 1983, January/February, p. 3.

7. Canine parvovirus was discovered in the mid-1970s. The symptoms are diarrhea (often bloody), vomiting, dehydration, depression and weight loss due to the inability of the intestinal tract to absorb nutrients. The condition is seen most commonly in puppies six to fourteen weeks of age. Before a vaccine was developed, the mortality rate was almost 100 percent. Kalu was a rare survivor.

8. The scene at the airport where the baggage handlers were afraid to go near the dog crate reminds me of the early description of traveling with a Tibetan mastiff on the Indian railroad, told by Robert Leighton in *The New Book of the Dog* (1907, p. 515): "Bhotean's journey through India was an expensive one, as he had to have a carriage to himself. He effectually cleared the platform at all stations where we stopped, and where he was given exercise..."

9. I was happy to learn that Ch. Bathsheba, the Rampur Hound, won first prize at another dog show in India later that year. For a glimpse of India's unique and beautiful indigenous dogs, see www.dogsindia.com and *The Indian Dog* by W.V. Soman (1963). Indigenous dogs were also featured in Theodore Baskaran's illustrated article, 'India's Canine Heritage' in *Swagat*, the Indian Airlines magazine of May 1985.

10. This account of the Gurung funeral grounds, where Kalu's ashes are buried, is based on my own travels and research, and on Stan Mumford's *Himalayan Dialogue* (1989), Judi Pettigrew's 'Parallel geographies' (1999),

and Liesl Messerschmidt, *et al*'s *Stories and Customs of Manang* (2004).
11. Liesl Messerschmidt, *et al*, *Stories and Customs of Manang* (2004, p.52).
12. Judi Pettigrew, 'Parallel geographies' (1999).

## Chapter 5

1. Ann Rohrer and Cathy J. Flamholtz, *The Tibetan Mastiff* (1989, p.26).
2. Irma Bailey, 'Dogs from the roof of the world' (1937), reprinted in Appendix 1, p.203 of this book.
3. J. Taring, 'Tibetan dogs' (1981).
4. Dan Taylor considers the Mount Kailash region of western Tibet to be the main provenance of KyiApsos ('Meet the KyiApso,' 1994). The report from eastern Tibet is by Juliette Cunliffe (personal communication, 2006). Sightings in the mountains of western Nepal are by me and in the eastern mountains by Major Dan James of the UK (personal communication, 1988).

    For a glimpse of the landscape and culture of western Tibet, *Sacred Landscape and Pilgrimage in Tibet: In Search of the Lost Kingdom of Bön* by Geshe Gelek Jinpa, *et al* (2005) and M.C. Goldstein and C.M. Beal, *Nomads of Western Tibet* (1990).
5. The reports of "sightings" of KyiApsos in America prior to the 1970s is from Judy Steffel (personal communication, 2002).
6. P. Gardner, *The Irish Wolfhound* (1931, pp.144-5).
7. Hilary Jupp, in 'Captain Graham and the resuscitation of the breed' (2005); see also Jan den Hoed (2001).
8. Andrew Wilson, *The Abode of Snow* (1886, pp.168-9).
9. Hedy Nouc wrote of her visit to the 1981 Dhauladhar dog show in 'Himalayan Mastiff—A Show at 3000 Meters' (1981).
10. Christian Ehrich (personal communication, 1983).
11. Nouc (1981).
12. I didn't know at the time that my friends Mel Goldstein and Dan Taylor had taken several KyiApsos to the USA from northern Humla, a few dozen miles north of my own encounter with them. Unfortunately, their bitch died before she could be bred. See Taylor, 'Meet the KyiApso' (1994). Prof. Melvyn C. Goldstein is a Tibetologist who teaches at Case Western University in Cleveland, Ohio, and Daniel Taylor is an educator and founder of a non-governmental organization called Future Generations, which has projects in Tibet.
13. Ann Rohrer and Cathy J. Flamholtz, *The Tibetan Mastiff* (1989, p.25).
14. The description is based on personal communications with the American KyiApso breeders Dan Taylor and Judy Steffel, and from the official standard of the Tibetan KyiApso Club (TKC, 1995). The opinions I have expressed may not reflect those of other KyiApso fanciers.
15. Judy Steffel (personal communication, January 2006).
16. *Pashmina* is a type of wool commonly used for making high quality woolen shawls. The term derives from the Persian *pashm*, the fine woolen under fur of certain Tibetan animals, particularly goats; but Tibetan

mastiffs and KyiApsos also have this fine fur, which is occasionally saved and used to weave small carpets.

17. Dan Taylor (personal communication, 1988) and Major Dan James (personal communication, 1988).

18. Steffel (personal communication, 2006). A parallel example of smoothies is occasionally reported in litters of purebred Lhasa Apsos. In a short article entitled 'What is a 'Prapso'?' (1997), Juliette Cunliffe writes that "'Prapso', or occasionally 'Perhapso', is the name given to Lhasa Apsos which are purebred but to all outward appearances closely resemble Tibetan Spaniels. One or more can be produced in a litter and it is perfectly possible that both parents are fully coated." In the late 1960s, she says, it was estimated that six percent of Lhasa Apsos had smooth coats, but that the number has probably decreased since then. "The incidence of Prapsos is a complex one, particularly because the smooth factor is displayed so infrequently and is such an irregular pattern." In conclusion she advocates taking steps to assure that the untypical coat patterns are eradicated in future generations. "Clearly it is now not considered wise to breed from Prapsos…"

    The same advice regarding the culling of KyiApso smoothies was given by the Tibetan KyiApso Club to its members.

19. The description of Suptu's behavior is from Major James (personal communication, 1988).

20. The high-strung nature of some of these big dogs, both KyiApsos and Tibetan mastiffs, is legendary. When kept in a yard, they become agitated if they cannot see the front door where strangers enter. Their anxiety to break loose when confined is also well known. My standard Tibetan mastiff, Kalu, would become highly agitated when left alone at home or placed overnight in a boarding kennel. Once, he attempted to chew his way through a wooden fence. Another time he successfully chewed his way through a chain link fence. And, he once showed his immense displeasure at being boarded at a commercial kennel by devouring a fiberglass food dish.

21. As early as 1933, the distinction between the "Tibet dog" (Tibetan mastiff) and the "Great Dog of Tibet" (thought to be the KyiApso) was made in a German dog book by Werner Mut, as reported in the *TMCA Sentinel* (vol. 4, 2001) by Jan den Hoed, a Dutch Tibetan mastiff enthusiast. Hoed writes that "In 1933 Dr. Werner Mut described in his book the existence of two Tibetan breeds, the Tibet Dog (Mastiff) and the Great Dog of Tibet, with a description which was consistent with the photo in the book of Mrs. Gardner. Dr. Werner Mut categorizes both dogs along with dogs such as the Newfoundlander, the Kuvasz, the Komondor, the Bergamasco and the Maremma." (I have been unable either to contact Jan den Hoed or to find any bibliographic reference to a W. Mut vis-à-vis Tibetan dogs.)

    Hoed's reference to "the book of Mrs. Gardner," is to Phyllis Gardner's *The Irish Wolfhound* (1931). The "Great Dog of Tibet" was first identified and discussed by George Graham in his book of that title (1939 [1885]). See also Edmund Hogan, *The Irish Wolfdog* (1887) and Hilary Jupp, 'Captain Graham and the resuscitation of the breed' (2005).

22. The quote about sharing characteristics is from a *Cambridge Dictionary* definition of a biological 'type.'
23. Jay N. Singh (personal communication, 1988).
24. Steffel (TMCA listserv, 2002).
25. Dan Taylor's conviction that the KyiApso is a distinct and separate breed is stated in several of his publications; e.g., see Taylor (1994).
26. Steffel (personal communication, 2006; all quotations from Steffel in this section are from this source).

## Chapter 6

1. Originally a larger group was planned, but some Canadian friends dropped out fearing snowfall, and the author's wife took ill the first day and returned to Kathmandu. That left the author and Judy Steffel to complete the journey with our two Tibetan friends, Dorjé and Lodenla. We were a congenial foursome, for it turned out that we agreed on almost every point that came up in discussion about the big dogs based on personal research and experience, from their place in history and old writings, to their natural temperament and traditional functions and looks, to their place in contemporary Tibetan society and their probable future. Dorjé and Lodenla were especially helpful during our trip, and we thank them for their patience and knowledgeable assistance.
2. Dorjé, our guide and translator, told Judy of a village that suffered from bears attacking the flocks at night because all the do-kyis had been sold off to buyers from the cities. So, in fact, they may not be entirely "ubiquitous".
3. Nyalam, as the 'Gateway to Hell,' is said to have been so named by the 11th century Tibetan saint Milarepa, who meditated in a cave a few miles north of the town. Nyalam is also spelled Nilam in some early accounts, and was also known in the past as Kuti, or Kutti, the last Tibetan frontier post before descending the Bhoté Kosi river gorge down out of Tibet into northern Nepal.
4. Ippolito Desideri, *An Account of Tibet: The Travels of Ippolito Desideri of Pistoia, SJ, 1712–1727*, edited by Filippo de Filippi (1932, p.312).
5. The Uighurs are Muslims from East Turkestan and, within China, from Urumqi, Kashgar, Yarkand, Khotan and other old cities on the historic Silk Route across Asia.
6. By one account, Lhoka's Comai County is considered to be the home of the Tibetan mastiff (Joy 2005). The 'tsang' in tsang-kyi, however, refers to the larger region formerly called Tsang (or more inclusively, Ü-Tsang), which was centered on Shigatse and covered a vastly larger area than just Lhoka. Technically, therefore, *any dog from Tsang* (large or small) is a tsang-kyi, and all tsang-kyi are do-kyi, 'tied' or guardian dogs.
7. Robert Ekvall, *Fields on the Hoof: Nexus of Tibetan Nomadic Pastoralism* (1968).
8. Steffel's observations here are from personal discussions while traveling in Tibet as well as from correspondence (2007) and miscellaneous postings on Tibetan mastiff Internet listservs. She bases her observations on considerable research including interviews and field observations during

our trip. For further discussion of the pros and cons of artificial selection for looks (color, size and other considerations such as sales, status and ego) see Raymond Coppinger & Lorna Coppinger, *Dogs* (2001, Chapter 7 and *passim*).

Steffel concludes her discussion with this observation about comparing very large dogs in Tibet with the presumably smaller (in some cases) Tibetan nomads' do-kyis: "Maybe the giant dogs are 'the true Tibetan mastiff' or the 'tsang-kyi' or whatever. Maybe the smaller dogs are 'the true do-kyi.' Palace guardians *vs.* livestock guardians. Temple dogs *vs.* nomads' dogs... But, dividing the 'types' into separate breeds in order to make them conform to written standards would cause harm to *all* of them, sooner or later, by reducing the gene pools. Not to mention that this would be (yet another) dramatic departure from the way the Tibetans created and maintained the dogs before we (outside of Tibet) got our hands on them."

9. The Coppinger references are from *Dogs* (2001, Ch. 7, especially pp.246-7).
10. 'FCI' is the Fédération Cynologique Internationale.
    The genetic relationship of red, golden and cream dogs is described in 'Coat color genetics in the Tibetan Mastiffs' by Charles Radcliffe and Matthew Taylor; reprinted here as Appendix 4, pp.221-6.
11. Ekvall's observation is from 'Role the dog in Tibetan nomadic society' (1963, p.168; see Appendix 2, pp.208-17 in this book).
12. The first two quotes are from K.R. Salmeri, *et al*, 'Gonadectomy in immature dogs: Effects on skeletal, physical, and behavioral development' (1991), and R. Bowen, 'Early sterilization in dogs and cats' (2004), respectively. See also: 'Spaying and castration—pros, cons, myths, and Dobermans' by Ione L. Smith (2000).
13. Landon's description of the "so-called" Tibetan mastiff is from *The Opening of Tibet* (1905, p.425).
14. One of the most seriously atypical bloodhound-like faces on a Tibetan Mastiff may be seen on page 25 of Li Qian's *China's Tibetan Mastiff* (2006), along with this startlingly outrageous caption: "This Tibetan Mastiff retains a great deal of the facial and head features of the ancient pure-blood Tibetan Mastiffs."
15. Judy Steffel's analogies and comparisons here are also from discussions during the Tibet trip and in follow-up correspondence.
16. 'Beefeater' is the popular nickname for the Yeomen of the Guard (and Yoemen Warders) established in the late 15th century to guard the Tower of London. Dressed in their bright scarlet and gold uniform and carrying a pike, the Yeomen are very imposing, like guardsmen (and guard dogs) should be.
17. We are not alone in our concern over the breeding of a new line of super-dogs and calling them the "true Tibetan mastiffs." In a recent exchange on the TMTalk2 Internet listserv over the descriptions of big dogs in Tibet by Robert Ekvall (1963), Mary Fischer (an avid Tibetan dog fancier) makes the following observations of alarm over what is happening to Tibetan mastiffs internationally. Her disquiet is similar to our own. "The use of the term 'Tsang Khyi' (a dog from Tsang) has been used to justify

the splitting of an already vanishingly small breed for purely political reasons." Then she notes that the "loose eyelid skin" (extreme haw) found in some of these "new" dogs has resulted in the breeding of "the most disgusting crossbreds and [passing] them off as purebred TMs..." She is also suspicious (as we are) about the breeding of Neapolitans, black-and-tan Coonhounds and/or Bloodhounds with Tibetan mastiffs to create a giant, jowly dog with extreme haw, "regardless of the loss of native character, soundness, structure, and function." When these issues are raised with the super-dog breeders, she hears them "remonstrate with anyone who can see through this charade: 'but it's authentic to the breed,' they say. Just look what the travelers said about the dogs." Fischer concludes that "The size mentioned has been used to justify 'breeding up' inferior stock, instead of eliminating it and breeding better animals via sound husbandry practices." We agree. (Excerpted from correspondence in 2007 by Mary Fischer to Kristi Sherling on the TMTALK2 listserv at http://pets.groups.yahoo.com/group/TMTALK2.)

18. The reports of extremely high prices being charged for these dogs in China are from several sources, including: 'First Tibetan mastiff exhibition in Jinjiang' (CTIC 2007), 'Hounded to extinction' (Li Jin 2005) and 'A conservationist campaigns to keep Tibet's mastiffs mean' (Blythe Yee 2006). See also 'Celebrities' pooch—Magnificent Tibetan mastiff' (Dong Zhen 2007) and 'Tibetan mastiffs may die out' (Chinanews 2006). These are only a few of many news articles on the subject.

    The highest price ever paid for a Tibetan mastiff or, for that matter, *any* dog, was 4 million Yuan, or the equivalent of over 585,000 US dollars, according to reports in the world press in September 2009. See the report by Lauren Cooper (2009) entitled 'Tibetan Mastiff becomes world's priciest dog: Chinese woman pays $600,000' (slightly exaggerated) online at www.dailyfinance.com/2009/09/12/tibetan-mastiff-becomes-worlds-priciest-dog-chinese-woman-pays. The dog named 'Yangtze River Number Two' is owned by a wealthy woman known only as Mrs Wang.

19. More to the point here: What is happening to traditional Tibetan culture at large? The Dalai Lama recently predicted that "If the present situation is the same in 15 years then I think Tibet is finished... Tibetan culture, Tibetan language and the Tibetan environment." A depressing thought.

20. Charles Bell, *Portrait of the Dalai Lama* (1946).

21. The story of the Younghusband military expedition that opened Tibet to the outside is summarized in Chapter 5, and at length in books by Perceval Landon (1905), Peter Fleming (1974 [1961]), Alan Ross Warwick (1962), Anthony Verrier (1991), Patrick French (1994), Charles Allen (2004), and Shubhi Sood (2005).

22. Stan Mumford, in *Himalayan Dialogue* (1989, p.74), reprinted as 'Shamanic sacrifice and Buddhist renunciation' (Mumford, 2004). See also Draghi, 'The stag and the hunting dog' (1982). The correct translation of Khyira Gonpo Dorjé (a mixture of Tibetan and Nepali) is 'The dog (*khyi*) and (*ra*) Gonpo Dorjé (the hunter's name).'

23. John Hedley, *Tramps in Dark Mongolia* (1910, pp.295-6).

24. Douglas Carruthers, *Unknown Mongolia* (1914, pp.227-8).

25. Irma Bailey, 'Dogs from the roof of the world' (1937), reprinted here as Appendix 1, pp.203-07.
26. A sighthound (sometimes called a gazehound) hunts by sight, as compared with dogs that track their prey by hearing or scent. The more well-known sighthound breeds are Saluki, Borzoi, Greyhound and Afghan hound.
27. The Khampa dog is also called the "Tibetan Hound" by W.V. Soman in *The Indian Dog* (1963, p.90), whose description of them is virtually identical to Heinrich Harrer's in *Seven Years in Tibet*, from whom it is obviously borrowed (without attribution): "It is a very swift, short-haired unusual type of dog, usually kept by nomad tribes known as Khampas. They are classified as unusual because they have less hair than any other Tibetan dog, lean, swift as a wind and rather ugly and cannot be easily described. They have not yet been identified or recognized by any of the associations or Kennel clubs."
28. While Buddhism prohibits taking of any life, it is common for Tibetan nomads to kill both domestic and wild animals for food. In one of the few studies of Tibetan hunting practices, Tony Huber, in an article entitled 'Antelope hunting in northern Tibet' (2004), describes Gertse pastoralists of Tibet's northern Changtang who "all admit to being Buddhists" but who regularly kill domestic livestock and engage in hunting wildlife. "All such killing," he writes, "is viewed in practical terms as necessary for survival." Only among elderly men who no longer participate in such strenuous activity does concern develop about taking life and its effects on one's ultimate salvation and afterlife or rebirth.
29. Robert Ekvall, 'Role of the dog in Tibetan nomadic society' (1963, p.167), reprinted here as Appendix 2, pp.208-17.
30. In this and the following three paragraphs on Tibetan hunting dogs (referring to the writings of Simsova, Sellar and Hally) all quotations and notations are from A.L. Wynyard, *Dogs of Tibet* (1982, pp.39-41).
31. See: www.hundeguiden.no/Hunderasene_Molosser/The_1st_Group/tibetan_hunting_dog.htm. See also: molosserdogs.com/modules.php?name=Molossers.
32. Melvyn C. Goldstein and Cynthia Beall, *Nomads of Western Tibet* (1990, p.125).
33. W.M. McGovern, *To Lhasa in Disguise: A Secret Expedition through Mysterious Tibet* (1924).
34. The *Wall Street Journal* article was written by Blythe Yee (2006).
    See Raymond and Lorna Coppinger's *Dogs* (2001) for a lengthy discussion of how domestic dogs revert back to feral scavengers when abandoned or ignored by humans and left to survive on their own.

## Chapter 7

1. A large number of wild dogs roaming the Canary Islands was reported by the Roman scholar Pliny in the 1st century CE. The dogs of the Canary Islands today are a large molosser type known as Perro de Presa Canario ('Canarian Dog of Prey' or 'Dogo Canaria' for short), established as a hunting breed in the 16th or 17th century.

2. *Nakyu* and, alternatively, *nagi* and *naghi* are words meaning 'dog' in Gurung, a Tibeto-Burman language spoken in the north central Nepal Himalayas. The prefix *na-* may derive from the Gurung term for 'nose,' a prominent canine feature. Gurung is spoken by a mountain ethnic group in Nepal for whom high altitude pastoralism is an important economic activity. The Gurung herd sheep and goats, but very early in their history of migration southward off the Tibetan plateau into the northern mountains of Nepal (hundreds of years ago) they herded yaks. Today, the northernmost Gurungs keep livestock guardian dogs of Tibetan origins, and some of them maintain ties with Tibetan traders and herdsmen over the border, from whom they periodically obtain new dogs. A recent archaeological study in the Nepal Himalayas sheds more light on their Tibetan origins: Christopher Evans (with J. Pettigrew, Y.K. Tamu and M. Turin), *Grounded Knowledge/Walking Land: Archaeological Research and Ethno-Historical Identity in Central Nepal* (2009).

3. Victor Mair's definition of cynic is from 'Canine Conundrums' (1998). A cynic is "A person who believes all people are motivated by selfishness; a person whose outlook is scornfully and often habitually negative; and a member of a sect of ancient Greek philosophers who believed virtue to be the only good and self-control to be the only means of achieving virtue" (*American Heritage Dictionary of the English Language*, 4th ed., 2000; www.bartleby.com). I am reminded of the dual themes of cynicism and conflict found in *Don Quixote* and *The Dialogue of the Dogs* by Miguel de Cervantes. In *Dialogue* Cervantes defines 'cynic' succinctly through a discussion between two canines, Scorpio to Berganza: ". . .gossip itself will call us cynics, which means gossiping dogs" (*Dialogue*, 2004 [1613], pp.xiii, 38).

4. The discussion of breed is based in part on D.P. Sponenberg's insightful article, 'Livestock guard dogs: What is a breed, and why does it matter?' (1998).

5. Since the Coppingers wrote this, dog genome studies have advanced to the stage that there is now commercially available testing which purports to distinguish the distinct breeds in mixed breed dogs. The science is new, however, and the results are debatable. It may well be that geneticists will discover the causes of canine type differentiation through *epi*genome studies, referring to heritable traits that do *not* involve changes to underlying DNA sequence.

6. Except for the blatantly racist overtones, Mrs G. Hayes got it right in 1932 when she wrote: "It is a curious and interesting fact that a primitive and uncultured [sic] people such as the Tibetans should be so devoted to their dogs and should actually breed and keep fairly true to type five or six different breeds. There is no Asiatic country of which the same can be said. . ." (quoted by Ann Lindsay Wynyard in *Dogs of Tibet*, 1982, p.34).

7. Several other terms for Tibetan dogs are noted by Robert Ekvall in 'Role of the dog in Tibetan nomadic society' (1963); reprinted here as an Appendix 2.

8. D.P. Sponenberg (1998) states that the term 'gentrification' was coined by David and Judy Nelson (no source). I have adapted it here in the context of modern modifications to big Tibetan-Himalayan dog breeds. A notable exception to the form of 'gentrification' that is focused on looks, however, is the standardization of sheep herding dogs such as border collies. While looks

are important in a purebred border collie, it is far more important that he *acts* like one, is able to function as a sheep herding canine, and excels in sheepdog trials. You can read about border collies in Donald McCaig's *Eminent Dogs, Dangerous Men* (1998) and *A Useful Dog* (2007), and in his novels *Nop's Trials* (1992) and *Nop's Hope* (1998). And be sure to read McCaig's *The Dog Wars: How the Border Collie Battled the American Kennel Club* (2007).

9. 'Haw' refers to inflammation of the loose lower eyelid in some animals. A distinct haw is often seen, for example, in Saint Bernards and Bloodhounds, but has recently begun showing up in an exaggerated form in Tibetan mastiffs bred to conform to a type of non-traditional gigantism that some contemporary breeders are promoting.

10. I have borrowed part of the heading to this section from Jan Libourel's 'The Molossus myth & other mastiff malarkey' (1993).

11. Philippe Touret (personal communication, 2007) alerted me to Raymond Triquet's wisdom about scientific inquiry and doubt. The original French reads: *"La conclusion est que le Doute étant l'essence de la science et de la recherche, il faut d'abord cesser d'affirmer sans preuves."*

12. The quotations and citations noted in this list are representative of the various writings, not exhaustive. Emphases in bold are mine.

13. Canine Information Library (2006), dogbreeds.bulldoginformation.com/tibetan-mastiff.html.

14. The full title of James Baillie Fraser's book is *Journal of a Tour through Part of the Snowy Range of the Himālā Mountains, and to the Sources of the Rivers Jumna and Ganges* (London, 1820). Today, though it is exceedingly scarce, a good copy may be purchased from a rare book dealer for $3,500 or more. All quotes from Fraser in these paragraphs are from his book (pp.176 and 354-5).

15. In fact, tigers were quite common where Fraser was trekking, in Kumaon, in the modern Indian state of Uttar Anchal. This became well known when Jim Corbett (1875–1955), a renowned British colonial hunter, published his popular books about tigers: *Man-Eaters of Kumaon* (1944), *The Temple Tiger and More Man-Eaters of Kumaon* (1954), and others. See also Neale Bates, 'Tracking man-eaters: The Jim Corbett story' (2008). Leopards are also commonly found in the Himalayan hills.

16. The term mastiff for big Tibetan dog also appears in C.R. Markham's 1876 translation of 'Brief account of the Kingdom of Tibet, by Fra Francesco Orasio della Penna di Billi, 1730.' Father Francesco referred to the dogs that pick over human corpses at Tibetan funeral grounds. In this delightful passage such dogs are referred to as "mastiffs": "...divers prayers having been recited... they take him [the corpse] thither... reciting divers things; and when they have arrived [at the funeral grounds] they celebrate other rites, placing the naked corpse on a great stone. Then a professed scholar... breaks the corpse to pieces with a great bar of iron, and distributes it among the dogs in presence of all the company. After the mastiffs are satisfied, the relations of the deceased gather up the most clean-picked and the largest bones, and make a bundle of them, throwing all that is left by the dogs into the neighbouring river, near which are the places set apart for this inhuman rite."

The original account of 1730 was in Italian. The use of the term "mastiff" in Markham's translation reflects *usage at the time of translation*, not when the original document was written. It is similar to the license others have taken by using the word "mastiff" in translations of Marco Polo's *Travels* (see Chapter 1).

17. It appears that Polo used the term *mastin* in the Old French language version of his *Travels*, dated 1298 CE. In the section that begins *"Tebet est une grandisime provence"* ("Tibet is a large province"), he goes on to describe *"grandismes chenz mastin qe sunt grant come asnes,"* which Henry Yule translated as "mastiff dogs as big as donkeys" (in the 1902 Yule-Cordier edition of *The Book of Marco Polo the Venetian*). In Old French, however, *mastin* simply means 'domesticated' (and not 'mastiff' as we interpret it, as a breed, today). Yule's substitution of the term 'mastiff' simply follows the custom of using the popular terminology of his day at a time when it was commonplace to call the big dogs of Tibet "mastiffs." My source for the Old French version of Polo's *Travels* (a few lines only) is the unique *Hobson-Jobson*, a dictionary published by Henry Yule and A.C. Burnell (2nd ed., 1902, p.918a), which gives as the source this all too brief and uncertain citation: "*Marco Polo*, Geog. Text. ch.cxvi. 1330."

18. See P. Savolainen, *et al*, 'Genetic evidence for an East Asian origin of domestic dogs' (2002). For review articles see Kate Dalke, 'Who is that doggy in the window:—Scientists trace the origin of dogs' (2002), and Karen E. Lange, 'Wolf to woof: The evolution of dogs' (2002).

19. A Molosser connection is hinted at in another version of Polo's *Travels* that, loosely interpreted, might be used to claim that Tibetan and Molosser dogs are closely related. Molosser mastiffs, however, have smooth short coats, quite unlike the shaggier Tibetan livestock guardian dogs, and the likelihood that Molossian mastiffs and Tibetan livestock guardian dogs are closely related is spurious. In the Latin translation of Polo's book (the so-called *Z Codex*, dated 1400 CE, discussed in the notes to Chapter 3), the same passage that describes the big dogs of "Thebet" as being as large as asses refers to them as *"magnos canos molosos,"* or "large *molosos* dogs" (literally, *molosos*-like dogs). This is no more than an offhanded comparison of the Tibetan dogs with the Greek dogs, and it further supports my contention (and that of other Polo interpreters) that writers like Polo and his translators favored analogies and descriptive terms with which their European readers would be familiar (much as we, today, might quite figuratively compare the immense size of something "as big as a barn" or its swiftness as "faster than a speeding bullet").

20. There is some confusion in the literature about the terms 'Bhutia' and 'Bhotia.' In the past, Tibetan border area dogs have been called by these terms after the ethnicity of some of the people who live there. The terms are one and the same and simply mean the people of Bhot ('Tibet') or of the Tibetan borderlands, and nothing more. Hence, some Tibetan border people (and their dogs) in India, Nepal and Bhutan are known as Bhutia, Bhotia, Bhotea or Bhotiya.

In some accounts, such as Fraser's (1820), Tibet itself is confusingly called 'Bhootan.' Some early writers seem perplexed by the similarity between these terms and the name of the country of Bhutan, and apparently think that the dogs, at least, are exclusively or originally from Bhutan. Ann Lindsay Wynyard, in *Dogs of Tibet*, for example, states that *"The real home for this dog is Bhutan*, but, like the Bhutia pony, it is found everywhere along the hills" (1982, pp.44-45; emphasis added). Did she, too, think Bhutan meant Tibet?

21. The thin line between types noted by Juliette Cunliffe is from her article 'Early history of the Tibetan Mastiff' in *Our Dogs* (2002). Cunliffe is the author of the breed book *Tibetan Mastiff* (2001, 2007). She previously edited the popular UK periodical, *Tibetan & Oriental Breeds International*.

22. Reflecting on Jacob's challenge, another comparison comes to mind: "It would be the same with the various sighthounds—Saluki, Sloughi, Afghan, etc. There are many regional variations depending upon climate, use, etc. One 'variety' may be consistent at the center of its range, but the 'varieties' blur at the edges of their ranges. I'd wager that this is true of virtually all landrace 'breeds'" (Judy Steffel, personal communication 2009).

23. Kristina Sherling, 'Do Khyi–Tsang Khyi Tibetan Mastiff' (2002).

24. Thomas Stevens, 'Some Asiatic dogs' (1890).

# Appendix 1

## Dogs from the Roof of the World
### Many Unusual Breeds Found in Tibet, the Strange Land that Lies in the Clouds

By Hon. Mrs. Eric Bailey*

Due to the airship and the radio, the world is growing smaller and smaller, and there is practically no country that is now a complete mystery to civilized man. Probably Tibet still remains the most mysterious, although the veil that hides this astonishingly interesting country is rapidly being torn aside, and we are learning more and more about these ancient people who live beyond the Himalayan mountains.

I have no intention of going into any great description of Tibet. I am as assuming that you know it is a strange land of quaint, little people, cheerful and kindly, yet hot tempered when roused. But I do want to talk about their dogs, which are commencing to be of interest to fanciers of both Great Britain and the United States.

Broadly speaking, Tibet has several distinct types of dogs. These have all been seen in either England or the United States. In addition to these major types, there are a number of other breeds—many hardly known out side of Tibet—that I will mention a little later in my story.

It quite often happens that names mean very little when they are transplanted planted in another country. As an example take the name Tibet, or, as it is sometimes written, "Thibet," is not generally used by Tibetans. They call their country Bod, and they call them selves Bod-pa, or "people of Bod."

According to the Encyclopedia Britannica, the word "Tibet" came to be used by Europeans because the great plateau with its uplands bordering the frontiers of China, Mongolia, and Kashmir, through which the traveler communicated with Bod, is called by the natives Tu-bhot, or "High Bod" or "Tibet," which designation, in the loose orthography of travelers, assumed a variety of forms. Therefore it is not at all surprising to find that the word "Apso" is the Tibetan name for any long-haired dog. It is a corruption of "Rapso" which means "goat-like."

These dogs are, in general appearance not unlike the small, long haired goats of the country.

---

* Originally published in *The American Kennel Gazette*, Vol. 54, No. 3, March 1, 1937.

The Tibetans have very little imagination in naming their dogs, and most dogs of the Apso breed, being the lion-dogs of Tibet, are called either Seng-tru or simply Apso. When we started our kennel we tried to keep to the idea of the young of carnivorous animals. Thus we got Tak-tru, the "young tiger"; Sik-tru, the "young leopard"; Sa-tru, the "young snow-leopard"; I-tru, the "young Lynx"; Chang-tru, the "young wolf," and so on. Later, we had to forsake the carnivorae for Tsi-tru, the "young mouse," and so forth.

Owing to this difficulty of finding sufficient names from the young of animals we finally used the names of Tibetan girls and goddesses for the bitches among our Apso dogs.

Speaking of names, mastiffs in Tibetan are called Do-Kyi, which means "a dog you can tie up." Do-Kyi are kept by every nomad and sheep- or yak-herd to guard the tents. Marco Polo, in the account of his journey into Tibet in the fourteenth century, mentions these dogs and remarks that they are as large as donkeys.

This has always been considered an absurd exaggeration on his part, but as the donkeys in Tibet are abnormally small this is not such an inaccuracy as one might think.

Besides guarding tents they are very often also used to guard houses.

To make them fierce, the people keep these mastiffs tied up all their lives from the moment they are about a month or two old.

One result is that the mastiffs in villages are sometimes rickety with twisted, deformed limbs, while the hind-legs especially may be poorly developed.

The nomad's or shepherd's dog has occasionally to move camp with his master and avoids this trouble in an exaggerated form, but in Tibet they are not active dogs except when actually carrying out the military precept that attack is often the best form of defense. They are, of course, quite unsuited to hunting game in any form.

As you approach a Tibetan encampment, the first sign of life is usually the barking of dogs. On this, the owners come out of their black yak hair tents and inspect the cause of the alarm.

They, or more usually their children, then see that all dog fastenings are secure, and often hold the dog down while it strains to reach the stranger.

Although fierce, as the result of being tied up from puppy hood, these mastiffs are very affectionate and good tempered with the people they know, and one often sees the smallest children handling and calling them off from their attempts to attack the intruder with perfect ease and safety.

These mastiffs, and sometimes the hunt hunting dogs also, wear a large fluffy collar of wool dyed bright red. This red collar can usually be seen in pictures of scenery by Tibetan artists. Tibetan mastiffs are usually black in color with tan points. One of the high officials of His Holiness the Tashi Lama once had some entirely black ones, of which he was very proud, but tan mark markings are more usual. Not infrequently red dogs are found in a litter.

The dog is very heavily built with a thick coat. The head is particularly heavy and the flews so pendant that the red of the eye is conspicuous. This, in a country like Tibet, subject to dust, wind and glare, often leads to diseased eyes owing to dirt and lack of care and cleanliness.

The whole head is large and heavy, but the heads of females are notably smaller and lighter than those of males. The Tibetans especially admire a deep-voiced bark.

In 1928 we imported five of these mastiffs. The best one was undoubtedly Tomtru (meaning "young bear"), a village dog.

Then there was Rakpa, whom we bought from a caravan of mules traveling in Tibet. He was a fine red dog which was given first prize for Foreign Dogs at the Kensington Show some years ago, and was also a winner at Cruft's and the Kennel Club shows. An imported bitch was the black and tan Gyan-dru.

These mastiffs seem quite impervious to cold. On a winter's day with a high wind and the temperature well below freezing, they will elect to lie out on a patch of snow if they can find one.

Tibetan mastiffs were shown at the Alexandra Palace in 1875. More than a quarter of a century elapsed before the breed again appeared, when one was brought to England from Lhasa after the Younghusband Mission in 1904. This dog can now be seen in a glass case at the Natural History Museum in London. The dog known in England and the United States as the Tibetan Spaniel has, as far as I know, no special name in Tibet. There seem to be more of them in the Chumbi Valley than in other parts of Tibet. Claude White, who was the first political officer in Sikkim, had a fine kennel of these dogs many years ago.

My husband got a very nice dog of this breed when in Lhasa with Sir Francis Younghusband's Expedition in 1904. This dog accompanied him on a journey of more than 1,000 miles through Tibet to Simla and the photograph on page 7 was taken on that expedition while crossing a high, snowy pass.

This dog was called "Lhasa," and was given to Mrs. Frank Wormald, who brought him to England in 1905. He was shown, and won prizes, and died at the age of 18.

I was given six of these dogs, some cream, black and red, but I parted with them and confined my attention to Apsos.

Besides the known breeds I have mentioned, there are two other very distinct breeds of Tibetan dogs. The first is the hunting dog, known in Tibet as Sha-Kyi - Kyi being the Tibetan for dog.

This dog is about the size of an Airedale. In color it is a creamy grey with a thick coat. The tail may be carried curled over the back, but also some times down. The head is long and is a smoky black shading into the creamy grey of the body. The ears hang forward.

The dog is used for killing game. He is taken on a leash to within sight of the game - Bharal (wild sheep) musk deer, serow, etc. - and slipped.

When the quarry is pursued, it adopts its natural defense against a wolf by getting into a cliff, where it turns to bay and attempts to butt the dog over the precipice.

This is where the quarry is wrong, for the dog does not go in and attempt to kill as a wolf would presumably do, but keeps barking in complete safety and distracts the attention of the quarry while the hunter comes up and shoots the animal at close quarters with his primitive matchlock.

These dogs are very keen sighted. I once saw a pair which spotted a herd of Ovis Ammon at a very great distance, but the owner would not slip them as he explained that there was no suitable place to which the wild sheep could be turned to bay!

I have attempted to keep these dogs. Those obtained when adult were tiresome in attacking strangers. A hunter once told me that they trained these dogs by tying the pup to the mother and letting her go after game when she forgot everything but her hunting. The puppy was dragged and bumped along after her. This made them fierce and keen!

Attempts to keep young ones were difficult as I found them delicate. There is no doubt, however, that this distinctive dog would be most attractive and if bred from imported parents would, like the mastiff, be a quiet house dog. There is no trace of wolf in them as the droop ears testify.

Another very distinctive breed in Tibet is the Kongbo dog, and has never been exported from Tibet, as far as I know. Kongbo is a province in South eastern Tibet. This dog is on the lines of a schnauzer, with small, prick ears. I have only seen two of these dogs, and both were reddish in color. I was once given one by a Tibetan friend. I did not keep him as he was very old and in bad health.

When in Tibet, I kept Finnish spitz, and Tibetans, on seeing these, would always point with surprise and say: "Kongkyi" (i.e. Kongbo dog). The Kongkyis I have seen are much heavier in build than the Finnish spitz, with coarse hair like a schnauzer and the ears are shorter.

Another very popular dog in Tibet is called Gya-Kyi, which means simply "Chinese dog." This is more like a smooth-haired Pekingese than anything else. I believe this dog to be the Chinese Pug.

This dog is, I think the same as, or akin to the Chinese ha-pa dog which was shown in England a few year ago. His Holiness the late Dalai Lama once gave me one of these dogs, which I kept for some time. It was a very nice, affectionate pet. This dog usually has a collar of colored cloth (often red) on which bells are sewn

# Appendix 1. Dogs from the Roof of the World

I have said that "Apso" in Tibetan means "any long-haired dog." Do-Kyi means a "tied dog" or mastiff. The late Dalai Lama, himself, kept many dogs, among them one described as "Do-Kyi-Apso," which may have been a cross between a Tibetan mastiff and the dog known in Tibet as the large Apso, which are called Tibetan Terriers in England. My husband took a photograph of this dog. It was the only one that he ever saw of this breed.

Tibetans have, of course, not standardized types. This results in no very specialized, or, as one might perhaps say, unnatural dogs, being bred there.

If you wish a dog for a pet in the house it should be small, lively and faithful; for hunting, fast and courageous; for a guard, large, strong and fierce on occasion.

However, breeding is not entirely haphazard as can be seen from the dogs which come from Tibet. After all, the points of breed were only standardized in England about a hundred years ago, when dog shows were first held.

As in other eastern countries, every Tibetan village contains stray dogs, which make a night's rest almost impossible. Some of these are quite fine dogs as indeed they must be to stand the rigorous winter climate.

On the Ling-Kor, or sacred road on which pilgrims circumambulate Lhasa, are numbers of dogs of all descriptions which are fed by the pilgrims as an act of piety. This has nothing to do with the dogs being sacred. All talk of sacred dogs bred only in monasteries is nonsense. But the Buddhist theory of reincarnation encourages kindness to animals, especially in a sacred place like Lhasa.

It must not be imagined that native dogs are the only ones seen in Tibet. Occasionally, one finds pure-breds from England or the Continent in the homes of Tibetans. I know that a former Dalai Lama kept English greyhounds, and his favorite dog, which was always with him in the house, was a dachshund.

This goes to prove that we all like things that are not native to our own soil. That is why you might be interested in owning one of the dogs that are indigenous to Tibet, that strangely interesting country which lies up in the clouds.

# Appendix 2

# Role of the Dog in Tibetan Nomadic Society

By Robert B. Ekvall*

The role of the dog in Tibetan nomadic society has many aspects, some of them are obvious, and what one would expect in any society—aid in hunting, watchdog function, etc.—and others which are quite the reverse of the obvious; which relate to social integration, and to influences affecting the development of personality within a culture. In any case the role of the dog is of surprising importance and is well worth some attention. Before discussing, however, the role of the dog in Tibetan nomadic society, the position of the dog in Tibetan society as a whole should be noted and some account should be taken of Tibetan attitudes in general toward dogs.

In many Asian Societies the dog is despised and quite literally is an outcast; barely tolerated as a scavenger. This is particularly so in Islamic societies where the unbeliever was stigmatized as a "dog" but was also true in traditional China where, except for the toy spaniels which became somewhat of status symbols and those dogs which were bred and fattened for food, the life of the dog was a miserable one. The Chinese have an extensive repertoire of invectives which begin with the word "dog," and Chinese living in the border areas between China and Tibet in moments of exasperation derive satisfaction in calling the Tibetans "dog headed barbarians."

Among the Tibetans, on the other hand, the dog is favored and even honored in traditions, in symbolism and in real life. The year of the dog is the year of good harvests and among the twelve animals by which the years are named the dog ranks with the tiger and the horse as the most auspicious. As an omen he is good and he is credited with the special sight—the ability to see ghosts and demons—and thus his outcries have peculiar significance. Mi La, the Tibetan poet-saint, told his disciple, Ras Chung, that the dogs the latter saw in his dream represented mKHaa aGro Ma (sky going females), or the celestial courtesans of Tantric doctrine.[1] In some of the religious dances only dog masks are permitted and in doctrinal discourse the neophyte is urged to be toward his spiritual leader like a dog is toward his master—faithful and loyal; with no touch of resentment no matter how harshly treated.

---

* Originally published in *Central Asiatic Journal*, 1963, v.8, pp.163-73. (Minor typological errors in the original have been corrected. /DM)

# Appendix 2. Role of the Dog in Tibetan Nomadic Society

For the earliest Chinese one set of barbarians, the Ti who were bracketed with the Ch'iang and later were incorporated into the Tibetan empire, were represented in ideographic writing by the symbols of dog and fire combined; and the Chinese chroniclers of the eighth century, in proper civilized scorn, wrote about the Tibetans that "when they honor their leaders they bark like dogs."[2] The Tibetans, on the other hand, see nothing amiss for the dog to be by the fire when he does not need to be on guard, and seem to feel no self-consciousness or embarrassment in explaining that Tibetans show their tongues in greeting just as dogs show their affection by licking the hands of their masters and those whom they like.[3] Nor do they feel demeaned by comparing themselves to dogs, for dogs are creatures whom they like and even respect.

Like all people who consider dogs individuals, and not just anonymous animals, the Tibetans give their dogs personal names. Some of these names are combinations of the word rGya, which can have any one of a number of meanings, chief of which is "expanse" or "big one," with color and other designations such as rGya Ser (big one yellow), rGya dMar (big one red), rGya Bo Mig bZHi (the big one eyes four), rGya sBrug (big one dragon), rGya dGe (big one virtue). Other names are names of animals followed by the diminutive of endearment PHrug (young one); such as Dom PHrug (black bear young one), sPang PHrug (wolf young one) and Seng PHrug (lion young one). But a great many of the dogs are called by religious names which in the religion oriented culture of the Tibetan Tibetans is a good indication of their status and how they are regarded. They may be called Sangs rGyas (the Buddhahood), Don aGrub (realization of causes)—which is a personal name of Buddha—, sGrol Ma (savior female), or the Tara goddess, and bKra SHis (blessing).

The attitude of liking and even respect is even more pronounced among the nomadic pastoralists of the Tibetan plateau. Without undue sentiment—although the womenfolk and children give the growing puppies much affection—they accord dogs recognition as companions and acknowledge their value as allies in meeting a common danger. They are cared for and treated very much as equals. When they are exercising their guard function they have clearly defined rights as to how they may and may not be treated: what weapons may be used against them and what weapons may not be used, unless indeed one is prepared to become embroiled with the owners. Stones, staves, whips, and weighted ropes, the butt end of spears and the backs of swords—even, with good wrist work, to the taking out of a tooth or giving of a toothache—may be used: but edged weapons, spear points, and above all firearms, even if only fired in the air, may not be used. The position and value of the dog is most clearly stated in the Tibetan proverb which says "The three most excellent possessions of the wilderness one are his gun, his horse and his dog."

The livestock which the Tibetan nomads rear and keep are, in the cycle of production "fields on the hoof" which are tended in expectation of a harvest in milk, meat, wool, hair, hides and the annual increase. Dogs are not part of these "fields on the hoof." Their flesh is not eaten. The Tibetans profess the utmost abhorrence of the very idea and enjoy throwing back at the Chinese an epithet which includes the word "dog"; calling them KHyii SHa Za mKHan (dog meat eaters). Dogs are not reared for their pelts or for sale, although a few skins of those which have died or been killed by enemies are offered for sale and occasionally a nomad will sell a dog, invariably getting moreover a good price, more as a favor than for gain. But it is also true that most of the dogs which are found throughout Tibet come from the land of the tents where greater availability of protein foods such as buttermilk, whey from cheese making, scraps of meat and the carcasses of stock which have died, favor the rearing of dogs. No nomad, however, wants the reputation of being a trader in dogs.

Only the relatively few dogs which are used in hunting make any direct contribution to the income of their owners. Nor do dogs contribute to mobility, for their use as draft animals as among the circumpolar peoples or as the travois pulling dogs of the Plains Indians, prior to the time the latter acquired the horse, has never been reported. At moving time all that is expected of dogs is that they should carry their own weight, and occasionally the puppies even must be packed on loads or stuffed into saddle bags.

Nevertheless, they are highly valued and well treated for themselves and for the obvious roles they fulfill. In those roles they are accomplices in the subsidiary subsistence techniques of hunting, and in the primary subsistence techniques of animal husbandry they are creators of security—precarious at times but greatly needed. In addition to these two obvious roles there are other roles which the dogs fulfill of which the Tibetans have slight, if any, appreciation yet which are of great importance. They function to create privacy and social distance in a situation where both privacy and social distance are a felt need. They help shape the behavior patterns of the children of the tents and thus influence character formation. And finally, although the dog is one of the "three excellent possessions of the wilderness one," yet as the peculiar pets of the womenfolk and as pitted against the horse, by the threat they pose to the mounted man, they counter balance the arrogance of the male on horseback.

The role of hunting dogs was of much greater importance in earlier periods of Tibetan history than at the present time. The general term for hunting dog is Ra KHyi (corral dog) for his function is to fence in weak game and to bring to bay and form a ring around big and strong game. Folk tales and legends which are over a thousand years old stress the use of rGya (dog(s)), sHa KHyi (meat dog(s)) and Ri KHyi (mountain dog(s)) in coursing the stag and surrounding game.[4] Marco Polo also speaks of "mastiffs as big as donkeys which are capital at seizing wild beasts (in particular the wild oxen)."[5] In the well-known encounter between *mGon Po rDo rJe* the hunter and Mi La, the poet-saint of the eleventh century, the Red Bitch figures prominently in the chase of the Black Stag, but eventually is converted to the gentle doctrine of seeking the benefit of all living creatures.[6]

In modern times the pervasive influence of that gentle doctrine, and Buddhist scruples against the taking of life have reinforced older fears of the aboriginal hunt gods, who object to poaching among their prized and jealously guarded herds of wild life, and consequently, hunting as a subsidiary subsistence technique, has declined in importance. Along with that decline the number of hunting dogs kept among the nomads has been much reduced and the importance of their role has lessened. In some areas a few rangy prick-eared dogs—mostly fawn or brindle—are kept for tracking stag or musk deer. They are called SHa KHyi (meat dog(s)) which is a somewhat general term for hunting dogs, and sometimes more specifically SHwa KHyi (stag dog(s)). They vary greatly in pelage, conformation and size for the breed is not at all pure. In a Golok tribe which I visited in the summer of 1940 I saw quite a number of slender greyhound type dogs which had certain resemblances to both borzois and salukis and came in rather unusual colors such as light fawn, ivory and pale ash grey. They were considered useless as watchdogs but, as their name Wa KHyi (fox dog) indicates, were kept for hunting foxes.

The watch dogs, called Srung KHyi (guard dog(s)) or sGo KHyi (door dog(s)), on the other hand, are extremely numerous among the nomadic pastoralists. There is no tent which does not have at least two, and chiefs or wealthy men may own twenty or more. In one small encampment of six tents where I visited for several weeks there were twenty-one dogs—averaging three and one half dogs per tent. They are two varieties or breeds: the true Tibetan mastiff—which is somewhat rare—and another equally ferocious and almost equally large mongrel. The latter undoubtedly has, among other strains, more or less mastiff blood.

The mastiffs, known variously as Sang KHyi (Sang (breed name) dog(s)) or gTsang KHyi (dogs of Tsang) constitute what Tibetans call a "bone line" and considerable efforts are made to keep the breed pure. Their possession is somewhat of a status symbol; it is very difficult to find one for sale; and if found the price is usually that of a good horse. They have the typical heavy muzzle, high domed head, hanging lips, the red of the eyelid showing and massive forequarters of the mastiff breed. The rather long tail is somewhat lightly feathered and carried in a loose curl. In color they are usually black—always so if considered pure bred—with tan trim on the face, neck and legs, usually some white on the throat and chest, and a tan spot over each eye; from which they get the name Mig bZHi Can (four eyed one(s)). As to size, the one I had weighed 160 pounds.

Their most distinctive characteristic is an incredibly heavy baying bark—much more like the sound of a fog horn than the outcry of any animals.

The mongrel watchdogs, which are by far the more numerous, are little, if any smaller than the mastiffs and quite as ferocious. But in their bark they do not have quite the foghorn quality of tone which distinguishes the true mastiff. In coloration they exhibit greater diversity; ranging from pure black to grey wolf color and even an occasional white one. Their pelage is somewhat longer than that of the mastiffs, they have somewhat wider and flatter heads with more pointed muzzles, and the heavily featured tail is carried in a tight curl over the back.

Both varieties are used only as watchdogs and although a dog may occasionally follow the herdsman for the day they are not trained as sheep and cattle dogs to drive and herd livestock. In an incidental fashion they act as scavengers—always hungrily on hand when the butchering is being done. Whenever the stock die in unusual numbers from disease or in the heavy snows of late spring the dogs feast to repletion. Their proper function in life, however, is as watchdogs, and as such they are savage and alert. Their mission in life, as defined by the Tibetans, is to "guard wealth against beasts of prey and thieves" and to the fulfillment of this mission they bring vigilance and ferocity. Indeed, they are so fierce that some of them are castrated to heighten their vigilance but also to lessen their ferocity and make them less likely to charge into a spear point or wildly swinging sword, and get hurt.

During the day they range the perimeter of the encampment as a pack, although the composition of the pack changes for dogs drop out and join it as it moves around. The dogs which belong to each tent keep fairly close to that tent, and any dog which gets too far away finds himself among more enemies than friends, and in trouble. They make it impossible for anyone to get near to or within the circle of the encampment without that approach being known; and unless the one who comes is very much of an expert in the only effective technique a horseman can use against the dogs, that approach is soon halted.

A man on foot may use a staff, throw stones and, at very close quarters, use the back of his sword; but the man on horseback must rely on whirling a rope with a weight on one end. The rope may be the long lead rope of the horse's bridle or any rope which is handy. The weight may be a wooden or horn whip handle, a tent peg, a bone or any other similar weight. The rope is whirled so the weight swings around to make a circle; and a twelve foot length—which is about the maximum which can be handled by a rider protecting himself and his mount—thus creates a sort of charmed circle, about twenty-four feet in diameter, of safety for himself and his horse as long as he can keep the weight swinging and the rope does not get tangled under his horse's neck or tail, or around the gun, with its long forked rest, which he carries on his back. Nor should he hit

# APPENDIX 2. ROLE OF THE DOG IN TIBETAN NOMADIC SOCIETY 213

too squarely one of the snapping, lunging dogs on the circumference of the circle, for then the rhythm of the swing will be broken. When coolly and efficiently done this maneuver will keep the dogs—except for the occasional maniac animal which rushes in regardless of rope, spear or sword—at bay indefinitely when riding through an encampment, or when waiting on the outskirts for reaction, permission or welcome from the members of the nearest tent.

All legitimate approach to tent or encampment is based, however, on the assumption or hope that eventually friends or someone will come to the rescue and stand off the dogs: thus permitting final arrival at the tent door. Once welcomed and accepted as guest the stranger must stay warily within the tent—on rare occasions even there me may be in danger of a sneak attack—and he can only move outside when he is escorted and protected by the owners of the dogs. When they escort him they too must change as he moves around the encampment for the dogs of each tent only recognize the members of their own tent family; barely tolerating even next door neighbors who must exercise caution in approaching the neighboring tent.

At night the encampment is given over to the dogs who redouble their vigilance and ring the tents with sound and fury, which dies down at times to querulous bickering only to break out into a roar of suspicion at any unusual sound or sign of movement, as they rush from place to place. The men who have their sleeping places on the perimeter of the encampment sustain an oddly symbiotic relationship with the dogs in the maintenance of this vigil: their occasional shouts, and once in a while a shot into the air, stimulate the dogs, and the dogs in turn keep the men in an uneasy state of half wakefulness. Thus together they build a defense—sensitive as a burglar alarm which moreover has teeth—against thieving and surprise attack.

Nor is the menace of the dogs all bark and no bite. On occasions all precautions fail and people get bitten; sometimes even killed. Three times during the eight years I lived in Amdo I learned of a rider who, for one reason or another, was not able to keep the protective rope whirling and the dogs, getting in close quarters, sought to hamstring the horse which panicked and bolted; bucking wildly. In each instance the rider was thrown—for one his saddle girth broke and for another his horse stumbled and fell—and falling in the midst of the dogs never got to his feet.

It is not only strangers who run such risks. Members of an encampment—both adults and children—may get severely bitten right in their own encampment. In my experience, next to sword wounds, dog bites were the most frequent occasion for requests for first-aid. Such accidents may be regretted and may arouse a certain amount of ill feeling, but an attack by dogs which results even in a death does not lead to a feud and reprisal, such as would follow the killing of anyone in a sudden brawl or by accidental gunfire.

The security thus created is not only a defense raised against attack and thievery from outside the community, but forestalls pilfering and casual or surreptitious borrowing within that community. In fair weather much of what the tent family owns is scattered in somewhat careless disarray outside the tent. Saddles, pads, clothing, wool which has been washed and skins which are being tanned are exposed to the sun to dry out; cheese and fuel are spread out in the process of making; and items of equipment which are being used are left near the tent or hung on the tent ropes, as the daily chores continue. Yet because of the watchfulness of the dogs nothing may be casually picked up or appropriated, and the community itself is spared much of bickering and suspicion between its members because possessions have gone astray.

In addition to this direct and obvious role of creating security—and incidentally keeping everyone more or less near his own tent—the dogs of the Tibetan nomads have a role in creating and maintaining pervasive, shifting zones of danger around the tents and throughout the encampment; thus ensuring privacy and ample social distance where both privacy and social distance are strongly felt needs.

To understand this need one must visualize the typical nomadic encampment. The black tents, with flapping curtains—seldom completely pegged down in summer—for walls are pitched on the common meadow within thirty to sixty yards of each other. The activities and doings of each tent family are very much open to constant visual surveillance by all members of the encampment who have the prying curiosity of a small society where news is scarce and gossip is at a premium. When I was camped with the nomads my sponsors and hosts again and again would ask to use binoculars and would sit by the hour bringing into closer focus and greater detail the doings of the encampment, and their pungent comments would have brought delight to the editor of a gossip column.

There are no walls or hedges: no hidden courtyards or private rooms. Each family lives in full view of every other family. Nor are there any barriers to approach, and were it not for the dogs visual surveillance from a distance could be supplemented by sudden visits and the nuisance of constant nearby loitering, for the members of the community could wander at will. But because of the dogs access to each tent is made difficult, and zones of danger are thus created which, in important aspects, are more effective than high walls or barred gates. The danger zones become factors making for social distance and any approach is a matter of permission: sought, granted and implemented by escort, service and protection. The permission again may be refused, or withheld by the simple expedient of neglecting to call off the dogs or seeming not to hear. In some respects this is much easier than not answering the doorbell. Even meeting in the area enclosed within the ring of tents cannot be casual or surreptitious, for it is advertized by the dogs, and is formalized by the common need to take measures for protection against them.

The privacy thus established is quite effective. I remember a nomad who had an attractive and somewhat provocative young wife with breasts of which she was obviously proud. One day he broke up their noontime visit with me by announcing with somewhat of a leer, "I am going to my little tent for I want to lie with *lHa Mo mTSHo*—she is not busy and it is warm and pleasant." To the suggestion that someone might come and interrupt them he answered, "That is alright, the dogs will keep us from being bothered." His assurance of complete privacy at noon was testimony to the vigilance of the five dogs who loitered and lived their lives around his tent.

In this creation of social distance and by the injection of the factor of danger the dogs also play a role in making a constantly sensed and known menace to life and limb one of the dimensions of living in the unfolding experience of the young child of the tents. From his earliest consciousness danger is sensed as one of the facts of life. He early learns to be wary; to stay on his own ground close to his own tent and among his own dogs; and he only moved away from that safe base under the protection of his elders. All of his play activities and the chores he learned to do—first handling sheep and then larger livestock—are carried on with wariness; never forgetting the menace of the dogs. As soon as he is able he learns to carry a fence slat or other stick, or a rope to swing each time he ventures beyond the immediate vicinity of his own tent.

This menace and the child's reaction to it soon harden into a pattern of behavior which effectively isolates him from the other children of the encampment. The invitingly open and—during the day—empty space in the center of the ring of tents can never be a common playground where children may run and play at will: there are always the dogs. Even a visit to a neighboring tent can never be casual and play with the child of that tent family can only take place—if indeed it takes place at all—with one eye on the dogs. Thus childhood for the relatively few children of the encampment is permeated by the experience of aloneness. Each child grows up alone yet not knowing that he is lonely and cut off from the give and take of association with other children in co-operative or competitive play.

With the limited mobility which comes with learning how to take precautions and protect himself he will have grown beyond the true playtime of childhood. He then meets other children in the more serious cooperation or competition of handling livestock; the girl catching and milking sheep and cows and learning how to make butter and cheese, and the boy driving sheep, cattle and horses where they should go, either in taking them to pasture or in rounding them up and tethering them at night. He meets his fellows in the context of adult responsibilities—guarding the herds and sharing a common fire for the day. Then only—a child no longer—does he begin the hard lesson of how to cooperate with others outside the immediate circle of his own family.

In the situation thus created the dogs play yet another role in relation to the children of the nomads. To a greater degree than any other animals—lambs, calves and colts—which for one reason or another at one time or another are kept close to the tent, the dogs become the playmates of the children. Babies just learning to crawl out of their fur and felt wrappings tumble amongst puppies and their attendant mothers, and find their earliest playmates among the dogs who seem to keep their love of play longer than other animals. I have seen children up to six and seven years old who romped with the dogs around the tent on their hands and feet with as much ease as they walked or ran erect, and seemed indeed to communicate with their canine playmates. From that association the child goes on; caring for and mastering beasts larger and stronger than himself. He begins by driving and occasionally riding sheep; then he learns to tether cattle—avoiding their horns and taking them by nose ropes or the foreleg; and eventually he handles horses and comes to the day when he finds himself on a horse—pulling hard on the reins—and in charge of the livestock he herds.

What influence these manifold situations and experiences which are created by the dogs have on the children of the nomads, and how they effect personality and character formation are problems for the child psychologist and student of personality in culture to analyze and assess with finality. It would, however, seem that these formative influences do not produce the conformist and the socially cooperative individual who is skilled in adjusting to the requirements of associative living with other individuals. Tentatively, it may be suggested that these situations, and the influence they exert, have played a part in making the nomad the independent, prickly individualist and aggressive fighter which he so frequently turns out to be.

In relation to this facet of the nomadic personality one more role of the dog, a role which is symbolic rather than real and yet which has aspects of reality, is of final interest. The dogs belong to the womenfolk of the tents in a somewhat special way. They are fed by the women and respond to that care with an affection which extends beyond the individual women who feed them to any and all women to such a degree that any woman—even a stranger—is in less danger from attack by them than is any man. The men, on the other hand, who are most in danger are those who are proudly mounted and carrying their guns, which are their most prized possessions, on their backs; for it is that very gun with its long forked rest which makes the trick of whirling the protection rope much more difficult. Once that rope is tangled and the protective it creates is lost, the horse seems to sense in anticipation the slashing hamstringing attack of the dogs and breaks into wild unmanageable panic. It is then that the horseman, who epitomizes the arrogance of the nomad—the mounted male intoxicated with the feel of a horse's power between his knees,—may be unhorsed to fall among the dogs. One of the tales about Gesar the national epic hero is that he was chased by dogs, his horse reared and he fell and was killed.

As an old and very wise Tibetan remarked, "When there are dogs which are the special animal of the womenfolk, the pride of the man is brought down to earth."

## Notes

1. According to sDe gZHung Rin Po CHe the story of the dream of *Ras Chung* and the interpretation is found in Vol. I of *CHag Med Ri CHe* which, however, is not available for folio citation.
2. Pelliot, P., *Histoire Ancienne du Tibet*, 1961, p.3.
3. This explanation was given by dDe gZHung Rin Po CHe, the most learned and candid of the three Tibetan scholars previously cited.
4. Thomas, F.W., *Ancient Folk-Literature from North-East Tibet*, 1957 pp.14, 88, 131.
5. Yule, Sir H., *Travels of Marco Polo*, Vol. II (1903), pp.49-50. The front piece of Vol. I, taken from the original Italian edition, shows the Tibetan mastiff which Marco Polo brought to Italy with him. [See Chapter 3 for the correct interpretation of the origin of frontispiece. /DM]
6. *RjE bTSun Mi Lai rNam THar Dang mGur aBum*, Vol II f.123-a.

# Appendix 3

## Adaptive Traits of Tibetan Mastiffs:
### Natural *vs.* Cultural, Landrace *vs.* Purebred

By Don Messerschmidt

| Physical and Behavioral Traits and Characteristics | Adaptation to Environment | | Natural Adaptation of Landrace Dogs | Cultural Adaptation of Landrace and Purebred Dogs |
|---|---|---|---|---|
| | **Natural:** Physical, zoological, climatological, etc., including 'pack' behaviors. **Cultural:** Based on human belief, preference, superstition, etc.; including guardian 'function' in landrace and 'looks' in purebred show dogs. | | | |
| Strong, powerful dog, massive frame, well boned and muscled, powerful hindquarters; never light, but agile afoot. (It's been observed that when two or more dogs work together, relative lightness and agility in one may balance out mass and power of the other, especially when confronting of attempting to corner or hamstring a predator. | Confronting large, powerful predator (wolf, leopard, bear, etc.) and human intruders. | | X | |
| Aggressive, fierce, territorial, protective. | Further protection against predators and intruders, especially of owner's livestock, territory and family. ('Appropriately' aggressive behavior is more adaptive—i.e., selectivity aggressive towards intruder, while maintaining gentleness with own stock and human companions. | | X | |
| Intelligent, independent, loyal; commands respect. (Able to make independent judgments/decisions/choices of behavior when out of range of human control.) | Companionship, loyalty to master and home or encampment (or other asset being guarded). | | X | X |
| Deep chest, enlarged lung capacity | Stamina and survival in high altitudes (low oxygen) zone. | | X | |

## Appendix 3. Adaptive Traits of Tibetan Mastiffs

| Trait | Function/Significance | | | | |
|---|---|---|---|---|---|
| Neck and shoulders heavily coated, with mane (enhanced in males); strong jaws and muzzle (extra long jaws for huge bite; eye teeth curve backwards). | Protection against predator attacks to the neck. | X | | | |
| Dense double coat, fairly long and thick, with heavy, woolly undercoat (shed in summer). | Protection against cold weather, biting wind and bites to the body (matted coat may provide even more protection). | X | | | |
| Medium to long tail curled over back; pendant ears. | Differentiates dogs from wolves at a distance; reduces the chance of being mistaken (or shot by mistake). | | | | X |
| Well-feathered tail. | Protects muzzle against snow, cold and wind while sleeping. | X | | | |
| Single annual estrus (like wolves). | Seasonal adaptation (livestock and herder encampment normally stationary during period of birth and weaning). | | X | X | |
| Dark color (primarily black or black-and-tan). | Dark color (primarily black or black-and-tan). | | X | X | |
| Tan spots over eyes (in black-and-tan dogs). | Signified supernatural ability to detect evil. | | | | X |
| White blaze on chest (occasional). | Signifies stout heart, bravery. | | | | X |
| Loud, sonorous bark. | Identifies and locates dog; designates territory; carries well over great distances. | | X | X | |
| Noble, dignified look. | A 'commanding presence' where the 1st level of 'force' is mere presence of an impressive enforcer; attractive and commanding presence on exhibition; greater marketability of breed. | | X | X | |
| Tendency to 'low normal' thyroid levels (without negative symptoms or consequences). | Maintains fitness on less or low quality nutrition. | X | | | |
| Long memory of friends and foes, location and landscape, terrain and structures. | A protective behavior (makes them better 'cops'), noticing and investigating changes or aberrations. | X | | | |
| Strong prey drive towards non-flock animals combined with scavenger instinct/behavior (i.e., able and willing to eat almost anything). | Less reliant on feeding by humans, greater survival during difficult times. | | X | X | |

| | | | |
|---|---|---|---|
| Strong pack instinct and structure. | Allows working together in packs; members may take better care of each other. (Aggressive attack by pack members on wounded pack members has also been observed.) | X | |
| Retains instinctive avoidance of breeding with close relatives (especially by bitches). | As in wild canids, inbreeding avoidance enhances the long-term viability of the breed. | X | |
| Instinct to regurgitate food for pups. | Increased viability and survivability of offspring. | X | |
| Strong maternal/protective instinct (in both genders). | Enhances and extends survival chances. (Primarily vis-à-vis other pack members, but observed to extend to human companions (included 'pack-like' in dog's social circle.) | X | |

# Appendix 4

# Coat Color Genetics in Tibetan Mastiffs

By Charles W. Radcliffe and Matthew J. Taylor*

Have you ever been to a show that sports Tibetan Mastiffs appearing in a variety of colors and asked yourself where these colors come from, or how a breeder can predict what colors will appear in a given litter? Well, hopefully, this article and some color photos** will begin to shed some light on this colorful subject. Before we get to the fun stuff, however, we will have to review a little genetics.

Most vertebrates, including dogs, have two copies of every chromosome (except sex chromosomes). Since genes make up the chromosomes, this means that there are two copies of every gene present in every cell of every dog. However, the two copies do not have to be identical. They can be different versions (alleles) of the same gene. For example one could be an allele for Black coat and one for Gold coat. Hmmm, this must be a Black and Gold spotted dog! No, because some alleles are dominant to others. Since Black is dominant to Gold, this dog will appear as solid Black.

Although the above example shows a simple relationship between two alleles located at one gene locus, coat color in dogs is ultimately quite complicated and not fully understood. Many genes, some of which have more than two allele choices, control the variety of colors and patterns seen in canine coat color. The many different genes interact by intricate rules, to create the final coat.

For the sake of discussion, each gene locus and all the alleles related to coat color are given alphabetical designations. Dominant alleles are shown in capital letters and recessive alleles are given in lower case letters. For the purposes of this article, only those loci and alleles seen in the Tibetan Mastiff will be discussed. Given that this breed is so colorful, that happens to be most of them.

Okay, back to our example. The gene locus involved in the example is the Agouti, or *A* locus. Three alleles of this gene are found in the Tibetan

---

*Originally published in the *TMCA Sentinel* v. 4, 2001; revised (2008) and reprinted here by permission of the authors and the Tibetan Mastiff Club of America.
**Images referred to in this article are to be found on the back cover of this book.

Mastiff: the allele for Black, designated *A*; the allele for Gold *Ay*; and the allele for Black and Tan, designated *at*. These alleles are given here in order of dominance, where Black is the most dominant and Black and Tan is the most recessive. Although there are three different alleles that are found in the Tibetan Mastiff breed, any given dog can only carry one or two of the three; two copies of the same allele, or two different alleles.

Why two? Well, every dog has two sets of chromosomes (they get one set from each parent) and each set of chromosomes contains one copy of each gene. Hence every dog has two copies of each gene, one from each parent. In our example, the dog is *A/Ay*. This is written so that we see both alleles, Black and Gold, with the most dominant allele, Black, written first. Remember, this dog appears Black, but so would a dog that is *A/A* or *A/at*. Thus, if a dog appears Black (image 1), we may have no idea what the second allele is until we breed that dog and find out what is passed to the offspring. Likewise, dogs that are *Ay/Ay* or *Ay/at* will appear Gold (image 3), as Gold is dominant to Black and Tan. Only when a dog is *at/at* does it appear Black and Tan (image 6).

What we call "Gold" actually ranges in color from Light Gold (image 2) to rich Reddish Gold (image 4), with the precise shade being determined by modifying genes. These modifying genes are called rufous polygenes and act like, well, a collection of checkers (lets say red and white checkers). For example, if a dog has mostly red checkers in its collection, it will appear Reddish Gold. If a dog has mostly white checkers, it will appear Light Gold, and if a dog has a balance of red and white checkers, it will appear Gold. Each puppy receives a handful of checkers from each parent, so Red Gold puppies are unlikely to come from two Light Gold parents and vice versa. These same genes work to determine the shade of tan on Black and Tan dogs. The dogs shown in images 5, 6, and 7, are all Black and Tan (determined by the *A* locus), but the tan varies considerably. So, as in the example above, Black and Red-Tan puppies are unlikely to come from parents that are Light Gold and/or Black and Tan (Light Gold Tan).

So far, the discussion has focused solely on dogs that have what is called full pigmentation. That is, their genetic makeup allows them to develop all colors to their fullest extent: pitch blacks and bright red-golds, golds, and light golds. There are some genes whose effect is to dilute the pigmentation. In other words, whatever *A* locus alleles the dog has will be expressed as expected, but the pigmentation will be diluted, even in the eyes and on the nose leather. In the Tibetan Mastiff, two different genes cause dilution effects. Each of these genes has only a dominant and one recessive allele.

The first of these genes is the Dilute, or *D* locus. The dominant allele *D* is necessary for full pigmentation. The effect of the recessive allele d when homozygous (two identical alleles in the same dog), is to dilute Black coat and nose and eye pigment toward Blue/Gray. Although this gene is completely separate from the *A* locus, there is an interaction that produces the final color of the dog. So, if a dog is *A/-*, *d/d* (where the dash represents any allele choice) it will appear Blue/Gray (image 8). If a dog is *at/at, d/d* then it will appear Blue/Gray and Tan (image 9).

The second gene that serves to dilute coat and pigment is the Brown, or *B* locus. This gene works the same way as the *D* locus. That is, when both copies of the *B* gene are recessive (*b*), in this case the Black coloration fades to Chocolate/Brown. Again, this gene is completely separate from the *A* locus, but the interaction of the two loci determines the final color of the dog. Thus, if a dog is *A/-, b/b* it will appear Chocolate/Brown (image 10). If a dog is *at/at, b/b* it will appear Chocolate/Brown and Tan (image 11). Again, one dominant allele *B* is necessary in any dog to get the normal black pigment. Any dog with black nose leather is not homozygous for any dilute gene. At least one dominant allele of both of these genes is required for the full black nose coloration.

Finally, a dog can receive a pair of recessive alleles at both the *D* and *B* loci. Although, for the moment, and to our best knowledge, the Western world has not produced a Tibetan Mastiff with this genetic makeup, it is thought that one would appear somewhat like a Weimaraner in color. The TMCA currently refers to this color as Double Dilute. Of course, this would dilute the black coat and pigment in Black and Tan dogs as well, so there could be dogs that can appear as Double Dilute and Tan *(at/at, b/b, d/d)*.

Gold dogs with dilution are a little more complicated. They still appear gold, but appear more washed-out than normal, and their nose leather may appear Blue/Gray, Chocolate/Brown, or some muddy combination of the two. These dogs are *Ay/-, d/d* (where the dash represents either *Ay* or *at*), *Ay/-, b/b*, or *Ay/-, b/b, d/d*. The TMCA refers to ALL of these colors as Gold Dilute (image 12), as determining exactly which dilutions are at work may be difficult.

So far, so good, and the above mechanisms are well established in the dog world, but what follows is not. There is one type of Black Tibetan Mastiff that is turning up more frequently in American litters. Breedings of two Black and Tan dogs (*at/at* mated with *at/at*) are yielding puppies that are all Black. This is completely unexpected! Since the allele for Black, **A**, is dominant to Black and Tan, *at*, neither of the parents could be carrying Black **A**, or that dog would appear Black itself. So, some other explanation must exist for these dogs to appear solid Black. Breeders have found that when bred, some of these dogs produce as if they were Black and Tan, not as Black. So, genetically, they are just what one would expect, *at/at*. One explanation for the conundrum would be that there is a recessive modifying gene present that completely masks the tan, yielding a Black dog. In some cases, a few tan hairs develop between the toes or under the tail, and over time may become more traditional in patterning. However, at birth, these Black pups are indistinguishable from the true Black colored puppies (*A/-*). The TMCA is now referring to these as recessive blacks. A recessive Black female is seen in image 13. It is also clear that this recessive masking gene can mask all the gold on genetically gold TM as well. The only way to tell whether a recessive black TM is really a masked black and tan or a masked gold is to breed it and see what you get. These recessive black dogs can also appear in matings of Gold dogs and the explanation is exactly the same. Although we do not know of any cases, we assume this masking gene would also affect dogs carrying the two dilute genes for blue and chocolate, thus converting genetically blue and tan and chocolate and tan dogs into blues and chocolates respectively.

For clarity, and to move us to the fun part, a couple of genetic word definitions will help; the first is phenotype. This refers to what is seen. A Tibetan Mastiff's color phenotype may be Black, for example, or Black and Tan. The second word is genotype, which describes the genetic recipe (our alphabet soup) carried by that dog. For example, a Tibetan Mastiff genotype may be *A/at* (which may appear as a Black phenotype), or *at/at* (which may appear as a Black and Tan phenotype).

Oh yeah, here is the fun part. Ready?

Let's look at an illustration using dogs that are phenotypically Black (not including the recessive blacks), that is, these are Black dogs with no tan points, although they may have markings like white on the chest (to be discussed in a future article). Keep in mind that Black dogs must have at least one copy of the dominant allele *A*. Now for this dog to be full-pigmented Black, it must also have at least one dominant allele *D* and at least one dominant allele *B*. If we do not know any more than how the dog appears, the genotype of this dog can then be written, *A/-*, *B/-*, *D/-* (as alleles represented by the dashes cannot change the phenotype—how the dog appears). So, we cannot tell by looking at the dog whether it is carrying any or all of the recessive alleles. Only breeding will reveal whether this dog is carrying the Gold allele (*Ay*), the Black and Tan allele (*at*), the recessive Blue/Grey allele (*d*), or the recessive Chocolate/Brown allele (*b*).

So, how does a breeder find out the exact genotypes of his/her breeding stock?

The first thing to remember is that each parent contributes to their offspring exactly one of the two alleles it is carrying for every gene in its chromosomes. This means that when the sperm containing one allele of each gene fertilizes the egg containing one allele of each gene, then the offspring will once again have two alleles for each gene. As an example, lets' say a breeder crosses a Black dog (image 1) with a Gold bitch (image 3). For simplicity sake, it is assumed that no recessive dilution alleles are carried by either parent. So, the Black sire has to be *A/-* because he is Black,

# Appendix 4. Coat Color Genetics in Tibetan Mastiffs

and the Gold dam has to be *Ay/-* because she is Gold. The result yielded 8 puppies: 4 Blacks, 2 Golds, and 2 Black and Tans.

Why are there more Blacks than anything else? Where did those Black and Tans come from? To answer these questions, the breeder reasons backwards. If the breeding yields any Black and Tan puppies at all, then considering that these puppies' genotype MUST be *at/at*, we then know that each parent must also carry **at**. So now the breeder knows exactly what the genotypes of the parents must have been. The Black parent was *A/at* and the Gold parent was *Ay/at*. To find the ratios of what the breeder should expect in the litter, a list of all combinations should be made.

The Black male can produce sperm containing the *A* allele, but he will produce an equal number with the **at** allele. The Gold female can produce eggs with the *Ay* allele and an equal number with the *at* allele. What follows is a list of all possible outcomes.

| | |
|---|---|
| *A* from father, *Ay* from mother | offspring will be *A/Ay* and appear Black |
| *A* from father, *at* from mother | offspring will be *A/at* and appear Black |
| *at* from father, *Ay* from mother | offspring will be *Ay/at* and appear Gold |
| *at* from father, *at* from mother | offspring will be *at/at* and appear Black and Tan |

Notice that the ratio is 2 Blacks to 1 Gold to 1 Black and Tan, which is the same as the 4:2:2 that was yielded in the litter. It is important to say that the 2:1:1 ratio is EXPECTED in the offspring, but not guaranteed. Statistical variation will determine what actually appears. Although many combinations are possible, most litters from these parents will yield colors near that ratio. A breeding between different Black and Gold parents could produce either all Black puppies, or half Black puppies and half Gold puppies. It is left to the reader to work out the genotypes necessary to produce these results.

As another example, our breeder crosses a Black and Tan male having medium toned tan points, with a Gold female like the one pictured in image 3. Remember that the tan points on this male appear as a medium tan because of the rufous polygenes (in this case, a balanced checker collection). Assuming the Gold female (also with a balanced checker collection) is *Ay/Ay*, and knowing that the male is *at/at*; what will the breeder get? The male can only produce sperm with at and the female can only produce eggs with *Ay*, so all of the offspring will be *Ay/at*, and appear gold, right? Well, yes and no. All the offspring will be *Ay/at* alright, but because the male had a pattern of rufous polygenes that determined his tan points should be medium toned (about equal numbers of red and white checkers, to return to our analogy) and the female the same medium toned Gold; the offspring could be anything from rich Red (having randomly received lots of red checkers from both parents) to Light Gold (having randomly received lots of white checkers from both parents).

There is some prejudice against Light Gold in some Tibetan Mastiff circles, mainly in Europe where they were unlucky enough to not have this gene included in the original few dogs that were imported there. (These cases where an original small group of founders do not include an accurate sample of the genes in the whole population are well known sampling errors and are referred to as founder effects in genetics.) This example above shows (and many breedings have confirmed it) that you will often get Light Gold puppies from parents of "acceptable" colors, carrying "acceptable" genes. All of these offspring in the above cross are genetically identical *Ay/at* with respect to the main color determinants, but they can exhibit the full range of gold colors. Only variation in the rufous polygenes accounts for the differences in color. This supports the inadvisability of having color prejudice against any color found Tibetan Mastiffs in Tibet (now China) or the rest of the range. It simply further limits an already limited gene pool while discriminating against otherwise sound dogs.

One last interesting point, it is theoretically possible (but unlikely in the extreme, you would probably need to produce over 100 puppies to see all of the statistically unlikely combinations) to breed a Black dog with a Gold dog and get all the possible colors discussed in this article. For this to happen, however, the genotypes of the parents would have to be *A/at, D/d, B/b* for the Black parent and *Ay/at, D/d, B/b* for the Gold parent. This cross will give Blacks, Golds, Black and Tans, Chocolate/Browns, Chocolate/Brown and Tans, Blue/Greys, Blue/Greys and Tans, Gold dilutes, Double Dilutes, and Double Dilutes and Tans (and recessive blacks if those recessive genes are also present). If the reader wants to work out the ratios, it is suggested he/she finds Punnet squares in an old genetics text and makes up one with 8 squares on a side. Good luck.

# Appendix 5

# The Tibetan Mastiff in Popular Fiction

The mythical huge size of Tibetan mastiffs has made it into pulp fiction. Physically (and spiritually) powerful and gigantic Tibetan dogs are featured in several Westerns by the popular novelist Louis L'Amour, for example. The first one showed up in *Treasure Mountain* (1972):

"I heard a low growl. Mister, if that dog wasn't half bear he was half of something that was big, and he was mean and ugly. He must have weighed two hundred and fifty pounds. He had a head like a bull mastiff and teeth that would give one of them dinnysouers a scare.

'It's all right, Neb,' Nell said. 'He's friendly.'

'If I wasn't,' I said, 'I'd start being. That's the biggest durned dog I ever did see.'

'He's big, all right'."

Another big one appears under mysterious circumstances in L'Amour's *Haunted Mesa* (1987), a Western with a science fiction twist. The dog's name is 'Chief', and he is able to move between this world and the next. About his size, Chief is described as "unusually large, weighing 160 pounds. The Tibetan mastiffs have been guard dogs for thousands of years, known to fight bears, tigers, or wild yaks, to attack anything invading their premises.... a proud, fierce dog, afraid of nothing..."

As a writer, Louis L'Amour had a powerful imagination and a good way with words. In his youth he traveled the world and lived a life of high adventure. He visited China and is said to have hung out with Tibetan bandits for part of the time; and there, it seems, he discovered the big dogs for himself.[1]

## Note

1. Louis L'Amour's Tibet connection is described in *The American West* by Larry Schweikart and Bradley J. Birzer (2002).

# Bibliography

Allen, Charles, 2004, *Duel in the Snows: The True Story of the Younghusband Mission to Lhasa*, London: J. Murray.

Andrews, Ed and Randy Kidd, 1982, 'Livestock protection dogs,' *The Mother Earth News*, January/February, pp.126-7.

Anker, C. and D. Roberts, 2001, *The Lost Explorer: Finding Mallory on Mt. Everest*, New York: Simon and Schuster.

Anonymous, 2007, 'Tibetan Mastiff'; URL: http://hi.tibetwindow.com/阿伽蓝/read_2387.html (June 14).

Armstrong, Karen, 2005, *A Short History of Myth*, New Delhi: Penguin Books.

Ayto, John, 1990, *Dictionary of Word Origins*, New York: Arcade Publishing/Little, Brown and Co.

Bailey, F.M., 1945 (1911), *China—Tibet—Assam: A Journey*, London: J. Cape.

Bailey, F.M., 1946, *Mission to Tashkent*, Oxford and New York: Oxford University Press.

Bailey, F.M., 1957, *No Passport to Tibet*, London, Hart-Davis.

Bailey, Irma (Hon. Mrs. Eric), 1937, 'Dogs from the roof of the world,' *American Kennel Gazette*, vol. 54, no. 3.

Baskaran, Theodore, 1985, 'India's canine heritage,' *Swagat* (Indian Airlines magazine), May, pp.43-48.

Bates, Neale, 2008, 'Tracking man-eaters: The Jim Corbett story,' *ECS magazine* (Kathmandu), July, pp.44-46.

Battuta, Ibn, 1994, *Travels of Ibn Battuta: AD 1325-1354* (4 vols.), H.A.R. Gibb (editor), London: The Hakluyt Society.

Beazley, Sir Charles R., 1897-1906, *The Dawn of Modern Geography: A History of Exploration and Geographical Science* (3 volumes), London: J. Murray.

Beckmann, Ludwig, 1880, (article) in *Leipziger Illustrierte Zeitung* (*Leipzig Illustrated Magazine*), quoted in Max Siber, 1995 (1897), *The Venerable Tibetan Mastiff*, Centreville, AL: OTR Publications; revised and edited by Cathy J. Flamholtz (editor) from *Der Tibethund* (*The Tibetan Dog*), Vienna: Paul Gerin.

Bell, Sir Charles, 1924, *Tibet Past and Present*, Oxford: The Clarendon Press.

Bell, Sir Charles, 1928, *The People of Tibet*, Oxford, UK: The Clarendon Press.

Bell, Sir Charles, 1931, *The Religion of Tibet*, Oxford, UK: Clarendon Press.

Bell, Sir Charles, 1946, *Portrait of the Dalai Lama*, London: Collins.

Bergreen, Laurence, 2007, *Marco Polo: From Venice to Xanadu*, New York: Alfred A. Knopf.

Berry, Scott, 1989, *A Stranger in Tibet: The Adventures of a Wandering Zen Monk*,

London: Collins, and Tokyo and New York: Kodansha International.

Bierce, Ambrose, 1911, *The Devil's Dictionary* (e-text version based on Aloysius West, April 15 1933); URL: www.alcyone.com/max/lit/devils/d.html.

Bird (Bishop), Isabella L., 1894, *Among the Tibetans*, New York and Chicago: F.H. Revell Co.

Bishop, Isabella Bird (see Bird, Isabella).

Blackwood, William and Sons, 1852, 'Nepaul,' *Blackwoods Edinburgh Magazine* (William Blackwood and Sons, Edinburgh and London), vol. 71, no. 441 (July), pp.86-98.

Blakeney, T.L., 1971, 'The First Steps Towards Mount Everest,' *Alpine Journal*, no. 76.

Bonvalot, Gabriel, 1891, *Across Thibet: Being a Translation of 'De Paris au Tonking a Travers le Tibet Inconnu, with Illustrations from Photographs taken by Prince Henry of Orleans* (translated by C.B. Pitman) (2 volumes), London: Cassell.

Boswell, James, 1900, *The Life of Samuel Johnson, LLD.* (3 volumes) (edited by Percy Fitzgerald), London: Swan Sonnenschein and Co., Ltd. p.256.

Bowen, R., 2004, 'Early sterilization in dogs and cats'; URL: www.vivo.colostate.edu/hbooks/pathphys/reprod/petpop/early.html.

Bower, Hamilton (Captain), 1893, *Diary of a Journey across Tibet*, London: Rivington, Percival and Co.

Brauen, Martin (compiler and editor), 2002, *Peter Aufschnaiter's Eight Years in Tibet*, Bangkok: Orchid Press.

Breashears, D. and A. Salkeld, 1999, *Last Climb: The Legendary Everest Expeditions of George Mallory*, Washington DC: National Geographic.

Brehm, Alfred E., 1895, *The Animals of the World: Brehm's Life of Animals*, Chicago: A.N. Marquis and Co.

Brown, Doug, 2007, 'The Bible delusion' (review of *Misquoting Jesus: The Story Behind Who Changed the Bible and Why*, by Bart D. Ehrman, New York: HarperOne (*Powells Review-a-Day*; URL: www.powells.com/review/2007_07_14.html).

Buckland, C.E., 1906, *Dictionary of Indian Biography*, London: Swan Sonnenschein and Co.

Budiansky, Stephen, 1999, 'The truth about dogs,' *The Atlantic Monthly*, *The Atlantic Online*, July; URL: www.theatlantic.com/issues/99jul/9907dogs1.htm.

Burrard, Sidney Gerald and Horace H. Hayden, 1933 (1907-8), *A Sketch of the Geography and Geology of the Himalaya Mountains and Tibet* (revised edition), Calcutta: Manager of Publications.

Burrard, Sidney Gerald, 1915, *Records of the Survey of India*, Vol. 8: *Exploration in Tibet and Neighbouring Nations 1879-1892*, Dehra Dun, India: Office of the Trigonometrical Survey.

Bush, H.W., 1908, 'The dogs of Tibet,' *The Kennel Encyclopedia*, London.

Buxton, Bill, 2005, *Dramatis Personae of the History and Exploration of the*

*Greater Himalaya, Karakoram, Pamirs, Hindu-Kush, Tibet, Afghanistan, High Tartary and Surrounding Territories, up to 1921*; URL: www.billbuxton.com/dramatis.html.

Byers, Alton C., 1985, 'Resource management in the arid Himalaya: Problems and prospective solutions,' *Contributions to Nepalese Studies* vol. 13, no. 3, pp.107-36.

Calvino, Italo, 1974, *Invisible Cities (Le Città Invisibili)* (translated by Helen and Kurt Wolff), Orlanda, FL: Harcourt/Harvest Books.

Cameron, Nigel, 1970, *Barbarians and Mandarins: Thirteen Centuries of Western Travellers in China*, Oxford, UK: Oxford University Press.

Camões, Luiz Vaz de, 1572, *Os Lusiadas (The Lusiads)*; URLs: http://lusiadas.gertrudes.com and www.gutenberg.org/etext/3333. Translated by Henry Yule and A.C. Burnell, 1903 (in *Hobson-Jobson: A Glossary of Colloquial Anglo-Indian Words and Phrases, and of Kindred Terms, Etymological, Historical*, London: Murray).

Campbell, Mary, 1988, *The Witness and the Other World*, Ithaca: Cornell University Press.

Canine Information Library, 2003-2009, 'Molosser breeds (Molosser dogs, Molossers, Mastiff breeds) (Molossoid breeds, Molossians, Molossi)'; URL: www.bulldoginformation.com/molossers-mastiff-type-dogs.html.

Canine Information Library, 2006, 'Tibetan Mastiff (Tibetan Dog, Tsang Khyi, Do Khyi or Phyu-khi)'; URL: http://dogbreeds.bulldoginformation.com/tibetan-mastiff.html.

Carey, William, 1983, *Travel and Adventure in Tibet, Including the Diary of Miss Annie R. Taylor's Journey from Tau-Chau to Ta-Chien-Lu through the Heart of the Forbidden Land*, Delhi: Mittal Publications.

Carruthers, Douglas, 1914, *Unknown Mongolia: A Record of Travel and Exploration in North-West Mongolia and Dzungari* (2 volumes), London: Hutchinson.

Cavenagh, Orfeur, 1851, *Rough Notes of the State of Nepal, its Government, Army, and Resources*, Calcutta (privately printed).

Cervantes, Miguel de, 2004 [1613], *The Dialogue of the Dog* (translated by William Rowlandson), London: Hesperus Press.

Chapman, F. Spencer, 1938, *Lhasa, the Holy City*, London: Chatto and Windus.

Chinanews, 2006, 'Tibetan mastiff may die out'; URL: http://english.cri.cn/2238/2006-1-24/64@295228.htm.

Christie, Agatha, 1970, *The Body in the Library*, London: Hamlyn.

Christopher, Tom, 2002, *In the Land of the Blue Poppies: The Collected Plant-Hunting Writings of Frank Kingdon-Ward*, New York: The Modern Library.

Cooper, Lauren, 2009, 'Tibetan Mastiff becomes world's priciest dog: Chinese woman pays $600,000'; URL: www.dailyfinance.com/2009/09/12/tibetan-mastiff-becomes-worlds-priciest-dog-chinese-woman-pays/ (September 12).

Coppinger, Raymond and Lorna Coppinger, 2001, *Dogs: A New Understanding* of Canine Origin, Behavior, and Evolution, Chicago: University of Chicago Press.

Corbett, Jim, 1944, Man-Eaters of Kumaon, Oxford, UK: Oxford University Press.

Corbett, Jim, 1954, The Temple Tiger and More Man-Eaters of Kumaon, Oxford, UK: Oxford University Press.

Cordier, Henri, 1920, Ser Marco Polo: Notes and Addenda to Sir Henry Yule's Edition, Containing the Results of Recent Research and Discovery, New York: C. Scribner's Sons.

Cotter, Eugene L., 1999, Roots of English; URL: http://ablemedia.com/ctcweb/showcase/roots.html#roots.

CTIC (China Tibet Information Center), 2007, 'First Tibetan mastiff exhibition in Jinjiang,' en.tibet.cn. (February), Beijing: China Tibet Information Center; URL: http://en.tibet.cn/en_index/hac/t20070202_206114.htm.

CTIC (China Tibet Information Center), 2005, 'Calling for protection of Tibetan mastiff'; URL: http://zt.tibet.cn/english/zt/unspoiledland/200402005126111120.htm.

Cunliffe, Juliette with Susan Elworthy, 2007, Tibetan Mastiff: A Special Rare-Breed Edition, A Comprehensive Owner's Guide, Freehold, NJ: Kennel Club Books, a division of BowTie, Inc.

Cunliffe, Juliette, 1997, 'What is a "Prapso"?,' Tibetan and Oriental Breeds International (previously Tibetan Breeds International Magazine), no. 9, p.30.

Cunliffe, Juliette, 2002, 'Early history of the Tibetan Mastiff,' Our Dogs (U.K.), Breed Feature, July, 12, p.31-32.

Cunliffe, Juliette, 2005, 'Special breeds from "The Roof of the World",' Dog World (UK), Special Supplement: Tibetan Breeds Showcase, September 9, p.3

Dalke, Kate, 2002, 'Who is that doggy in the window?—Scientists trace the origin of dogs,' Genome News Network; URL: www.genomenewsnetwork.org/articles/11_02/dog.shtml.

Darwin, Charles, 1868, The Variation of Plants and Animals Under Domestication, 2 vols., London: John Murray (in The Complete Work of Charles Darwin Online); URL: http://Darwin-online.org.uk.

Das, Sarat Chandra, 1902, A Journey to Lhasa and Central Tibet, London: J. Murray.

David-Neel, Alexandra, 1936, Tibetan Journey, London: John Lane.

David-Neel, Alexandra, 2005 (1927), My Journey to Lhasa (translated from Voyage d'une Parisienne à Lhassa), New York: Harper Perennial.

Dawson, Christopher, 1955, The Mongol Mission, New York: Sheed and Ward.

Denman, Earl, 1954, Alone to Everest, London: Collins.

Desideri, Ippolito, SJ, 1932, An Account of Tibet: The Travels of Ippolito Desideri of Pistoia, SJ, 1712-1727, Filippo de Filippi (editor), London: George Routledge and Sons.

Dong Zhen, 2007, 'Celebrities' pooch—magnificent Tibetan mastiff,'

Shanghai Daily, Wednesday 23 May; URL: www.shanghaidaily.com/sp/article/2007/200705/20070523/article_316793.htm.

Draghi, Paul Alexander, 1982, 'The stag and the hunting dog: A Bhutanese religious dance and its Tibetan source,' *Journal of Popular Culture* vol. 16, no. 1, pp.169-75.

Dunham, Mikel, 2004, *Buddha's Warriors : The Story of the CIA-Backed Tibetan Freedom Fighters, the Chinese Invasion, and the Ultimate Fall of Tibet*, New York: J.P. Tarcher.

Durand, Dana B., 1939, (Review of) *Marco Polo: The Description of the World, Vol. II [A Transcription of Z: The Latin Codex in the Cathedral Library at Toledo]*, by A.C. Moule and Paul Pelliot, editors, *Isis* vol. 30, no. 1 (February), pp.103-9.

Eckfeld, Tonia, 2005, *Imperial Tombs in Tang China, 618-907: The Politics of Paradise*, London, New York: Routledge.

Edel, Leon, 1984 (1959), *Writing Lives: Principia Biographica*, New York and London: W.W. Norton and Co.

Ehrman, Bart D., 2005, *Misquoting Jesus: The Story Behind Who Changed the Bible and Why*, New York: Harper San Francisco.

Eiselin, Max, 1961, *The Ascent of Dhaulagiri*, London: Oxford University Press.

Ekvall, Robert B. and James F. Downs, 1987, *Tibetan Pilgrimage*, Tokyo: Institute for the Study of the Languages and Cultures of Asia and Africa.

Ekvall, Robert B., 1963, 'Role of the dog in Tibetan nomadic society,' *Central Asiatic Journal* 8:163-75.

Ekvall, Robert B., 1964, *Religious Observances in Tibet: Patterns and Function*, Chicago and London: University of Chicago Press.

Ekvall, Robert B., 1968, *Fields on the Hoof: Nexus of Tibetan Nomadic Pastoralism*, Prospect Heights, IL: Waveland Press.

Evans, Christopher (with Judith Pettigrew, Yarjung Kromchaī Tamu and Mark Turin), 2009, *Grounding Knowledge/Walking Land: Archaeological Research and Ethno-Historical Identity in Central Nepal*, Cambridge, UK: McDonald Institute for Archaeological Research, University of Cambridge.

Ewans, Martin (editor), 2003, *The Great Game: Britain and Russia in Central Asia*, London: Routledge.

Fischer, Mary, 2007, Listserv correspondence from M. Fischer to K. Sherling re: 'Quote from Ekvall,' 7 September 2007; URL: http://pets.groups.yahoo.com/group/TMTALK2.

Fleming, Peter, 1974 (1961), *Bayonets to Lhasa: The First Full Account of the British Invasion of Tibet in 1904*, Westport, CN: Greenwood Press.

Fraser, James Baillie, 1820, *Journal of a Tour through Part of the Snowy Range of the Himālā Mountains and to the Source of the Rivers Jumna and Ganges*, London: Rodwell and Martin.

French, Patrick, 1994, *Younghusband: The Last Great Imperial Adventurer*, London: HarperCollins.

Fürer-Haimendorf, Christoph von, 1975, *Himalayan Traders: Life in Highland*

*Nepal*, London: J. Murray.

Gardner, Phyllis, 1931, *The Irish Wolfhound: A Short Historical Sketch*, Dundalk, Ireland: Dundalgan Press.

Gee, E.P., 1964, *The Wildlife of India*, London: Collins.

Gentry, Anthea, Juliet Clutton-Brock and Colin P. Groves, 2004, 'The naming of wild animal species and their domestic derivatives,' *Journal of Archaeological Science* 31: 645-51.

Gibb, H.A.R., 1958, *Travels [of Ibn Battutu]*, *A.D. 1335-1354*, Cambridge, UK: Cambridge University Press for the Hakluyt Society.

Gill, William, 1880, *The River of the Golden Sand: The Narrative of a Journey through China and Eastern Tibet to Burmah*, 2 vols., London: John Murray.

Goldstein, M.C. and C.M. Beall, 1990, *Nomads of Western Tibet, The Survival of a Way of Life*, Berkeley and Los Angeles: University of California Press.

Graham, G.A., 1939 (1885), *The Irish Wolfhound*, Dublin: The Irish Wolfhound Club of Ireland.

Greenblatt, Stephen, 1991, *Marvellous Possessions*, Oxford: Clarendon Press.

Hally, Will, 1932, 'Northern Tibetan hunting dog,' *Our Dogs* (UK), December 16.

Hancock, David, 1983, 'The big dog from Tibet,' *Kennel Gazette* (UK), vol. 114, no. 1233 (January), pp.4-5.

Harrer, H., 1957 (1953), *Seven Years in Tibet* (1st edition), The Adventure Library Edition, London: Rupert Hart-Davis.

Harrer, H., 1985, *Return to Tibet* (translated by Ewald Osers), New York: Schocken Books.

Haw, Stephen G., 2006, *Marco Polo in China: A Venetian in the Realm of Khubilai Khan*, Oxon, UK and New York: Routledge.

Hawley, E., 2004, *The Himalayan Database: The Expedition Archives of Elizabeth Hawley* (with Richard Salisbury, creator), New York: American Alpine Club; URL: www.himalayandatabase.com.

Hedin, Sven, 1899, *Through Asia*, New York and London: Harper and Brothers.

Hedin, Sven, 1901, *Adventures in Tibet*, Boston and Chicago, United society of Christian Endeavor.

Hedin, Sven, 1903, *Central Asia and Tibet: Towards the Holy City of Lhasa*, 2 vols., London: Hurst and B.

Hedin, Sven, 1909, *Trans-Himalaya: Discoveries and Adventures in Tibet*, 2 vols., London: MacMillan and Co., Ltd.

Hedin, Sven, 1925, *My Life as an Explorer*, Garden City and New York: Garden City Publishing Co.

Hedley, J., 1910, *Tramps in Dark Mongolia*, London: T. Fisher Unwin.

Hemmleb, J., Johnson, L.A., Simonson, E.R. and W.E. Nothdurft, 1999, *Ghosts of Everest: The Search for Mallory and Irvine*, Seattle: Mountaineers Books.

Hermanns, M., 1949, *Die Nomaden von Tibet*, Wien: Verlag Herold

Herzog, Maurice, 1952, *Annapurna: The First Conquest of an 8000-Metre Peak*, translated by Nea Morin and Janet Adam Smith, London: J. Cape.

Hodgson, Brian H., 1832, 'On the Mammalian of Nepal,' *Journal of the Asiatic Society of Bengal*, vol. 1, pp.335-49.

Hodgson, Brian H., 1833, 'Description of the wild dog of the Himalaya,' *Asiatic Researches* 18(2):221-37.

Hoed, Jan den, 2001, 'The great dog of Tibet,' *TMCA Sentinel* 4 (Tibetan Mastiff Club of America); URL: www.tmcamerica.org/TMCA/Sentinel4/the_great_dog_of_tibet_by_jan_den_hoed_doc.htm.

Hogan, Edmund, 1939 (1887), *The Irish Wolfdog; The Irish Wolfhound*, Dublin: The Irish Wolfhound Club of Ireland.

Holdich, Thomas H., 1906, *Tibet, the Mysterious*, London: Alston Rivers, Ltd., and New York: F.A. Stokes Co.

Hooker, Joseph Dalton, 1854, *Himalayan Journals: Or, Notes of a Naturalist in Bengal, the Sikkim and Nepal Himalayas, the Khasia Mountains, &c.*, 2 vols., London: J. Murray (reprinted in 1980, New Delhi: Today and Tomorrow Printers and Publishers).

Hopkirk, Peter, 1980, *Foreign Devils on the Silk Road: The Search for the Lost Cities and Treasures of Chinese Central Asia*, London: John Murray, Ltd.

Hopkirk, Peter, 1982, *Trespassers on the Roof of the World: The Secret Exploration of Tibet*, London: John Murray, Ltd.

Hopkirk, Peter, 1984, *Setting the East Ablaze: Lenin's Dream of an Empire in Asia*, London: John Murray, Ltd.

Hopkirk, Peter, 1992, *The Great Game: The Struggle for Supremacy in Central Asia*, New York: Kodansha International.

Hopkirk, Peter, 1995, *On Secret Service East of Constantinople: The Plot to Bring Down the British Empire*, Oxford, UK and New York: Oxford University Press.

Hopkirk, Peter, 2001, *Quest for Kim: In Search of Kipling's Great Game*, Oxford, UK: Oxford University Press.

Houston, C.S. and Bates, R.H., 1954, *K2, The Savage Mountain*, New York: McGraw Hill.

Howard-Bury, C.K., 1922, *Mount Everest: The Reconnaissance 1921*, London: E. Arnold and Co.

Huber, Tony, 2004, 'Antelope hunting in northern Tibet: Cultural adaptations to wildlife behaviour,' pp.5-17 in A. Boesi and F. Cardi (editors), *Wildlife and Plants in Traditional and Modern Tibet: Conceptions, Exploitation and Conservation*, Memoire della Societa Italiana di Scienze Naturali vol. XXXII, no. 1, pp.5-17.

Huc, Evariste Régis and Joseph Gabet, 1998 (1852), *Travels in Tartary, Thibet, and China During the Years 1844-1846*, translated from French by W. Hazlitt (*Souvenirs d'un voyage dans la Tartarie, le Thibet, et la Chine années 1844, 1845 et 1846*), 2 vols., New Delhi : Asian Educational Services.

Hummel, Siegbert, 1961, 'Der Hund in der religiösen Vorstellungswelt des

Tibetans,' *Paideuma, vol.* 6, n.8 (November 1958), pp.500-509; vol. 7, no. 7 (July 1961), pp.352-361.

Hunter, William Wilson, 1886, *The Indian Empire: Its People, History and Products*, London: Trübner and Co.

Hutchinson, W., 1934-35, *Hutchinson's Dog Encyclopædia*, 3 vols., London: Hutchinson.

Isseman, Maurice and Stewart Weaver, 2008, *Fallen Giants: A History of Himalayan Mountaineering from the Age of Empire to the Age of Extremes*, New Haven: Yale University Press.

Jackson, D.P., 1976, 'The early history of Lo (Mustang) and Ngari,' *Contributions to Nepalese Studies* (Tribhuvan University, Nepal) vol. 4, no.1, pp.39-56.

Jackson, D.P., 1978, 'Notes on the history of Serib, and nearby places in the upper Kali Gandaki,' *Kailash (Journal of Himalayan Studies)* 6(3):195-227.

Jackson, D.P., 1984, *The Mollas of Mustang: Historical, Religious and Oratorical Traditions of the Nepalese-Tibetan Borderland*, Dharamsala: Library of Tibetan Works and Archives.

Jackson, R., Ahlborn, G., Ale, S., Gurung, D., Gurung, M., and U.K. Yadav, 1994, 'Reducing livestock depredation in the Nepalese Himalaya,' pp.241-7 in R.M. Timm and A.C. Crabb (editors), *Proceedings of the 17th Vertebrate Pest Conference*, Davis, CA: University of California/Davis.

Jackson, Rodney, 1990, *Threatened Wildlife, Crop and Wildlife Depredation and Grazing in the Makalu-Barun Conservation Area*, The Makalu-Barun Conservation Project Working Paper Publication Series, Report 12, Kathmandu: Government of Nepal/Department of National Parks and Wildlife Conservation, and Woodlands Mountain Institute

Jacob, L., 1901, 'Indian breeds,' pp.49-55 in *Dog Owner's Annual 1901*, London: Dean and Son.

Jardine, W. (editor) with Charles Hamilton Smith, 1843, in *The Naturalist's Library*, Vol. 5: *Mammalia, Dogs* Part 2, by C.H. Smith, Edinburgh: W.H. Lizars

Jinpa, Geshe Gelek, Charles Ramble and Carroll Dunham (photographs by Thomas L. Kelly), 2005, *Sacred Landscape and Pilgrimage in Tibet: In Search of the Lost Kingdom of Bön*, New York and London: Abbeville Press.

Johnston, R.F., 1908, *From Peking to Mandalay: A Journey from North China to Burma through Tibetan Souch Uan and Yunnan*, London: J. Murray. Joy, 2005, 'Mastiff makes a big industry', *en.tibet.cn*. (December), Beijing: China Tibet Information Center; URL: http://info.tibet.cn/en/news/tin/t20051208_76985.htm.

Jupp, Hilary, 2005, 'Captain Graham and the resuscitation of the breed,' in 'Irish Wolfhound History,' *Irish Wolfhound Web Ring*; URL: www.irishwolfhounds.org/graham.htm.

Kawaguchi, Ekai, 1909, *Three Years in Tibet*, Madras, Benares and London: Theosophical Publishing Society (reprinted 2005, Bangkok: Orchid Press).

Keay, John, 2000, *The Great Arc: The Dramatic Tale of How India was Mapped and Everest was Named*, London: HarperCollins.

Keyes, Ralph, 2006, *The Quote Verifier: Who Said What, Where, and When*, New York: St. Martin's Griffin.

Kingdon-Ward, Frank, 1910, *On the Road to Tibet*, Shanghai: Shanghai Mercury Ltd.

Kingdon-Ward, Frank, 1913, *Land of the Blue Poppy*, Cambridge, UK: Cambridge University Press.

Kingdon-Ward, Frank, 1923, *Mystery Rivers of Tibet: A Description of the Little-Known Land where Asia's Mightiest Rivers Gallop in Harness through the Narrow Gateway of Tibet, Its Peoples, Fauna, and Flora*, London: Seely Service and Co.

Kingdon-Ward, Frank, 1934, *A Plant Hunter in Tibet*, London: Jonathan Cape (reprinted 2006, Bangkok: Orchid Press).

Kingdon-Ward, Frank, 1990, *Himalayan Enchantment, Frank Kingdon-Ward: An Anthology* (chosen and edited by John Whitehead), London: Serindia Publications.

Kipling, John Lockwood, 1891, *Beast and Man in India: A Popular Sketch of Indian Animals in Their Relations with the People*, London: Macmillan.

Kipling, Rudyard, 1901, *Kim*, London, Macmillan and Co.

Kirkpatrick, William, 1969 (1811), *An Account of the Kingdom of Nepaul: Being the Substance of Observations Made During a Mission to that Country in the Year 1793*, London: W. Bulmer. Reprint edition, Delhi: Manjusri Publishing House.

Knowledgerush.com, 2003, 'Dog'; URL: http://knowledgerush.com/kr/encyclopedia/Dogs.

Komroff, Manuel, 1926, *The Travels of Marco Polo*, New York: Boni and Liveright, Inc.

Komroff, Manuel (editor), 1928, *Contemporaries of Marco Polo*, New York: Boni and Liveright.

Konradsen, Kåre, 2002, 'Molosserworld's origin and history of the Molossers' (Molosserworld); URL: www.moloss.com/001/ori.

Kostova, Elizabeth, 2005, *The Historian*, New York: Little, Brown.

Krassnig, Adolf, 2006, 'Do Khyi—mythos und wirklichkeit—killing legends!,' Kaukasen-*Blättle (Das Sprachrohr der Hirtenhunde)*, translated from German by Barbara Neidereiter; URLs: http://pets.groups.yahoo.com/group/TMPreservation, and www.hirtenhunde-liptak.de/blaettle/pages/rundum/2006-09-dokhyi.html (in 2 parts).

Kreitner, Gustav, 1881, *Im fernen Osten: Reisen des Grafen Bela Széchenyí* (*In the Far East: Travels with Count Bela Széchenyi*), Vienna: Hölder.

L'Amour, Louis, 1972, *Treasure Mountain*, New York: Bantam Books.

L'Amour, Louis, 1987, *Haunted Mesa*, New York: Bantam Books

Lal, Mukundi, 1981, 'Tibetan dogs,' *Journal of the Federation of Kennel Clubs of India* 4(4):17-19.

Landon, Perceval, 1905, *The Opening of Tibet: An Account of Lhasa and the Country* and People of Central Tibet and of the Progress of the Mission Sent There by the English Government in the Year 1903-4, New York: Doubleday Page and Co.

Landor, A. Henry Savage, 1898, *In the Forbidden Land*, 2 vols., London: Heinemann.

Landor, A. Henry Savage, 1910, *An Explorer's Adventures in Tibet*, New York: Harper Brothers.

Lange, Karen E., 'Wolf to woof: The evolution of dogs,' *National Geographic Magazine*, January, pp.4-31.

Langkavel, Bernard, 1882, in *Die Neue Deutsche Jagdzeitung* (*The New German Hunter's Magazine*) (quoted in Max Siber, 1995 (1897), *The Venerable Tibetan Mastiff*, Centreville, AL: OTR Publications (revised), Cathy J. Flamholtz (editor) from *Der Tibethund* (*The Tibetan Dog*), Vienna: Paul Gerin).

Larner, John, 1999, *Marco Polo and the Discovery of the World*, New Haven, CN: Yale University Press.

Latham, Ronald (editor and translator), 1958, *The Travels of Marco Polo*, Harmondsworth, UK: Penguin.

Leighton, Robert, 1907, *The New Book of the Dog: A Comprehensive Natural History of British Dogs and Their Foreign Relatives, with Chapters on Law, Breeding, Kennel Management, and Veterinary Treatment*, 2 vols., London: Cassell and Co., Ltd.

Lewis, Jon E. (editor), 2003, *The Mammoth Book of Eyewitness Everest*, New York: Carroll and Graf.

Lewontin, R.C., 1991, 'Facts and the factitious in natural sciences,' *Critical Inquiry*, vol. 18, no. 1, pp.140-153.

Li Jin, 2005, 'Hounded to extinction,' *China Daily/Beijing Weekend* (Beijing, August 4, 2005); URL: www.chinadaily.com.cn/english/doc/2005-08/05/content_466494.htm.

Li Qian, 2006, *China's Tibetan Mastiff*, Beijing: Foreign Language Press.

Libourel, Jan, 1993, 'The Molossus myth and other mastiff malarkey: A critical look at the ancient history of the bull breeds,' *Bulldog Review* 4(2, #14).

Macfarlane, Robert, 2003, *Mountains of the Mind: A History of a Fascination*, London: Granta Books.

Mair, Victor H., 1998, 'Canine conundrums: Eurasian dog ancestor myths in historical and ethnic perspective,' *Sino-Platonic Papers* 87.

Markham, C.R. (editor), 1999 (1876), *Narratives of the Mission of George Bogle to Tibet and of the Journey of Thomas Manning to Lhasa*, New Delhi: Asian Educational Services (London: Trubner).

Marsden, William (translator), 1818, *The Travels of Marco Polo, a Venetian, in the Thirteenth Century: Being a Description, by that Early Traveller, of Remarkable Places and Things, in the Eastern Parts of the World*, translated from the Italian, with notes, London, Longman, Hurst, Rees, Orme and Brown (reprinted in 1908 in an Everyman Library edition).

Martin, William Charles Linnaeus, 1845, *The History of the Dog*, London: Knight's Weekly Volume.

Matthiessen, Peter, 1978, *The Snow Leopard*, New York: The Viking Press.

Mazuchelli, Nina, 1876, *The Indian Alps and How We Crossed Them: Being a Narrative of Two Years' Residence in the Eastern Himalaya and Two Months' Tour into the Interior*, New York: Dodd, Mead and Co.

McCaig, Donald, 1992, *Nop's Trials*, Guilford, CT: The Lyons Press.

McCaig, Donald, 1998, *Eminent Dogs, Dangerous Men: Searching Through Scotland for a Border Collie*, Guilford, CT: The Lyons Press.

McCaig, Donald, 1998, *Nop's Hope*, Guilford, CT: The Lyons Press.

McCaig, Donald, 2007, *A Useful Dog*, Charlottesville, VA: University of Virginia Press.

McCaig, Donald, 2007, *The Dog Wars: How the Border Collie Battled the American Kennel Club*, Hillsborough, NJ: Outrun Press and Exeter, UK: RevaluatioBooks.

McGovern, William Montgomery, 1924, *To Lhasa in Disguise: A Secret Expedition through Mysterious Tibet*, New York and London, The Century Co.

McGranahan, Carole, 2006, 'Tibet's Cold War: The CIA and the Chushi Gangdrug Resistance, 1956-1974,' *Journal of Cold War Studies* vol. 8, n.3, pp.102-130.

McRae, Michael, 2002, *The Siege of Shangri-La: The Quest for Tibet's Sacred Hidden Paradise*, New York: Broadway Books.

Menzies, Gavin, 2003, *1421: The Year China Discovered America*, New York: Harper/Perennial.

Messerschmidt, Don, 1974, 'Gurung shepherds of the Lamjung Himalaya,' *Objets et Mondes* (Paris) vol.14, pp.307-16.

Messerschmidt, Don with Naresh Jang Gurung, 1974, 'Parallel trade and innovation in central Nepal: Gurung and Thakali Subbas compared,' pp.197-221 in Christoph von Fürer-Haimendorf (editor), *Contributions to the Anthropology of Nepal*, London: Aris and Phillips.

Messerschmidt, Liesl, Tsering Dolma Gurung and Frances Klatzel, 2004, *Stories and Customs of Manang, as Told by the Lamas and Elders of Manang*, Kathmandu: Mera Publications.

Messner, Reinhold, 2002, *The Second Death of George Mallory: The Enigma and Spirit of Mount Everest* (translated by Tim Carruthers), New York: St. Martin's Griffin.

Meyer, K.E. and S.B. Brysac, 1999, *Tournament of Shadows: The Great Game and Race for Empire in Central Asia*, Washington DC: Counterpoint.

Miller, Luree, 1976, *On Top of the World: Five Women Explorers in Tibet*, New York: Paddington Press.

Millington, Powell (Mark Synge), 1905, *To Lhassa at Last*, London: Smith, Elder.

mnteverest.net, 2003, 'Quotes from Everest'; URL: www.mnteverest.net/quote.html.

Moorcroft, William and George Trebeck, 1841, *Travels in the Himalayan Provinces of Hindustan and the Punjab in Ladakh and Kashmir; in Peshawar,*

## Publications on Tibetan Dogs by the Author

**Messerschmidt, Don**

2008 'Women explorers and the dogs they saw in Tibet and the Himalayas,' *TMCA Sentinel* (Tibetan Mastiff Club of America), no. 11 (July), pp.11-12.

2008 'Trekking high in the monsoon, and the shepherds of Lamjung Himal,' *ECS Nepal* (Kathmandu), September, pp.56-60.

2008 'Bhote kukur & the Nepaul dog: Early tributes to the big dogs of the mountains by explorers, missionaries, diplomats and spies,' *ECS Nepal* (Kathmandu), July, pp.52-54.

2007 'Big Dogs of Tibet and the Marco Polo Effect,' *Proceedings of the First International Conference on 'Aboriginal Dog Breeds as Part of Biodiversity and of the Cultural Heritage of Humankind* held in Almaty, Kazakhstan, September 10-15, 2007, sponsored by the Russian Branch of Primitive and Aboriginal Dog Society (R-PADS). Published in *PADS Newletter*, no. 17 (2008). URL: www.pads.ru.

2006 'What's a Bhote kukur? Debate goes on,' *The Himalayan Times* (Kathmandu): *Sunday Perspectives*, June 25, p.4.

2005 'Ghost dogs: To be seen, not heard,' *The Himalayan Times* (Kathmandu): *Sunday Perspectives*, November 27, p.4.

1988 'The bearded Tibetan mastiff of the trans-Himalaya,' *Dog World* (USA), October, pp.14, 130, 133, 134, 160, 178, 180; November, pp.124, 126, 137; and December, pp.130-133.

1988 'On the derivation of 'Emodus' and 'Saipal' kennels,' *ATMA Gazette: Notes from Emodus*, July, vol. 12, no. 4, p.8.

1987 'The mysterious mastiffs of Tibet: Finding the dogs of Kesang Camp,' *Dog World* (USA) vol. 72, no.1, pp.12-13, 135-6, and no. 2, pp.91-92, 94.

1987 'On varieties of TM dogs (Or: Travelers' tales—fact or fancy?),' *ATMA Gazette: Notes from Emodus*, March, vol. 1, no. 2, pp.21-23.

1986 'Tibetan goats, sheep and Humli dogs,' *ATMA Gazette: Notes from Emodus*, May, vol. 10, no. 3, pp.15-16.

1986 'On Tibetan mastiff size,' Notes from Emodus, *ATMA Gazette*, January, vol. 10, no. 1, p.15.

1986 'On dogs in Tibet,' Notes from Emodus, *ATMA Gazette*, September, vol. 10, no. 5, pp.20, 22.

1986 'Kalu and the kat,' Notes from Emodus, *ATMA Gazette*, March, vol. 10, no. 2, pp.20-21.

1985 'TMs in the snow,' Notes from Emodus, *ATMA Gazette*, March, vol. 8, no. 4, pp.14-15.

1985 'TM dogs at work in the Himalaya,' *ATMA Gazette: Notes from Emodus*, May, vol. 9, no. 3.

| | |
|---|---|
| 1985 | "Tall tails', Notes from Emodus,' *ATMA Gazette*, October, vol. 9, no. 2. |
| 1985 | 'Putting on the dog at the Delhi dog show,' *Dog World* (USA) vol. 70, no. 2, pp.22, 145-50. |
| 1983 | 'The Tibetan mastiff: Canine sentinels of the range,' *Rangelands* (Society of Range Management) vol. 5, no. 4, pp.172-4. |
| 1983 | 'On Tibetan mastiff size and sound,' *Dog World* (USA), May, pp.152, 192. |
| 1983 | 'On Tibetan mastiff color, coat and collar,' *Dog World* (USA), September, pp.142-4. Reprinted in *Indian Kennel Trust*, April-June 1984, p.14. |
| 1983 | 'On the "ideal" Tibetan mastiff,' *ATMA Gazette: Notes from Emodus*, January/February, pp.2-3. |
| 1982 | 'Qualities of a Tibetan mastiff,' *Dog World* (USA), September, pp.136-7. |
| 1973 | 'Himalayan highlands,' *Summit*, June. pp.3-9. |

*Kabul, Kunduz and Bokhara 1819-1825* (edited by H.H. Wilson), London: John Murray.

Moule, A.C. and Paul Pelliot (editors), 1935, *Marco Polo: The Description of the World*, vol. II: *A Transcription of Z: The Latin Codex in the Cathedral Library at Toledo*, London: George Routledge & Sons.

Mumford, Stan R., 1989, *Himalayan Dialogue: Tibetan Lamas and Gurung Shamans in Nepal*, Madison: University of Wisconsin Press.

Mumford, Stan R., 2004, 'Shamanic sacrifice and Buddhist renunciation,' pp.256-71 in Andrei Znamenski (editor), *Shamanism: Critical Concepts in Sociology*, London: Routledge.

Murray, Hugh, 1852, *The Travels of Marco Polo, Greatly Amended and Enlarged from Valuable Early Manuscripts Recently Published by the French Society of Geography, and in Italy by Count Baldelli Boni, with Copious Notes, Illustrating the Routes and Observations of the Author and Comparing them with Those of More Recent Travellers*, New York: Harper & Brothers.

NASA, 1999, 'New Height for Mt. Everest: 29,035 ft.,' *NASA Observatorium: Observation of the Week* (December 8, 1999); URL: http://observe.arc.nasa.gov/nasa/ootw/1999/ootw_991208/ob991208.html.

Nelson, D., Laban, P., Shrestha, B.C., and G.P. Kandel, 1980, *The Major Ecological Land Units and Their Watershed Condition in Nepal*. Kathmandu: Government of Nepal/Department of Soil Conservation and Watershed Management, UN Food and Agriculture Organization and UN Development Program.

Nitartha, 2006, *Nitartha Online Tibetan English Dictionary*; URL: www.nitartha.com.

Noel, J.B.L., 1927, *Through Tibet to Everest*, London: Edward Arnold.

Norbu, J., 1999, *The Mandala of Sherlock Holmes: Based on the Reminiscences of Hurree Chunder Mookerjee C.I.E., F.R.S., F.R.G.S., Rai Bahadur*, New Delhi:

HarperCollins India.

Norbu, Thubten Jigme and Colin M. Turnbull, 1968, *Tibet*, New York: Simon and Schuster.

Nouc, Hedy, 1981, 'Himalayan Mastiff—A Show at 3000 Meters,' *Tibetan Mastiff Club of America Quarterly* (translated by Bob Walker).

Oliff, D., 1999, *The Ultimate Book of Mastiff Breeds*, New York: Howell Books.

Panwar, A.S., 1983, 'Himalayan Mastif or Bhotia,' excerpted from pp.7-8 of 'Can India be put on world dog map?,' *Journal of the Federation of Kennel Clubs of India* 6(2):5-9.

Pauthier, M., 1865, *Le Livre de Marco Polo*, Paris: M.G. Pauthier.

Peissel, Michel, 1968, *Mustang: The Forbidden Kingdom*, London: Harvill P.

Peissel, Michel, 1997, *The Last Barbarians: The Discovery of the Mekong in Tibet*, New York: Henry Holt and Co.

Pettigrew, Judith, 1999, 'Parallel geographies: Ritual and political values of a shamanic soul journey,' pp.247-70 in B. Bickel and M. Gaenszle (editors), *Himalayan Space: Cultural Horizons and Practices*, Zürich: Volkerkundemuseum.

Polo, Marco, 1298, *Il Milione*, Italy (see translations by Wright 1880, Yule 1903, and others).

Pradhanang, Tirtha Bahadur, n.d., 'An autobiography of Sagarmatha's height'; URL: www.nepalhomepage.com/himalaya/sagarmatha.html.

Quinn, Diana, 1994, 'The search for the Tibetan KyiApso'; URL: www.muddypaws.com.

Quinn, Diana, 2000, 'The Tibetan KyiApso'; URL:www.muddypaws.com/kyiapso.html.

Radcliffe, Charles W. and Matthew J. Taylor, 2001, 'Coat color genetics in the Tibetan Mastiffs,' *TMCA Sentinel* vol. 4 (Tibetan Mastiff Club of America); revised 2008.

Rana, Amar, 1976, 'The (Tibetan) Himalayan Mastiff,' *The Tibetan Terrier Association Year Book* (England), pp.30-32.

Rigg, Robin, 2001, *Livestock Guarding Dogs: Their Current Use World Wide*, IUCN/SSC Canid Specialist Group, Occasional Paper no. 1; URL:www.canids.org/occasionalpapers.

Rijnhart, Susie Carson, 1904, *With the Tibetans in Tent and Temple: Narrative of Four Years; Residence on the Tibetan Border, and of a Journey into the Far Interior*, London: Oliphant, Anderson and Ferrier.

Rockhill, William Woodville, 1891, *The Land of the Lamas: Notes of a Journey through China, Mongolia and Tibet*, London: Longmans.

Rockhill, William Woodville, 1894, *Diary of a Journey through Mongolia and Thibet in 1891 and 1892*, Washington, Smithsonian Institution.

Rockhill, William Woodville (translator), 1894, *The Life of the Buddha, and the Early History of His Order, Derived from Tibetan Works in the Bkah-hgyur and Bstan-hgyur, Followed by Notices on the Early History of Tibet and*

*Khoten*, London: Trübner and Co.

Roerich, Nicholas, 1929, *Heart of Asia*, New York: Roerich Museum Press.

Rohrer, Ann and Cathy J. Flamholtz, 1989, *The Tibetan Mastiff: Legendary Guardian of the Himalayas*, Ft. Payne, AL : OTR Publications.

Ruttledge, Hugh, 1934, *Everest 1933*, London: Hodder and Stoughton Ltd.

Ruttledge, Hugh, 1937, *Everest: The Unfinished Adventure*, London: Hodder and Stoughton, Ltd.

Ruysbroek, Willem van, 1900 (13th century), *The Journey of William of Rubruck to the Eastern Parts of the World, 1253-55, as Narrated by Himself, with Two Accounts of the Earlier Journey of John of Pian de Carpine* (translated from the Latin and edited with Introduction by William Woodville Rockhill), London: The Hakluyt Society.

Salmeri, K.R., Bloomberg, M.S., Scruggs S.L. and V. Shille, 1991, 'Gondectomy in immature dogs: Effects on skeletal, physical and behavioral development,' *Journal of American Veterinary Medical Association* vol. 198, pp.1193-1203; URL: www.ncbi.nlm.nih.gov/pubmed/2045340.

Savolainen, P., Y.P. Zhang, J. Lou, J. Lundeberg and T. Leitner, 2002, 'Genetic evidence for an East Asian origin of domestic dogs,' *Science* (November), vol. 22, no. 3, pp.1610-13.

Schafer, Edward G., 1963, *The Golden Peaches of Samarkand: A Study of T'ang Exotics*, Berkeley: University of California Press.

Schaller, George B., 1977, *Mountain Monarchs*: Wild Sheep and Goats of the Himalaya, Chicago: University of Chicago Press.

Schweikart, Larry and Bradley J. Birzer, 2002, *The American West*, New York: Wiley and Sons.

Sellar, A.J., 1980, 'Kattuk the Tibetan hunting dog,' in *Tibetan Terrier Association 1980 Yearbook*, Herts, UK: Tibetan Terrier Association.

Shanklin, Eugenia, 1985, 'Sustenance and symbol: Anthropological studies of domesticated animals,' *Annual Review of Anthropology 1985*, vol. 14, pp.375-403.

Shaumian, Tatiana, 2000, *Tibet: The Great Game and Tsarist Russia*, Oxford, UK: Oxford University Press.

Shebbeare, E.O., 1934, 'Transport,' Part III in Ruttledge, H., 1934, *Everest 1933*, London: Hodder and Stoughton Ltd.

Sherling, Kristina, 1987, 'Will varieties be necessary in the Tibetan mastiff,' *ATMA Gazette*, vol. XI, no. 1, p.15.

Sherling, Kristina, 2002, 'Do-Khyi—Tsang Khyi Tibetan Mastiff'; URL: www.moloss.com/001/breed/def/d007/.

Sherling, Kristina, 2002, 'Molosser World: Do Khyi—Tsang Khyi, Tibetan Mastiffs'; URL: www.moloss.com/001/breed/def/2007.

Siber, M., 1995 (1897), *The Venerable Tibetan Mastiff*, Centreville, AL: OTR Publications. Revised and edited in translation by Cathy J. Flamholtz, from *Der Tibethund*, Vienna: Paul Gerin.

Simsova, Sylva, 1979, *Tibetan and Related Dog Breeds: A Guide to their History*,

Welwyn, Hertfordshire, U.K.: Tibetan Terrier Association.

Smethurst, Paul, 2001, 'Travel writing: Writing the East—Marco Polo and Sir John Mandeville' (ENGL-2045), Hong Kong: Department of English, University of Hong Kong; URL: www.hku.hk/english.

Smethurst, Paul, 2005, Introduction,' pp.xxvii-xli, in Marco Polo, *The Travels of Marco Polo*, New York: Barnes and Noble/Library of Essential Reading Series (based on the translation by William Marsden, 1908 [1818]).

Smith, A. Croxton, 1931, *About Our Dogs*, London: Ward Lock.

Smith, Thomas, 1852, *Narrative of a Five Years' Residence at Nepaul*, 2 vols., London: Colburn and Co. (reprinted as *The Nepal Years: Narrative of a Five Years' Residence at Nepaul*, Delhi: Cosmos, 2003).

Smith, Ione L., 2000, 'Spaying and castration—Pros, cons, myths, and Dobermans,' East Tennessee Doberman Rescue; URL: http://etdr.doberinfo.com/health/spayinfo.html.

Snellgrove, David, 1961, *Himalayan Pilgrimage: A Study of Tibetan Religion by a Traveller Through Western Nepal*, Oxford, UK: Bruno Cassirer (reprinted 2010, Bangkok: Orchid Press).

Snellgrove, David, 2000, *Asian Commitment: Travels and Studies in the Indian Sub-Continent and South-East Asia*, Bangkok: Orchid Press.

Snellgrove, David and Hugh Richardson, 1968, *A Cultural History of Tibet*, New York: Frederick A. Praeger (reprinted 2003 Bangkok: Orchid Press).

Soman, W.V., 1963, *The Indian Dog*, Bombay: Popular Prakashan.

Sood, Shubhi, 2005, *Younghusband: Troubled Campaign*, New Delhi: India Research Press.

Speake, Jennifer, 2003, *Literature of Travel and Exploration: An Encyclopedia*, London and New York: Routledge.

Sponenberg, D. Phillip, 2006 (1998), 'Livestock guard dogs: What is a breed, and why does it matter,' *Akbash Sentinel* 44:13-17; URL: www.beautdogs.com/Sponenberg.htm.

Steffel, Judy, 1997(?), 'The Tibetan KyiApso,' *NetpetMagazine.com*; URL: http://netpet.batw.net.

Steffel, Judy, 2002, *TMCA Discussion Group* (listserv of the Tibetan Mastiff Club of America (April 10).

Stephanitz, Max von, 1923, *The German Shepherd Dog in Word and Picture*, Jena: A Kämpfe.

Stevens, Thomas, 1890, 'Some Asiatic dogs,' pp.314-19 in Mary Mapes Dodge (editor), *St. Nicholas: An Illustrated Magazine for Young Folks* (New York: The Century Co., and London: T. Fisher Unwin) vol. XVII, no. 1.

Stewart, Jules, 2006, *Spying for the Raj: The Pundits And the Mapping of the Himalaya*, Gloucestershire, UK: Sutton Publishing.

Stonehenge (John Henry Walsh), 1878, *The Dogs of the British Islands, Being a Series of Articles on the Points of their Various Breeds, and the Treatment of the Diseases to which they are Subject* (reprinted from 'The Field' newspaper), London: 'The Field' Office.

Strang, Paul D. and James M. Giffen, 1981, *The Complete Great Pyrenees*, New York: Howell Book House.

Strebel, Richard, 1904, 'Die Tibetaner Dogge,' in *Die Deutschen Hunde (The German Dog)*, Vienna: Paul Gerin, pp.183-204 (translated to English by Christina McFarland and complied by Peter Rees-Jones for the Tibetan Mastiff Club of Great Britain, 1990, pp.5-7).

Subedi, Abhi, 1999, *Ekai Kawaguchi: The Trespassing Insider*, Kathmandu: Mandala Book Point.

Swan, Lawrence W., 2000, *Tales of the Himalaya: Adventures of a Naturalist*, La Crescenta, CA: Mountain N' Air Books.

Swart, Sandra, 2003, 'Dogs and dogma: A discussion of the socio-political construction of Southern African dog "breeds" as a window into social history,' *South African Historical Journal* vol. 48; URL: http://academic.sun.ac.za/history/dokumente/dogsanddogma.pdf.

Synge, Mark; see Powell Millington.

Széchenyí, Count Bela, 1882, (article in) *Die Neue Deutsche Jagdzeitung (The New German Hunter's Magazine)* vol. 14, nos. 18,19, quoted in Richard Strebel, 1904, 'Die Tibetaner Dogge,' in *Die Deutschen Hunde (The German Dog)*, Vienna: Paul Gerin, pp.183-204 (translated to English by Christina McFarland and complied by Peter Rees-Jones for the Tibetan Mastiff Club of Great Britain, 1990, pp.5-7).

TABL (Tourism Administration Bureau of Lhoka), 2000, *Lhoka in Tibet*, Lhasa: China Travel and Tourism Press.

Tafel, A., 1914, *Meine Tibetreise*, Stuttgart: Union Deutscher Verlagsgesellschaft.

Taring, Jigme, 1981, 'Tibetan dogs,' *The Tibet Journal* vol. 6, no. 2.

Taylor, Daniel, 1994, 'Meet the KyiApso,' *Tibetan Breeds International Magazine*, no. 5, pp.19-21.

Taylor, Daniel, n.d. (1980?), 'Thumdru—The long haired immigré,' (unpublished essay)

Taylor, Daniel, 1992, 'What is a KyiApso,' *KyiApso News!*, vol. 2, n.1 (reprinted in *KyiApso News!*, 1998, vol. 7, n.4).

Taylor, Daniel, 1992, 'A question of identity: The Tibetan KyiApso or, a bearded Tibetan mastiff?,' *KyiApso News!*, vol. 2, n.1.

Thomas, F.W., 1957, *Ancient Folk-Literature from North-Eastern Tibet*, Berlin: Akademie Verlag (pp.14, 88, 131).

TKC, 1995, *Official Standard, Tibetan KyiApso Club* (Annex A to Club By-Laws, Ratified 28 December 1995), USA: Tibetan KyiApso Club.

Tucci, Giuseppe and E. Ghersi, 1935, *Secrets of Tibet, Being the Chronicle of the Tucci Scientific Expedition to Western Tibet (1933)*, London and Glasgow, Blackie and Son, Ltd.

Tucci, Giuseppe, 1967, *Tibet: Land of Snows*, London: Elek Books Limited, and Calcutta: Oxford and IBH Publishing Co.

Tucci, Giuseppe, 1973, *Transhimalaya*, London: Barrie and Jenkins.

Turner, Samuel, 1800, *An Account of An Embassy to the Court of the Teshoo Lama in Tibet--Containing a Narrative of a Journey through Bootan and Part of Tibet*, London: W. Bulmer and Co. / Messrs G. and W. Nicol (reprint edition, 1971, New Delhi, Manjushri Publishing House).

Tylor, E.B., 1964 (1878), *Researches Into the Early History of Mankind and the Development of Civilization*, Chicago: University of Chicago Press, Phoenix Books.

Unsworth, Walt, 1989, *Everest* (2nd revised and enlarged edition), Seattle: Cloudcap.

Verrier, Anthony, 1991, *Francis Younghusband and the Great Game*, London: J. Cape.

Vinding, Michael, 1988, 'A history of the Thak Khola Valley, Nepal', Kailash 14(3-4):167-211.

Vinding, Michael, 1998, *The Thakali: A Himalayan Ethnography*, London: Serindia Publications.

Waddell, L. Austine, 1895, *The Buddhism of Tibet or Lamaism, with its Mystic Cults, Symbolism and Mythology, and in its Relation to Indian Buddhism*, 2nd edition, London: W.H. Allen and Co. Ltd.

Waddell, L. Austine, 1906, *Among the Himalayas* (2nd edition), Philadelphia: J.B. Lippincott Company.

Waddell, L. Austine, 1906, *Lhasa and Its Mysteries: With a Record of the Expedition of 1903-4*, London: Methuen.

Waller, Derek J., 1990, *Pundits: British Exploration of Tibet and Central Asia*, Lexington, KY: University of Kentucky Press.

Walsh, John Henry, see: 'Stonehenge'.

Warwick, Alan Ross, 1962, *With Younghusband in Tibet*, London: F. Muller.

Washburn, Bradford and Lew Freedman, 2005, *Bradford Washburn: An Extraordinary Life*, Portland, OR: WestWinds Press.

Whelpton, John, 2005, *A History of Nepal*, Cambridge, UK: Cambridge University Press.

Wilson, Andrew, 1979 [1886], *The Abode of Snow: Observations on a Tour form Chinese Tibet to the Indian Caucasus, Through the Upper Valleys of the Himalaya*, Kathmandu: Ratna Pustak Bhandar, Bibliotheca Himalayica Series I, no. 12) [New York: G.P. Putnam Sons].

Wood, Frances, 1995, *Did Marco Polo Go to China?*, London: Secker and Warburg.

Wood, Frances, 2004, *The Silk Road: Two Thousand Years in the Heart of Asia*, Berkeley and Los Angeles: University of California Press.

Workman, Fanny Bullock and William Hunter Workman, 1900, *The Ice World of Himalaya: The Peaks and Passes of Ladakh, Nubra Suru and Baltistan* (2nd edition), New York: Cassell and Company.

Workman, Fanny Bullock and William Hunter Workman, 1917, *Two Summers in the Ice-Wilds of Eastern Karakoram; The Exploration of Nineteen Hundred*

*Square Miles of Mountain and Glacier*, New York, E.P. Dutton.

Workman, Fanny Bullock, 1908, *Ice-Bound Heights of the Mustagh; An Account of Two Seasons of Pioneer Exploration in the Baltistan Himálaya*, New York, C. Scribner's Sons.

Workman, Fanny Bullock, 1909, *Peaks and Glaciers of Nun Kun; A Record of Pioneer-Exploration and Mountaineering in the Punjab Himalaya*, London, Constable.

Wright, Thomas (editor), 1948 (1880), *The Travels of Marco Polo, a Venetian*, Garden City, NY: Doubleday.

Wynn, M.B., 1886, *The History of the Mastiff*, London: William Loxley.

Wynyard, Ann Lindsay, 1982, *Dogs of Tibet and the History of the Tibetan Spaniel*, Leicester, UK: Book World Rugby.

Yavorskaya, G., n.d., 'The Mongolian sheep-dog—the most ancient breed'; URL: www.mongoldog.ru.

Yee, Blythe (with J. Qin and C. Rong), 2006, 'A conservationist campaigns to keep Tibet's mastiffs mean,' *The Wall Street Journal Online*, March 10: A1; URL: http://online.wsj.com/article/SB114194472948594095.html.

Youatt, William, 1857 (1852), *The Dog: A Nineteenth-Century Dog-Lover's Manual, a Combination of the Essential and the Esoteric*, Philadelphia: Blanchard and Lea.

Yule, Henry and A.C. Burnell, 1903, *Hobson-Jobson: A Glossary of Colloquial Anglo-Indian Words and Phrases, and of Kindred Terms, Etymological, Historical*, London: Murray (reprinted in 1985 as *Hobson-Jobson: The Anglo-Indian Dictionary*, London: Routledge and Kegan Paul); URL: http://dsal.uchicago.edu/dictionaries/hobsonjobson.

Yule, Henry, 1913–16, *Cathay and the Way Thither*, London: Hakluyt Society.

Yule, Henry, 1903, *The Book of Marco Polo the Venetian: Concerning the Kingdoms and Marvels of the East* (translated and edited with notes by Henry Yule), (3rd ed., rev.), London: Henri Cordier (reprinted in 2 vols., London: J. Murray, 1975).

'Z Codex,' see Dana B. Durand (1939).

# Index

## A
Acharya, Babu Ram 38
Aguirre, Gustavo 18, 185(*n*44)
Alexander the Great 15, 166
Allen, Charles 36, 188(*n*32,33), 197(*n*21), 228
Amdo, Yamdo (region in eastern Tibet) 24, 135, 140, 213,
American Peace Corps xi, xiv, 51, 52, 57, 82, 97(*n*), 190(*n*4),
Andrade, António 22-23, 186(*n*9)
Andrews, Ed 60, 190(*n*5), 228
Anker, C. 189(*n*44), 228
Annapurna 47, 51, 52*ff.*, 74, 81, 111, 190(*n*2), 234
Anonymous 181(*n*19), 182(*n*23), 228
Anthropology, anthropologists xi, xii, xiii, 4, 18, 25, 80, 82, 95, 97, 105, 114, 117, 137-8, 140, 154, 159, 178(*n*2), 238, 242,
Armstrong, Karen xi, 3, 228
Aufschnaiter, Peter 22, 156, 185-6(*n*7), 229; *see also* Heinrich Harrer
Ayto, John 161, 228

## B
Bailey, Frederick M. ('Eric') 27, 187(*n*20), 228
Bailey, Irma (Hon. Mrs Bailey) vii, viii, x, 27, 106, 107, 156, 158, 171, 187(*n*20), 193(*n*2), 198(*n*25), 203, 228
Bajhang (district in northwestern Nepal) 84, 111, 114-5, 118, 192(*n*4); *see also* Singh, J.N.; Thalar
Bas reliefs of dogs at Assyrian Ninevah 166
Baskaran, Theodore 192(*n*9), 228
Bates, Neale 200(*n*15), 228
Bates, R.H. (Bob) and Gail 62, 234
Battuta, Ibn 5, 180(*n*12), 228
Beale, Cynthia x, 80, 157, 159, 182(*n*24), 191(*n*17,19), 198(*n*32), 233
Beazley, Sir Charles R. 180(*n*11), 228
Beckmann, Ludwig 185(*n*5), 228
Beefeater (Yoeman of the Guard) 144-5, 196(*n*16)
Bell, Sir Charles 25, 151, 179(*n*6), 187(*n*25), 197(*n*20), 229
Bergreen, Laurence 180(*n*11), 182-3(*n*22,29), 228
Berry, Scott 43, 189(*n*47), 228-9
Bhoté Kosi (river in Nepal) 136, 195(*n*3),
Bhutan, Bhootan; Bhutanese xiii, 19, 20, 82, 108, 128, 133, 140, 142, 168, 175, 201-2(*n*20), 232
Bible 15, 181(*n*16), 194(*n*40), 229, 232
Bierce, Ambrose 162, 229
Bird (Bishop), Isabella L. 27, 171, 187(*n*19), 229
Birzer, Bradley J. 227, 242
Blackwood, William 16, 185(*n*43), 229; *see also* Kirkpatrick Effect
Blakeney, T.L. 189(*n*42), 229
Blue dog; *see* Dog bite charm
Bogle, George 19, 23, 185(*n*3), 237
Bone line 211; *see also* Dog physical features: Bone

Bonvalot, Gabriel 25, 154, 171, 186(*n*14), 229
*Book of Travels* (Benjamin ben Jonah, of Tudela) 5
Boswell, James 14, 183(*n*36), 229, 246
Bowen, R. 196(*n*121), 229
Bower, Hamilton 24-25, 186(*n*14), 229
Brandt, Phil 54
Brauen, Margin 185-6(*n*7), 229
Breashears, D. 189(*n*44), 229
Breed standards 106-32 *passim*, 130, 193(*n*14), 244; *see also* Show dogs
Breed, 162-65, 175 and *passim*; *see also* Dog breeds
Brehm, A.E. 24, 186(*n*12), 229
Brown, Doug 6, 181(*n*16), 184(*n*40), 229
Bruce, Charles 39-40
Brysac, S.B. 188(*n*38), 238
Buckland, C.E. 229
Budiansky, Stephen 185(*n*44), 229
Burrard, Sidney G. 25, 82, 186(*n*16), 192(*n*1), 229
Bush, H.W. 37, 171, 175, 188(*n*36), 229
Buxton, Bill 186(*n*10), 229-30
Byers, Alton 74*ff.*, 157, 182(*n*24), 191(*n*13), 230

## C

Calvino, Italo 14, 180-181(*n*14), 183(*n*13), 230
Cameron, Nigel 180(*n*11), 230
Camões, Luiz Vaz de 83, 192(*n*2), 230
Campbell, Mary 180-181(*n*14), 230
Canary Islands (Insula Canaria, Island of Dogs); *see* Dog breeds, European: Perro de Presa Canario
Canine Information Library 184(*n*37), 200(*n*13), 230
Capuchin (Catholic order, mission) 23
Carey, William 187(*n*19), 230
Carruthers, Douglas 155, 197(*n*24), 230
Cavenagh, Orfeur 16-18, 185(*n*43), 230; *see also* Kirkpatrick Effect
Cenni, Qinto 11
Cervantes, Miguel de 199(*n*3), 230
Challenge Certificate (CC) 89, 93
Changtang (northwest region of Tibet) 54, 135, 140, 190(*n*28)
Chapman, F. Spencer 230
China 1, 5-13 *passim*, 21, 23, 28, 31, 35, 45, 65, 68, 143-8 *passim*, 164, 167, 173, 179(*n*6), 180-187 *passim*, 195-7 *passim*, 203, 208, 226-44 *passim*
Chinanews 197(*n*18), 230
Chokgyel Lama (Chokgya) 78-81
Chomolungma; *see* Mount Everest
Christie, Agatha 6, 181(*n*15), 230
Christopher, Tom 187(*n*22), 230
Chushi Gandrug 65, 190-91(*n*8,9), 238; *see also* CIA
CIA (U.S. Central Intelligence Agency) 65-68, 190-191(*n*9), 232, 238; *see also* Chushi Gandrug
Clemens, Samuel; *see* Twain, Mark
Cooper, Lauren 197(*n*18), 230
Coppinger, Raymond and Lorna 139-40, 162, 195-6(*n*8,9), 198(*n*34), 199(*n*5), 230-131

INDEX 249

Corbett, Jim 200(*n*15), 228, 231
Cordier, Henri x, 11, 167, 181-2(*n*20), 201(*n*17), 231, 246
Corte del Milione (the Polos' Court of the Millionaires) 13; *see also Il Milione*; Polo
Cotter, Eugene L. 161-2, 231
CTIC (China Tibet Information Center) 182(*n*27), 197(*n*18), 231
Cuiju (Chinese Province) 2-3, 10
Cultural Revolution 131, 148
Cunliffe, Juliette 36-37, 175, 179(*n*7), 182(*n*26), 188(*n*35), 191(*n*18), 193(*n*4), 194(*n*18), 202(*n*21), 231
Curzon, Lord, (Viceroy of India 1898-1905) 36, 40
Cynic (word origin) 14, 161-2, 166, 199(*n*3)
Cynology (root word: cyn), cynologists xiii, 7, 14, 161-2, 166, 167, 199(*n*3); *see also* Dog, word origin

# D
D'Orville, Albert 23, 186(*n*10)
Dalai Lama viii, 22, 36, 64, 106, 107, 151-2, 178(*n*1), 187(*n*25), 197(*n*19,20), 206-7, 228
Dalke, Kate 201(*n*18), 231
Darwin, Charles 23, 171, 231
Das, Sarat Chandra 24, 186(*n*13), 231
David-Neel, Alexandra 26, 171, 187(*n*18,19), 231
Dawson, Christopher 180(*n*), 231
Denman, Earl 231
Desideri, Ippolito 19, 23, 136, 167, 180(*n*13), 185(*n*2), 195(*n*4), 231
Dhauladhar mountains (Himalayan region in India) 110-114, 193; *see also* Dog shows: Dhauladhar
Dhaulagiri 47, 51, 54-57 *passim*, 69, 74, 190(*n*3), 232
Dhole (Asiatic wild dog) 61
Dingri (Tingri) 38, 133
Dog (word origin) 106, 161*ff*., 199(*n*2,3); *see also* Cynology
　Bark, voice, growl, etc. 7-8, 19, 21, 24-33 *passim*, 41-48 *passim*, 55, 60-64 *passim*, 68-71 *passim*, 79, 80, 88-90 *passim*, 98, 100-03 *passim*, 110, 117, 119, 124, 125, 130-131, 141, 147, 151, 156, 159, 160, 162, 173, 184(*n*41), 187(*n*25), 201, 204-5, 209, 212, 213, 219, 227, 237; *see also* Ghostlike sounds
　Behavior and physical features; temperament, function; adaptive traits (Appendix 3)
　Bite, nip, snap 26, 42, 89, 92, 100, 121, 145, 160, 213, 219; *see also* Dog bite charm
　Collar (spiked, leather), wool ruff ix, 19, 21, 24, 28-29, 37, 55, 58, 60, 101, 110, 111, 128, 149-51, 158, 204, 206, 240
　Companionship, friendliness, gentleness, affection; loyalty, faithfulness xiii, 4, 10-12, 19, 21, 30-34, 40-43 *passim*, 45-51, 61-64, 68-73, 78, 82, 86-89 *passim*, 92, 97-104, 109-10, 114-8 *passim*, 120, 126, 127, 132, 141, 165, 208, 209, 218-20; *see also* Dog behavior: Play
　Courage 160, 173, 207
　Diet, food xiv, 20, 42-43, 45, 48-49, 70-73, 194, 200(*n*16), 210, 219
　Dignity, nobility, pride 19*ff*., 24, 30, 51, 71, 73, 82, 85-86, 103, 126, 135, 153, 160, 165, 169, 177, 219, 227
　Ghostlike, ghost sounds, phantom dogs 29, 105, 208, 239; *see also* Sacred
　Guard dogs, guardianship, protection, watchfulness, vigilance (functional temperament) *passim*
　　Sentry dogs, sentinels 8, 77, 82, 92, 142, 144, 176, 184(*n*41), 240; *see also* Beefeater; War dogs

War dogs, dogs of war 8, 64-73, 77, 172; *see also* Threat, aggressiveness
Hunting; *see* Dog breeds, Asian: Hunting dogs
Loyalty; *see* Companionship
Personality traits *passim*; *see also* Dog behavior: Companionship; Threat
  Fearlessness; *see* Threat, aggressiveness
  Independence 43, 130, 145, 218
  Intelligence, cleverness 12, 58, 97-105, 109, 130, 155, 158, 218
  Playfulness 32-33, 86-89 *passim*, 100, 109, 110, 125-7 *passim*, 147, 215-6
Physical features
  Beard, bearded, shaggy face; muzzle vii, viii, 20, 23, 44, 71-72, 88, 106-132, 135, 142, 147, 148-52 *passim*, 155, 166, 170-73 *passim*, 211-2, 218, 239, 244; *see also* Lips, jowls
  Castration, sterilization, gondectomy 143, 196($n$12), 212, 242, 243; *see also* Size: Bone
  Coat color 28, 59, 66, 67, 71-72, 74, 79, 113-14, 118, 121, 123, 131, 135, 139, 141-2, 144, 149-56 *passim*, 160, 168, 171-4 *passim*, 181($n$17), 184($n$41), 195-96($n$8,10), 204-6 *passim*, 209-12 *passim*, 219, 221-6 *passim* (Appendix 4), 240, 241
    Undercoat 72, 169, 172, 219
  Dew claws 31, 170
  Ears xii, 20, 23, 61, 99, 123, 131, 139, 155, 156, 170-171, 173, 179($n$), 206, 219
  Estrus cycle 124, 219
  Eyes (color), eyesight 20, 23, 28, 33-34, 42, 45, 55, 72, 86, 89, 91, 98, 103-5, 111, 121, 123, 126, 131, 135, 138, 144, 145, 146, 149, 156, 160, 170-171, 173, 182($n$27), 189($n$46), 196-7($n$17), 200($n$9), 205, 206, 208, 209, 211, 219, 222
    Eye spots (tan/red/tawny/brown markings; 'four eyes', 'spirit eyes') 20, 72, 131, 138, 145, 170-171, 208, 209, 219, 211; *see also* Ghostlike
    Haw ('red eye', bloodshot; disease, inflammation) 20, 28, 33, 122-3, 144, 145-6, 196-7($n$17), 200($n$9), 211, 205
  Lips, flews, jowls 20, 28, 71, 121, 142-5 *passim*, 165, 170-171, 173, 176, 182($n$27), 205, 211
  Size (*incl.* gigantism) ix, 2-3, 4, 6-9, 11, 16-17, 24, 28, 31, 42, 65, 70-72, 74, 81, 88, 92, 106, 108, 110, 121, 123, 130, 137, 139-45 *passim*, 150, 155-7, 162, 165-75 *passim*, 181($n$17,20), 182($n$27), 184($n$41), 187($n$21), 195-7($n$8,17), 200, 201($n$17,19), 205-6, 211, 218, 226
    Bone 37, 71, 85, 121, 141, 143, 155, 211, 218; *see also* Bone line; Physical features: Castration
    Dimorphism, gender differences (size, speed) 28, 79, 205; *see also* Appendix 4
    Donkey-size 1-18 *passim*, 19, 137, 165, 167, 178-9(Ch1$n$4), 181($n$17,20), 184($n$41), 181-2($n$17,21), 201($n$17,19), 204, 211
    Head 20-28 *passim*, 55, 71, 74, 92, 121, 123, 141, 142, 144, 147, 154-6 *passim*, 170-73 *passim*, 196($n$14), 205-6, 211-2, 227
    Weight 20, 28, 55, 71, 74, 80, 121, 123, 131-2, 139, 141-2, 145, 173, 192($n$7), 205, 206, 210, 211, 227
  Tail 15, 20-21, 23, 24, 30, 31, 41, 45, 55, 60, 72, 74, 102, 109, 121-3, 130, 131, 135, 147, 155, 156, 160, 162, 168, 171, 205, 211-2, 219, 223, 240
  Whelp, whelping 73, 88, 109, 124, 135,
Sacred, ritual, spiritual; funerals; souls of monks xii, 94-96, 120, 192-3($n$10), 200$n$16), 207, 227; *see also* Ghostlike
Threat, aggressiveness, ferocity, fearlessness, attack; fighting 4, 10, 11, 19, 21, 25-27, 28-30, 31-2, 36, 43-45, 48, 55, 60-77 *passim*, 81, 92, 98, 100-101, 103, 106, 110-111, 124-8 *passim*, 135, 141, 143-4, 147, 167-71 *passim*, 173, 177, 187($n$22), 191($n$11), 204-18 *passim*, 227; *see also* Bark; Courage; War dogs; *and compare* Companionship, friendliness...

INDEX 251

Challenge, charge, 'rush' 29, 33, 45, 60, 69, 110, 135, 143, 190(*n*5), 212-3
Tied, chained, caged, fettered, leashed 8, 10, 19, 20-33 *passim*, 44, 55, 58-59, 66-72, 77, 81, 89-90, 92, 100, 102, 106, 110, 128, 135, 136, 144-5, 147, 151, 154, 156, 158, 160, 164, 195(*n*6), 204, 206, 207
Training, trainability 41, 67, 77, 86, 92, 99, 100-101, 105, 158, 190(*n*13); *see also* War dogs
Voice; *see* Dog behavior: Bark
Voice like a lion 7-8, 184(*n*41)
Dog bite charm viii, 25-26, 77
Dog breed clubs; *see* Kennel clubs, breed clubs; Dog kennels; Dog shows
Dog breed standards, standardization 64, 71, 72, 82, 106-32 *passim*, 135, 138, 139, 141-2, 145, 147, 162-63, 164, 170-171, 173, 175, 184, 193(*n*14), 194, 196, 199-200(*n*8), 207, 244; *see also* Breed; Tibetan KyiApso; Tibetan mastiff
Dog breeds, types
  *Asian breeds: Himalayan, Tibetan (& other Chinese) & C. Asian (incl. Mongolia)*
    Apso types
      Bearded (*or* shaggy) Tibetan mastiff 20, 36, 106-32 *passim*, 135, 148, 149, 152, 166, 239, 244
      Humli kukur (Humli KyiApso, Humli dog) 108, 111, 114, 117, 123, 239
      Kinnauri kutta (Kinnaur dog) viii107, 108, 111-3, 123
      KyiApso (Apso Do-Kyi, Apso Mastiff, Do-Kyi-Apso, Tibetan KyiApso, Tibetan Khyi Apso) viii, ix, 20, 32, 82, 105-32 *passim*, 133, 135, 146, 148-52 *passim*, 165, 166, 193-4(*n*4,5,12,14,16,18,20,21), 195(*n*25), 241-3
      Lhasa Apso 81, 84, 92, 107, 108, 135, 194(*n*18)
    Bara Benghali dog 112
    Bhoté (Bhoté kukur: 'Tibetan dog' in Nepali) (*var.* Bhotea, Bhotean, Bhotia, Bhotiya, Bhutia) ix, 53, 66, 74, 239
    Bhotea (Bhotea kutta: 'Tibetan dog' in Hindi) vii, 92, 168, 201-2(*n*20)
    Bischur 170,
    Chinese ha-pa 205
    Chinese Pug 206
    Do-kyi, Do-khyi ('door dog', 'tied dog' in Tibetan) *passim*
    Do-Kyi-Apso; *see* KyiApso
    Gaddi kutta 112; *see also* Gaddi shepherds; Dog breeds, Asian: Sheepdogs
    Great Dog of Tibet 107, 109, 194(*n*21)
    Gya-Kyi 206
    Himalayan mountain dog (Himalayan sheepdog, Tibetan mastiff) ix, 59, 62, 63, 66, 111, 112, 159, 166, 175-6, 184(*n*41)
    Himalayan sheepdog; *see* Himalayan mountain dog; Sheepdogs
    Hunting dogs
      Khampa dog (Tibetan hound) 198(*n*27)
      Kunlun hound 158, 159; *see below*: Kunlun mountain dog
      Mongolian hunting dog 155
      Nepal (shikari) ix, 62, 74, 79, 80, 182(*n*24), 197(*n*22)
      Northern Tibetan hunting dog 155
      Sha-kyi, SHa KHyi (hunting dog, meat dog) 153ff.
      SHwa KHyi (stag dog) 157, 211
      Tibet, Mongolia (sha-kyi, sheuke) ix, 2, 4, 8, 9, 10, 12, 14, 22, 28, 74, 79-80, 99, 129, 133, 153-60 *passim*, 169, 179, 182(*n*24), 190(*n*17,18), 197(*n*22), 198(*n*28,30,31), 204-7 *passim*, 208-17 *passim*, 232-34, 242
      Wa KHyi (fox dog) 158, 211
    Kongkyi (Kongbo dog) 206

Kunlun mountain dog 158
Lahauli kutta 112; *see also* Dog breeds, Asian: Sheepdogs
Landrace; *see* Breed
Lhasa Apso; *see* Dog breeds, Asian: Apso types
Lhasa terrier; *see* Dog breed standards, Asian: Tibetan terrier
Lhasan (Lassa) 170
*Mastivus Thibetanus* 179(*n*19)
Mongolian mastiff 175
Mongolian sheepdog; *see* Dog breeds, Asian: Sheepdogs
Mustang (red) dog 31, 81, 110, 123, 128, 140-141, 157, 170-171, 188(*n*30)
Nepaul dog, Nepal dog (Tibetan mastiff) 15-17, 19, 170-171, 239
Phö-kyi, Phyu-kyi 135, 141, 142, 164, 230
Prapso ('Perhapso') 194(*n*18), 231
Pug 92, 206
Ra KHyi ('corral dog')
rGya ('dog') 209-10
Ri KHyi (mountain dog) 210
Sang KHyi ('Sang dog') 211
sGo KHyi ('door dog') 211
Sheepdogs, Himalayan sheepdog 28, 112, 175-6 *passim* (Bhutia, Bhotia, Mountain dog; Tibetan mastiff)
    Bhutia sheepdog 175, 201-2(*n*20)
    Mongolian sheepdog (Mongolian mastiff) 175, 191(*n*170), 246
    Tibet sheepdog (Wolfdog) 107, 109
Srung KHyi ('guard dog') 211
Tibetan mastiff *passim*
Tibetan spaniel 37, 53, 66, 69, 92, 108, 135, 172, 181(*n*18), 193(*n*18), 205, 246
Tibetan terrier 37, 81, 92, 107, 108, 113, 124, 129, 135, 149, 181(*n*18), 241-3
Toy spaniel 208
Tsang-kyi (gTsang KHyi) viii, 27, 135-53 *passim*, 142, 160, 164, 165, 166, 195(*n*6), 196(*n*8,17), 202(*n*23), 211, 230, 242-3
Wolfdog 198; *see above*: Great Dog of Tibet; Irish Wolfhound
Yun-kyi ('untied dog') 8, 164
*European/American & other:*
    Airedale 156, 205
    Alsatian; *see* German shepherd
    Bloodhound 20, 144, 196(*n*14), 196-7(*n*17), 200(*n*9)
    Borzoi 80, 157, 198(*n*26), 199, 211
    Boxer 174
    Bull terrier 84
    Bulldog 16, 17, 127, 174, 184(*n*37), 200(*n*13), 230, 237
    Bullmastiff 174
    Canarian Dog; *see* Perro de Presa Canario
    Collie 93, 199-200(*n*8)
        Border 238
        Scotch 31
        Welsh 36
    Coonhound 196-7(*n*17)
    Dachshund 207
    Doberman 91, 93, 195, 121, 243
    Dogo Canaria; *see* Perro de Presa Canario

Dogue de Bordeaux 166, 174
English bulldog; *see* Bulldog)
English greyhound (*see* Greyhound)
English mastiff 130, 169, 170, 171, 177
English Newfoundland (*see Newfoundland*)
English sheepdog 109
Finnish spitz 206
French bulldog 93
Gazehound (*see* Hunting dogs: Sighthound)
German shepherd (Alsatian) 67, 73, 91, 93, 121, 160, 243
Golden Retriever 155
Great Dane 84, 92, 174
Greyhound 80, 157, 198($n$26), 207, 211
Hunting dogs of Britain 174
Irish setter 93, 101
Irish Wolfhound (Irish Wolfdog) 6, 109, 146, 193, 194($n$21), 233, 234, 235
Japanese spaniel 84
Labrador retriever 92, 155
Leonberger 146
Molosser, molloso 14, 145, 173, 174-5, 178-9(ch1$n$4), 184($n$37,39), 198($n$31,1), 201($n$19), 230, 236, 243
Neapolitan mastiff 174, 196-7($n$17)
Newfoundland 16-17, 20, 21, 194($n$21),
Perro de Presa Canario (Dogo Canaria, Canarian Dog of Prey) 161, 198($n$1)
Pit Bull 103
Pomeranian 92
Russian sheepdog 109
Saint Bernard 6, 92, 105, 175, 177, 200($n$9)
Schnauzer 206
Sheepdog (generic); *see* Collie
Spaniel (*incl.* Toy Spaniel) 93, 208; *see also* Dog breeds, Asian: Japanese spaniel; Tibetan spaniel
Swiss cattle dog 175
Wolfhound; *see* Irish Wolfhound; *see also* Wolf; Wolfdog
S & SW Asia/India:
Afghan hound 92, 198($n$26), 202($n$22)
Chippiparai 93
Hunting dogs of India (South Asia) 93, 169
Hunting hounds 80
Maratha 93
Pariah (Pi-dog) 93, 110, 116
Pashmi 93
Pekingese 92, 206
Pi-dog; *see* Pariah
Poligar 93
Rajapalayam 93
Rampur hound 92, 93, 192($n$9)
Saluki 80, 92, 153, 158, 160, 198($n$26), 202($n$22), 211
Sighthound 80, 153, 156, 191($n$18), 198(26), 202
Sindh 93
Sloughi 202($n$22)

Vaghari 93
Dog kennels, by name
    Emodus (Nepal) 82, 83, 86, 93, 238-9, 133, 171, 179(*n*4), 240-241
    Gangaling (Lhasa) ix, 151-3
    Kesang Camp (USA) 181(*n*20), 184(*n*41); *see also* Kesang Guerrilla Camp
    Saipal (Nepal) 82, 84-86, 239; *see also* Dogs named: Saipal Baron (Kalu)
Dog mask 72, 208
Dog origins (domestication; origin of Tibetan mastiff and KyiApso) 6, 14, 52, 106, 108, 111, 128-32, 137-45 *passim*, 163-4, 166, 184(*n*37), 190(*n*1), 199 201(*n*18), 202, 230, 236, 242; *see also* Dog (word origin); Mastiff (word origin); Wolf
Dog prices, selling dogs 3, 37, 81, 84, 118, 120, 135, 136, 142, 146, 149-50, 152, 158, 165, 169, 178, 196, 197(*n*18), 210, 211, 230
Dog shows; *see also* Kennel clubs, breed clubs; Show dogs
    Alexander Palace show (UK) 204
    China 146
    Crufts 27, 204; *see also* Kennel clubs
    Delhi Dog Show 89-94, 239
    Dhauladhar Dog Show (informal) viii, 110-114, 192(*n*9), 240
    Kennel Club shows (UK) 27, 36, 204
    Kensington Show 27, 204
    Nepal Kennel Club Dog Show 89
Dog training; *see* Dog behavior: Training
Dogs named
    Alubari mastiff ('Potatofield Dog') viii, 51-57 *passim*, 58
    Amjo, Amjo Gipu ('Big Ears') viii, xii, 61-68 *passim*
    Apo ix, 123
    Baby Puppy 34
    Badger 123, 125
    Bathsheba 93, 192(*n*9)
    Bhalu ('Bear') (*née* Emodus Boris KC) viii, 96, 97-105 *passim*,
    Bhoté ('Tibetan') 85, 129; *see also* Dog breeds, Asia: Bhoté kukur
    Bhotean viii, 36-37, 167, 192(*n*8),
    Bhotoo 85-86
    Bhout 167
    bKra SHis ('Blessing')
    Bounty viii, 84, 86, 88, 97(*n*)
    Dianga 20-21
    Djeóla 30-31
    Dom PHrug ('Black Bear Young One') 208
    Don aGrub ('Realization of Causes', a personal name of Buddha) 207
    Dzong ('Fortress') 129
    Emodus Boris KC; *see* Bhalu
    Emodus dogs; *see* Dogs named: Kalu; Bhalu
        Funeral 96
    I-tru ('Young Lynx') 204
    Kali ('Blacky') 71-73
    Kalu ('Blacky') (Ch. Saipal Baron of Emodus) vii, viii, xii, 58, Ch.4 *passim*, 161, 192-3(*n*7,10), 194(*n*20), 239
    Kang Rinpoche (another name for Mount Kailash) 120
    Kang; *see* Kang Rinpoche
    Kattuk 158
    Lata ('Mute') viii, 71-72

Lhasa 188(*n*34), 205
Lishi 123
Mig bZHi Can ('Four-Eyed One') 211
Mindu 120, 129
Minhsingh 123
Nako (Nako-Wallah) (named after the village of Nako) 109-10
Ngao 4-5, 167, 179-80(*n*10)
Nying-kar ('White Heart') 45-51 *passim*, 190(*n*50)
PHrug ('Young One') 209
Police-ie 40, 42-43, 189(*n*46)
Rakpa 27, 205
rGya Bo Mig bZHi ('Big One Eyes Four') 209
rGya dGe ('Big One Virtue') 209
rGya dMar ('Big One Red') 209
rGya sBrug ('Big One Dragon') 209
rGya Ser ('Big One Yellow') 209
Saipal Baron of Emodus; *see* Kalu
Saipal dogs; *see* Dogs named: Kalu; Bounty; Tü-bo; Dog kennels: Saipal
Sangs rGyas ('Buddhahood') 209
San-San 135-7, 153
Sa-tru ('Young Snow Leopard') 204
Seng PHrug ('Lion Young One') 209
Sengtru 204
sGrol Ma ('Savior Female', the goddess Tara) 209
Shakya ('sha-kyi' hunting dog?) 155
Sheroo 86
Sik-tru ('Young Leopard') 204
Singdru ('Little Lion') 121, 123, 125, 128, 129
SPang PHrug ('wolf young one') 209
Suptu 123-7 *passim*, 194(*n*19)
Takkar (a place name) viii, 32-35
Tak-tru ('Young Tiger') 204
Tara ('Goddess') 109
Tashi 120, 129
Tendrup ('Bear') viii, 40-42, 189(*n*45)
Thumdru ('Little Bear') ix, 121-8 *passim*, 244
Tiger 72-73, 85
Tomtru ('Young Bear') 27, 205
Tü-bo 85
Violet 125
Wangdu (Wangdi) 65, 68, 73
Wolf 109
Yollchi ('Him Who Was Picked Up On The Road') 32
Dogs, landrace; *see* Breed
Dogs, purebred; *see* Breed; Dog breed standards
Do-kyi ('Tied Dog') (Tibetan mastiff) *passim*
Dolpa (district in Nepal) ix, 74-81, 157
Dong Zhen 1979(*n*18), 231-2
Dougall, Major W. 36-37
Downs, James F. 232
Draghi, Paul Alexander 197(*n*22), 232
Dunham, Mikel 190-191(*n*9), 232

Durand, Dana 178-9(Ch1*n*4), 182(*n*22), 232, 246
During, Stuart 54

## E
East India Company 15
Eckfeld, Tonia 180(*n*11), 232
Edel, Leon 232
Ehrich, Christian x, xiii, 107, 111, 113, 193(*n*10)
Ehrman, Bart D. 181(*n*16), 229, 232
Eiselin, Max 190(*n*3), 232
Ekvall, Robert vii, xiii, 10, 12, 25, 76-80 *passim*, 138, 143, 157, 172, 192(*n*24), 186(*n*15), 191(*n*11,12,14,17), 195(*n*7), 196-97(*n*11,17), 198(*n*29), 199(*n*7), 208, 232
Elworthy, Susan 179(*n*7), 182(*n*26), 231
Emodus; *see* Kennel names
Evans, Christopher 199(*n*2), 232
Everest, Sir George; *see* Mount Everest
Ewans, Martin 188(*n*32), 232

## F
Fischer, Mary 196-97(*n*17), 232
Flamholtz, Cathy J. 3, 106, 108, 121, 179-80(*n*7,8,10), 181(*n*18), 185(*n*5), 193(*n*1,13), 228, 237, 242-3
Fleming, Peter 197(*n*21), 232
Franciscan (Catholic order, mission) 5, 186(*n*10)
Fraser, James Baillie 170, 168, 175, 200(*n*14,15), 201-2(*n*20), 232
Freedman Lew 189(*n*39), 245
French, Patrick 188(*n*32), 197(*n*12), 232
Fürer-Haimendorf, Christoph von 232, 238

## G
Gaddi shepherds 111-114; *see also* Dog breeds, Asian: Gaddi kutta; Sheepdogs
Gandhi, Indira 93
Gardner, Phyllis 107, 193(*n*6), 194(*n*21), 233
Gee, E.P. 233
Genghis Khan; *see* Khan
Gentry, Anthea 184-5(*n*42), 233
Ghosts; *see* Dog behavior: Ghost dogs
Gibb, II.A.R. 180(*n*12), 228, 233
Giffen, James M. 179-80(*n*10), 244
Gigantism; *see* Dog behavior... attributes: Size
Gill, William 20, 185(*n*4), 233
Goldstein, Melvyn C. x, 80, 108, 121, 157, 159, 182(*n*24), 191(*n*17,19), 193(*n*12), 198(*n*32), 233
Graham, George A. viii, 107, 109, 193(*n*7), 194(*n*21), 233, 235
Great Game (Tournament of Shadows) 36, 187(*n*20), 188(*n*32), 232, 234, 238, 242, 245
Great Wall of Pisang (Oblé, Swarga Danda, Paungda Danda, 'Mountain of Heaven') 94*ff.*; *see also* Gurung; Manang
Greenblatt, Stephen 181-2(*n*14), 233
Grewal, Amarjeet Singh 112
Grueber, John 23, 186(*n*10)

Gugé (Gu-gé; Zhang-Zhung) (ancient Tibetan kingdom) 54, 111,
Guizhou (Chinese Province); *see* Cuiju
Gurung (Tamu) (ethnic group, shepherds of Nepal), Gurung language 58-61, 94-96 *passim*, 162, 192-3(*n*10), 199(*n*2), 238, 240
Gurung, Naresh 58, 60, 238
Gyantse 133, 135, 153, 155, 158

# H

HAGAR (cartoon) ix, 177
Hally, Will 155-6, 198(*n*30), 233
Han Dynasty viii, 4, 5, 179(*n*9)
Hancock, David 176, 233
Harrer, Heinrich 22, 156, 185-6(*n*7), 198(*n*27), 233; *see also* Aufschnaiter
Haw ('red eye') 121, 142, 144-5, 147, 165, 197, 200
Haw, Stephen 13, 180(*n*11), 182-83(*n*29), 233
Hawley, Elizabeth 190(*n*4), 233
Hayden, Horace H. 25, 82, 186(*n*16), 192(*n*1), 229
Hedin, Sven x, 12, 30, 31-35, 188(*n*31), 233
Hedley, John 45, 155, 189(*n*49), 197(*n*23), 233
Helambu (mountainous region of Nepal) xii, 61*ff*.
Hemmleb, J., 189(*n*44), 233
Hermanns, M. 158, 233
Herzog, Maurice 53, 190(*n*2), 234
Hillary, Sir Edmund 40, 189(*n*44)
Himachal Pradesh (state in India Himalayas) 111-2, 168; *see also* Uttar Pradesh
Hodgson, Brian H. 39, 168, 169-71, 234
Hoed, Jan den 107, 193(*n*7), 234
Hogan, Edmund 194(*n*21), 234
Holdich, Thomas Hungerford 28, 171, 187(*n*22), 234
Holliday, Eric 172
Hooker, Joseph Dalton 23-24, 168, 171-2, 186(*n*11), 234
Hopkirk, Peter 187(*n*19), 188-9(*n*32,38), 234
House of Lords Guesthouse (New Delhi) 90
Houston, C.S. 234
Howard-Bury, C.K. 189(*n*40), 234
Huber, Tony 198(*n*28), 234
Huc, L'Abbé Evariste 23, 186(*n*9), 234
Humla (Humla-Jumla; Humla traders) viii, 111, 114-28, 193(*n*12); *see also* Jumla; Dog breeds, Asian: Humli kukur
Hummel, Siegbert 186(*n*15), 234
Hunan (Chinese Province) 3, 10
Hunter, William Wilson 171, 235
Hunting dogs; *see* Dog breeds, Asian: Hunting dogs; S & SW Asia: Hunting hounds; Europe/America: Hunting dogs of Britain
Hutchinson, W. 173, 235

# I

Ide, Jennifer 126-8
*Il Milione* ('The Million') 2, 13, 183, 241; *see also* Polo
Isseman, Maurice 189(*n*41), 235

## J

Jackson, D.P. 190(*n*7), 235
Jackson, R. (Rodney) 191(*n*15), 235
Jacob, L. 175, 202(*n*22), 235
James, Major Dan 124-8,193(*n*4), 194(*n*17,19); *see also* Dogs named: Suptu
Jardine, William 171, 235
Jesus 181(*n*16), 229, 232
Jewish merchants, *see* Radhanites
Jinpa, Geshe Gelek 193(*n*4), 235
Johnson, Elliott viii, 102
Johnson, Samuel 14, 183(*n*36), 229
Johnston, R.F. 29, 187(*n*25), 235
Jomsom (Dzongsam) 65, 81, *see also* Mustang District; Thak Khola
Jumla (northern border district in Nepal) 78; *see also* Humla (Humla-Juml)
Jupp, Hilary 109, 193(*n*7), 194(*n*21), 235

## K

Kachár (hillside, high hills) 169-70
Kali Gandaki (river in Nepal) 51, 52, 235; *see also* Mustang; Thak Khola
Kalimpong (a town in West Bengal, India) 41-42
Kang La (pass), Kangla Himal 47-48
Kashmir 31, 187-8(*n*28), 203, 240, 114-28
Kathmandu viii, xii, xiv, xv, 39, 47, 49-51, 61-68 *passim*, 81, 81, 84-86, 89, 93, 95, 114, 118, 120, 136, 146, 148, 153, 169-71, 178(*n*4), 195(*n*1)
Kawaguchi, Ekai 43-44, 189(*n*47), 235, 244
Keay, John 188(*n*38), 236
Kennel clubs, breed clubs; *see also* Dog kennels
    American Kennel Club (AKC) 131, 164, 199-200(*n*8), 238
    American Tibetan Mastiff Association (ATMA) xv, 178(N4), 184(*n*41), 192(*n*6), 203, 239-40, 242
    Delhi Kennel Club 90
    Fédération Cynologique Internationale (FCI) 71, 93, 141, 196(*n*10)
    Federation of Kennel Clubs of India 236, 241
    Irish Wolfhound Club of Great Britain 109
    Kennel Club of India 92, 93
    Nepal Kennel Club xii, 82, 85, 86, 89, 93, 97(*n*)
    The Kennel Club (Great Britain), UK Kennel Club 27, 93, 173, 205
    Tibetan Breeds Association 27
    Tibetan KyiApso Club (TKC) xv, 108-32 *passim*, 193(*n*14), 194(*n*18), 244
    Tibetan Mastiff Club of America (TMCA) xii, xv, 194(*n*21), 221*ff*., 243
Kesang Guerrilla Camp (Khampa camp) viii, 64*ff*., 184(*n*41), 239
Keyes, Ralph 185(*n*45), 235
Kham (region in eastern Tibet) 22, 65, 135, 140, 156, 190(*n*8); *see also* Khampa
Khampa (men of Kham, Tibetan resistance fighters) viii, xiii, 22, 64-73 *passim*, 77, 140, 153, 156, 157, 190(*n*8), 191(*n*13), 198(*n*27), *see also* Chushi Gandrug
Khampa dog, *see* Dog breeds: Asian
Khan (Genghis and Kublai, the 'Great Khan') 1,4,8,12, 13, 14, 173, 179, 182
Khumbu; *see* Solu-Khumbu, Mount Everest
Kingdon-Ward, Francis 25, 28, 186(*n*16), 187(*n*22), 230, 236
Kinnaur (Himalayan region in India) viii, 107-8, 111-4, 123, 128
Kipling, John Lockwood 176, 236

Kipling, Rudyard 36, 185(*n*45), 188(*n*32), 234, 236
Kirkpatrick Effect ix, 15-17
Kirkpatrick, William ix, 15-17, 18, 64, 199(*n*7), 185(*n*43), 236
Knowledgerush.com 236
Kodari (Nepalese town on Tibet border) 136, 153
Komroff, Manuel 179(*n*5), 180(*n*11), 236
Konradsen, Kåre 14-15, 166, 175, 184(*n*39), 236
Kostova, Elizabeth 1, 178(Ch1*n*1), 236
Krassnig, Adolf 179-80(*n*10,11), 183(*n*31), 184-5(*n*42), 236
Kreitner, Gustav 21, 185(*n*6), 236; *see also* Széchenyi
Kublai Khan; *see* Khan
Kuti, Kutti; *see* Nyalam
Kwei-chau (Chinese Province); *see* Cuiju
Kyirong ('Happy Valley', Tibet) xii, 86

## L

L'Amour, Louis 227, 236
Lake Manasarovar 108, 111, 114, 119, 128
Lal, Mukundi 176, 236
Lamjung Himal (Nepal) viii, 58-59, 238-9
Landon, Perceval 143-4, 196(*n*13), 197(*n*21), 237
Landor, A. Henry Savage 25, 186(*n*14), 237
Landrace dogs; *see* Breed
Landseer, Edwin ix, 169
Lange, Karen E. 201(*n*18), 237
Langkavel, Bernard 237
Latham, Ronald 9, 178(Ch1*n*2), 237
Leighton, Robert 167, 179-80(*n*10), 192(*n*8), 237
Leone; *see* Lion, Tiger
Lewis, Jon E. 189(*n*43), 237
Lewontin, R.C. 237
Lhasa (Lhassa, Lassa) 5, 19, 22-27 *passim*, 35-39 *passim*, 84, 92, 108, 114, 119, 128, 133-7 *passim*, 140, 142, 146-52, 153, 156, 164-5, 169-72, 185(*n*1,7), 186(*n*8,10,13,16), 187(*n*18,19,20), 188(*n*32,33), 198(*n*33), 205, 207, 228, 230-233, 236-7, 245
    Lhasa Apso; *see* Dog breeds, Asian: Apso types
    Lhasa dog market ix, 135, 141, 144, 146-52; *see also* Dog kennels: Gangaling
Lhoka (a prefecture in southeastern Tibet) 133, 137-50 *passim*, 153, 195(*n*6), 244
Li Jin 197(*n*18), 237
Li Qian x, 182(*n*23,27), 187(*n*21), 196(*n*14), 237
Libourel, Jan 200(*n*10), 237
Lion (Leone), lion-dog, lion voice 2, 7-8, 10, 121, 181-2(*n*20,21), 184(*n*41), 204, 209
Livestock (herds, herders, goats, sheep, yak) *passim*; *see also* Nomads
Lo Manthang; *see* Mustang

## M

Macfarlane, Robert 52, 237
Mair, Victor 162, 199(*n*3), 237
Mallory, George 40, 189(*n*44), 228-9, 233, 238; *see also* Mount Everest
Manang (Nyeshang) (district in northern Nepal) 47-51 *passim*, 94-96 *passim*, 111, 154, 192-3(*n*10,11), 238
Manning, Thomas 23, 237

Marco Polo Effect 1(*n*), 3, 14, 184(*n*40), 195(*n*40), 239; *see also* Kirkpatrick Effect
Marco Polo; *see* Polo
Markham, C.R. 185(*n*3), 200-201(*n*16), 237
Marple, Miss; *see* Christie, Agatha
Marsden, William 237, 243
Martin, William Charles Linnaeus 168, 171, 238
Mastiff (word origin; definitions) 166-72, 170, 172-6, 184(*n*37), 190(*n*1); *see also* Dog origins
Matthiessen, Peter 44-45, 77, 190(*n*48), 238
McGovern, William Montgomery 25, 160, 198(*n*33), 238
Messerschmidt
    Don vii, 108, 218, 238, 239-40
    Hans xiii, 86-88, 93-94
    Kareen xiii, 94
    Liesl xiii, 86-88, 93-94, 192-3(*n*10,11), 238
Messner, Reinhold 189(*n*44), 238
Meyer, K.E. 188(*n*38), 238
Milarepa, Mila, Mi-la (Tibetan saint; stag and hunting dog legend) 80, 154, 195(*n*3), 208, 211
Milione, Messer Marco; *see* Polo; *Il Milione*
Miller, Luree 27, 187(*n*19), 238
Millington, Powell (Mark Synge) 188(*n*33), 240, 244
*Mithé* ('jungle man', Yeti) 79
Mnteverest.net 44, 240
Molosser; *see* Dog breeds, European: Molosser
Monasteries, stupas, shrines, temples (gomba) 4, 8, 22, 47-48, 53, 55, 60, 67, 70, 73, 76, 86, 95, 113, 124, 133-5, 139, 143, 144-5, 147-8, 153, 154, 160, 163, 187(*n*19), 195-6 (*n*8), 207, 241
    Jokhang temple (Lhasa) 148,
    Kumbum Stupa (Gyantse) 135, 153
    Pelkhor Monastery (Gyantse) 135, 153
    Rongphu Monastery (near Mt Everest) 133-5
    Sa Yi Gompa ('Temple of the Dead') 95, 144
    Temple dogs 195-6(*n*8)
    Temple lions 182(*n*21)
Mongol Empire 192; *see also* Mongolia
Mongol Peace 1
Mongolia, Mongol(s) 5, 10, 13, 24, 31, 36, 43, 45, 155, 192, 203, 180, 190, 197, 230-233, 241-2
Mongolian dogs 10, 45, 155, 174, 190(*n*49), 191(*n*17), 246; *see also* Mongolia
Moorcroft, William 29, 187-8(*n*28), 240
Morley, John 40
Morrison, Bruce x, xiii, 51, 52-57 *passim*, 97ff.
Moule, A.C. 178(Ch1*n*4), 232, 240
Mount Everest viii, x, 29, 38-43, 51, 58, 85, 133-4, 187(*n*24,27), 188-9(*n*37-46), 190(*n*3), 228-45 *passim*
Mount Kailash ix, 108-32 *passim*, 133-60 *passim*, 193(*n*4)
Mumford, Stan R. 192-3(*n*10), 197(*n*22), 240
Murray, Hugh 180(*n*11), 240
Mustang (district in northern Nepal) (Moostang; Lo Manthang; former principality; Raja of Mustang) ix, 47, 51-54 *passim*, 64, 66, 68, 74-81, 110, 111, 123, 128, 141, 157, 170, 188(*n*30), 199(*n*7,9), 191(*n*10), 235, 241

Mustang dog; *see* Dog breeds, Asian
Myth, myth-making, mythology, mythos xi, 3, 8-14 *passim*, 25, 30, 70, 94, 108, 113, 161, 165, 166*ff*., 179-80(*n*10), 182(*n*26), 196(*n*12), 200(*n*10), 227, 228, 236, 237, 243, 245; *see also* Singha (mythological lion)

## N

Nangpa La (pass) xii, 85
Nar and Phu (villages in Manang District, Nepal) 47-48, 49
NASA (U.S. National Aeronautics and Space Administration) 39, 189(*n*39), 240
National Volunteer Defense Army (Tibet); *see* Chushi Gandrug; Khampa
Nelson, D. 191(*n*10), 240-241
Nelson, David and Judy 199-200(*n*8)
Nepal Army 65, 68-73, 77; *see also* Kesang Guerrilla Camp
Ngao; *see* Dogs named
Niligiri peaks 54, 66, 70; *see also* Annapurna; Mustang District; Thak Khola
Nitartha 241
Noel, John Baptist Lucius 29, 187(*n*24,27), 241
Nomads (nomadic life) vii, ix, x, xiii, 8, 10, 12, 19, 21-44 *passim*, 76, 78, 79-81, 92, 107, 111, 119-20, 124, 128, 131, 135-60 *passim*, 163, 165, 179-80(*n*10), 182(*n*24), 186(*n*15), 191(*n*12,14,17,19), 195-6(*n*7,8,11), 198(*n*27,28,29), 199(*n*7), 204, 232, 233, 234
Norbu, J. (Jamyang) 188(*n*32), 241
Norbu, Thubten Jigme 191(*n*11), 241
Norbulingka (Dalai Lama's Summer Palace) 106, 151; *see also* Dalai Lama
Nouc, Hedy xiii, 108, 111-3, 193(*n*9,11), 241
Nyalam (Kuti, Kutti) (trading town in southern Tibet) 133, 135*ff*., 153, 195(*n*3)

## O

Obedience training; *see* Dog training
Oliff, Douglas 187(*n*25), 241

## P

PADS (Primitive and Aboriginal Dog Society) 1(*n*), 239
Pamir mountains 32, 230
Panwar, A.S. 241
Parvovirus 86, 192(*n*7)
Pauthier, M. 179-80(*n*10), 241
Peace Corps; *see* American Peace Corps
Peissel, Michel xii, 77, 178(*n*3), 191(*n*13), 241
Pelliott, Paul 178-9(Ch1*n*4), 217(*n*2), 232, 240
Pettigrew, Judith 95, 192-3(*n*10,12), 199(*n*2), 232, 241
Phu (village in Manang District, Nepal); *see* Nar and Phu
Pilgrims, pilgrimage 12, 21-22, 29, 43-51 *passim*, 52, 45, 47, 119-20, 148, 154, 187-9 (*n*28,38), 190(*n*50), 193(*n*40, 207, 232, 235, 243
Polo
    Maffeo viii, 1, 11
    Marco viii, ix, x, xi, Ch.1 *passim*, 19, 28, 137, 145, 161, 165, 167, 178-201(*notes passim*), 204, 210, 217(*n*5), 228-46 *passim*; *see also* Il Milione
    Nicolo viii, 1, 11
Pradhanang, Tirtha Bahadur 188(*n*37), 241
Pundits 188-9(*n*38)
Purebred dogs; *see* Breed

## Q

Qian, Li; *see* Li Qian
Quinn, Diana 118, 119-20, 128, 241

## R

Radcliffe, Charles W. vii, xiii, 184(*n*41), 196(*n*19), 221-7 (Appendix 4), 241
Radhanites (Jewish merchant travelers) 5
Ramusio, Giovanni Battista 11, 183(*n*34)
Rana, Amar 175, 241
Richardson, Hugh x, 241
Rigg, Robin 241
Rijnhart, Susie Carson 27, 241
Roberts, D. 189(*n*44), 228
Rockhill, William Woodville 24, 171, 241, 241-2
Roerich, Nicholas 241
Rohrer, Ann v, xiii, 3, 106, 108, 121, 178(*n*4), 179(*n*7,8), 181(*n*18), 193(*n*1,13), 242
Rustichello de Pisa 2
Ruttledge, Hugh viii, x, 40-42, 189(*n*45,46), 242
Ruysbroek, Willem van 180(*n*11), 241

## S

Sagarmatha; *see* Mount Everest
Saipal Himal 73, 84, 114
Saipal; *see* Kennel names
Salkeld, A. 189(*n*44), 229
Salmeri, K.R. 196(*n*12), 241
Satlej (Sutlej) river 35, 113
Savolainen, P. 201(*n*18), 242
Schafer, Edward G. 180(*n*11), 242
Schaller, George B. 44-45, 77, 191(*n*15), 241
Schweikart, Larry 227, 242
Sellar, A.J. 158, 198(*n*30), 242
Shanklin, Eugenia xii, 178(*n*2), 242
Shaumian, Tatiana 188(*n*32), 242
Shebbeare, E.O. 42-43, 189(*n*46), 242,
Sheep (herds, herders) *passim; see also* Livestock; Nomads
Sherling, Kristina 166, 176, 196-7(*n*17), 202(*n*23), 232, 242-3
Sherpa (Sherpani) 29, 30, 45, 51, 58, 60, 61*ff*, 63, 81, 85, 190(*n*50),
Shigatse 133, 147, 151, 153, 195(*n*6)
Shipki La (pass) 30, 34, 113
Show dogs, breed for show 3, 82, 106-32 *passim*, 135, 138, 140, 141, 142, 145, 156, 162-3, 164, 165, 175, 178(*n*4), 184(*n*41), 218, 221-6; *see also* Breed; Dog shows; Kennel clubs; Breed standard
Siber, Max 179-80(*n*10), 185(*n*5), 228, 237, 243
Silk Road 1, 5, 170(*n*11), 187(*n*19), 233, 244
Simsova, Sylva 158< 198(*n*30), 243
Singh, Singha
    Chapal 53-54
    Jay N. v, viii, xii, xiii, 82, 84-86, 108, 115, 130, 192(*n*5), 195
    Kishen 188-89(*n*38)
    Nain 188-89(*n*38)

Raghab Narayan (Raja of Thalar) 84
Singha (mythological lion) 8
Smethurst, Paul 13, 14, 178(Ch1n3), 180-181(n14), 182(n22), 183(n31,32,34), 243
Smith, A. Croxton 167, 175, 243
Smith, Charles Hamilton 179-81, 235
Smith, Ione 196(n12), 243
Smith, Thomas 16-18, 185(n43), 243; *see also* Kirkpatrick Effect
Snellgrove, David x, 45-51, 190(n50), 243
Society of Range Management (USA) xv, 184(n41), 239
Solu-Khumbu xii, 51, 85
Soman, W.V. 192(n9), 198(n27), 243
Sood, Shubhi 197(n21), 243
Speake, Jennifer 180(n4,12), 243
Sponenberg, D. Phillip 199-200(n8), 243
Stag and hunting dog legend; *see* Milarepa
Steffel, Judy x, 123-32 *passim*, 133-60 *passim*, 193(n5,14,15), 194(n18), 195-96(n24,26,1,8,15), 202(n22), 243
Stephanitz, Max von 243
Sterilization; *see* Castration
Stevens, Thomas 202(n24), 243
Stewart, Jules 188-9(n38), 243
Stonehenge; *see* John Henry Walsh)
Strang, Paul 179-80(n10), 244
Subedi, Abhi 244
Swan, Laurence 29-30, 187(n26), 244
Swart, Sandra 244Swiss, Switzerland 57, 62-64, 174, 190(n3,4)
Synge, Mark; *see* Powell Millington
Széchenyí, Count Bela 20, 60, 185(n5,6), 236, 244; *see also* Gustav Kreitner

# T

TABL (Tourism Administration Bureau of Lhoka) 244
Tafel, A. 244
Tamu, Yarjung Kromchai 199(n2), 232
Taring, Jigme Wangchuk xi,106, 178(n1), 193(n3), 244
Tashi Lama 28, 204
Taylor, Daniel (Dan'l) xiii, x, 107, 108, 118, 119, 119-32 *passim*, 244
Taylor, Matthew J. vii, xiii, 184(n41), 196(n19), 221-7 (Appendix 4), 241
Thak Khola, Thakali (ethnic group) viii, 47, 51, 52ff., 74ff., 190(n4), 238, 245; *see also* Mustang; Tukché
Thalar (former principality, Nepal) 84-85, 115, 192(n4); *see also* Bajhang; Singh, J.N.; Singha, Raghab (Raja of Thalar)
Thomas, F.W. 217(n4), 244
Tibet Expeditionary Force; *see* Sir Francis Edward Younghusband)
Tibetan KyiApso; *see* Dog breeds, Asia: Apso types
Tibetan mastiff *passim*
  Breed standards 64, 71, 72, 82, 121, 125, 126, 138, 141-2, 145, 147, 164, 165, 184(n41), 207
  Gigantism, *or* super-mastiff; *see* Dog behavior: Size
Tibetan resistance; *see* Chusi Gandrug; Khampa
Tibetan Theory 14; *see also* Konradsen
Tiger 8, 10, 12, 146, 168, 181-2(n20), 182(n27), 200(n15), 208, 227, 231; *see also* Dogs named; Lion

Tilichho peak 66, 69
Tingri; *see* Dingri
Tinker La (Tinker-Lipu La) (pass) 65, 73, 84
TKC (Tibetan KyiApso Club); *see* Kennel clubs, breed clubs
Tomb guard dog (Han Dynasty) viii, 4, 5
Tournament of Shadows; *see* Great Game
Training; *see* Dog training
Triquet, Raymond 166, 200($n$11)
Tucci, Giuseppe 25, 28, 186($n$16), 187($n$23), 244-5
Tukché (Tukucha) (village, peak in northern Nepal) 53-57 *passim*, 66, 81, 190($n$4); *see also* Mustang; Thak Khola
Turin, Mark 199($n$2), 232
Turnbull, Colin M. 191($n$11), 241
Turner, Samuel 19, 167, 185($n$3), 245
Twain, Mark (Samuel Clemens) 18, 185($n$45)
Tylor, E.B. xi, 3, 245

**U**
Unsworth, Walt 189($n$41), 243
Uttar Anchal (state in India Himalayas) 168, 200($n$15)
Uttar Pradesh (state in India Himalayas) 93; *see also* Himachal Pradesh

**V**
Venice viii, 1, 10, 11, 13, 182
Verrier, Anthony 197($n$21), 245
Vinding, Michael 190($n$7), 245

**W**
Waddell, Laurence Austine x,19, 25-26, 39, 195($n$1), 196($n$16,17), 199($n$38), 245
Waller, Derik J. 288-9($n$38), 245
Walsh, John Henry ('Stonehenge') 171, 244, 245
Wangdi (Wangdu) (Tibetan resistance leader) 1909($n$9); *see also* Dogs named; Kesang Guerrilla Camp; Khampa
Warwick, Alan Ross 197($n$21), 245
Washburn, Bradford 189($n$39), 245
Weaver, Stewart 189($n$41), 235
Whelpton, John 192($n$4), 245
Wiki (Wikipedia) Effect 15
Wild dog; *see* Dhole
Wilson, Andrew 12, 30-31, 107, 109-10, 171, 175, 188($n$29,30), 193($n$8), 245
Wolf, wolves 12, 19, 28, 29, 33, 60-61, 78, 124, 131, 156, 174, 182($n$27), 201($n$18), 204, 206, 209, 212, 218-9, 237
Wolfdog; *see* Dog breeds, European: Irish wolfhound
Wood, Frances 180($n$11), 245
Workman, Fanny Bullock 27, 187($n$19), 245-6
Workman, William Hunter 187($n$19), 245-6
Wright, Thomas 7, 8-9, 167, 180-81($n$14,19), 241, 246
Wuwang (Wou-Wang; Chinese emporer, Han Dynasty) 5, 167, 174-5($n$10)
Wynn, M.B. 7, 14, 181($n$19), 184($n$38), 246
Wynyard, Ann Lindsay 172, 198($n$30), 199($n$6), 201-2($n$20), 246